Reading Essentials
An Interactive Student Workbook

gpscience.com

New York, New York · Columbus, Ohio · Chicago, Illinois · Peoria, Illinois · Woodland Hills, California

Glencoe Science

To the Student

In today's world, knowing science is important for thinking critically, solving problems, and making decisions. But understanding science sometimes can be a challenge.

Reading Essentials takes the stress out of reading, learning, and understanding science. This book covers important concepts in science, offers ideas for how to learn the information, and helps you review what you have learned.

In each chapter:

- **Before You Read** sparks your interest in what you'll learn and relates it to your world.
- **Read to Learn** describes important science concepts with words and graphics. Next to the text you can find a variety of study tips and ideas for organizing and learning information:
  - The **Study Coach** offers tips for getting the main ideas out of the text.
  - **Foldables™ Study Organizers** help you divide the information into smaller, easier-to-remember concepts.
  - **Reading Checks** ask questions about key concepts. The questions are placed so you know whether you understand the material.
  - **Think It Over** elements help you consider the material in-depth, giving you an opportunity to use your critical-thinking skills.
  - **Picture This** questions specifically relate to the art and graphics used with the text. You'll find questions to get you actively involved in illustrating the concepts you read about.
  - **Applying Math** reinforces the connection between math and science.
- Use **After You Read** to review key terms and answer questions about what you have learned. The **Mini Glossary** can assist you with science vocabulary. Review questions focus on the key concepts to help you evaluate your learning.

See for yourself. *Reading Essentials* makes science easy to understand and enjoyable.

The McGraw·Hill Companies

Copyright © by the McGraw-Hill Companies, Inc. All rights reserved. Except as permitted under the United States Copyright Act, no part of this publication may be reproduced or distributed in any form or by any means, or stored in a database or retrieval system, without the prior written permission of the publisher.

Send all inquiries to:
Glencoe/McGraw-Hill
8787 Orion Place
Columbus, OH 43240

ISBN 0-07-866089-0
Printed in the United States of America
17 18 19 20 045 14 13 12 11 10 09

Table of Contents

To the Student ii

Chapter 1	The Nature of Science	.2
Chapter 2	Motion	.20
Chapter 3	Forces	.36
Chapter 4	Energy	.52
Chapter 5	Work and Machines	.66
Chapter 6	Thermal Energy	.84
Chapter 7	Electricity	.102
Chapter 8	Magnetism and Its Uses	.120
Chapter 9	Energy Sources	.136
Chapter 10	Waves	.156
Chapter 11	Sound	.174
Chapter 12	Electromagnetic Waves	.194
Chapter 13	Light	.212
Chapter 14	Mirrors and Lenses	.232
Chapter 15	Classification of Matter	.252
Chapter 16	Solids, Liquids, and Gases	.264
Chapter 17	Properties of Atoms and the Periodic Table	.282
Chapter 18	Radioactivity and Nuclear Reactions	.300
Chapter 19	Elements and Their Properties	.322
Chapter 20	Chemical Bonds	.338
Chapter 21	Chemical Reactions	.360
Chapter 22	Solutions	.378
Chapter 23	Acids, Bases, and Salts	.398
Chapter 24	Organic Compounds	.412
Chapter 25	New Materials Through Chemistry	.432

chapter 1 The Nature of Science

section 1 The Methods of Science

What You'll Learn
- how scientists solve problems
- why scientists use variables
- compare and contrast science and technology

Mark the Text

Make Flash Cards Highlight each heading that is a question. Use a different color of marker to highlight the answers to the questions.

FOLDABLES

A **Build Vocabulary** Make the following vocabulary Foldable to help you study and learn key terms, which are always bold, from this section. You will need to make more than one vocabulary Foldable.

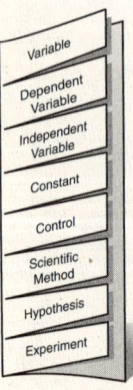

Before You Read

When you hear the word *scientist*, what comes to mind? Brainstorm a list of words that describe a scientist. Write them on the lines below.

_____

_____

Read to Learn

What is science?

Science is not just a subject in school. Science is a way of studying the world. The word *science* comes from a Latin word that means "knowledge." Science is a way to learn or gain knowledge by observing and investigating.

Nature follows a set of rules. The rules for understanding how the human body works are complicated. The rules for understanding the pattern of the Earth spinning once every 24 hours on its axis are simpler. Science is the study of the rules of nature.

What are the major categories of science?

Science covers many different topics. These topics fall under three main categories: (1) life science, (2) Earth science, and (3) physical science. Life science is the study of living things. Earth science is the study of Earth and outer space. Physical science is the study of matter and energy. In this textbook, you will study physical science. You will also learn how these three main categories sometimes overlap.

How does science explain nature?

Science helps us understand the natural world. Scientists use investigations to get new information. Technology has helped scientists learn more about the world. Sometimes, this new information causes scientific explanations to change.

2   CHAPTER 1 The Nature of Science

What are investigations?

Scientists learn new information by doing investigations. There are three types of investigations: observing and recording, experimenting, and building a model. Scientists might use one, two, or all three types to get new information about how nature works.

What is a variable?

An experiment might test three fertilizers to see which one helps plants grow biggest. In this experiment, there are two variables being studied. A **variable** is a quantity that can have more than one value. The height of the plants is a variable because it can have a range of values. The variable measured at the end of the experiment is called the **dependent variable** because it depends on what happened in the experiment. The height of the plants is the dependent variable because that is what is measured at the end of the experiment. The fertilizer is also a variable because the researcher is using three different kinds. A variable that is controlled and changed by the researcher is called the **independent variable**. In an experiment, there should be only one independent variable.

Why are constants and controls important?

To keep an investigation fair, all other factors must remain the same for each experiment or trial. A factor that does not change in an experiment is a **constant**. In the fertilizer experiment, the constants are the amount of water and sunlight the plants get and the kind of soil they are planted in. All of these constants are kept the same for the three types of fertilizer being tested. One plant is a control. Researchers use a **control** to compare the results of the experiment against. The control plant in the fertilizer experiment does not receive any fertilizer but does receive the same amount of water and sunlight and is planted in the same type of soil as the other plants. This helps scientists understand whether the fertilizer is having an effect on the plants or whether they would grow that way in regular conditions. The table shows possible results from the experiment testing the three fertilizers.

Think it Over

1. **Infer** Why is a control important in an experiment?

   _____
   _____
   _____

Picture This

2. **Make and Use Tables** In the table, fill in each blank with one of these terms: *constant, independent,* or *dependent*.

Constants and Variables	Plant 1 (control)	Plant 2	Plant 3	Plant 4
Fertilizer (_____ variable)	None	Mix A	Mix B	Mix C
Water (_____)	5 mL/day	5 mL/day	5 mL/day	5 mL/day
Soil (_____)	loamy	loamy	loamy	loamy
Growth (_____ variable)	5.2 cm	5 cm	10 cm	6 cm

Think it Over

3. Draw Conclusions Why might steps in scientific methods be skipped or changed?

_____
_____

Picture This

4. Interpret Why are there two arrows leading to different parts of the chart at the bottom?

_____
_____

Scientific Methods

A set of steps used in an investigation is called a **scientific method**. A scientist follows these steps when doing an investigation. These steps guide the scientist. Some steps may be repeated. Other steps may be skipped. Even if the steps change, the scientist is still using a scientific method. The flowchart below shows one way to use scientific methods to solve a problem. The arrows show where to go after each step is completed.

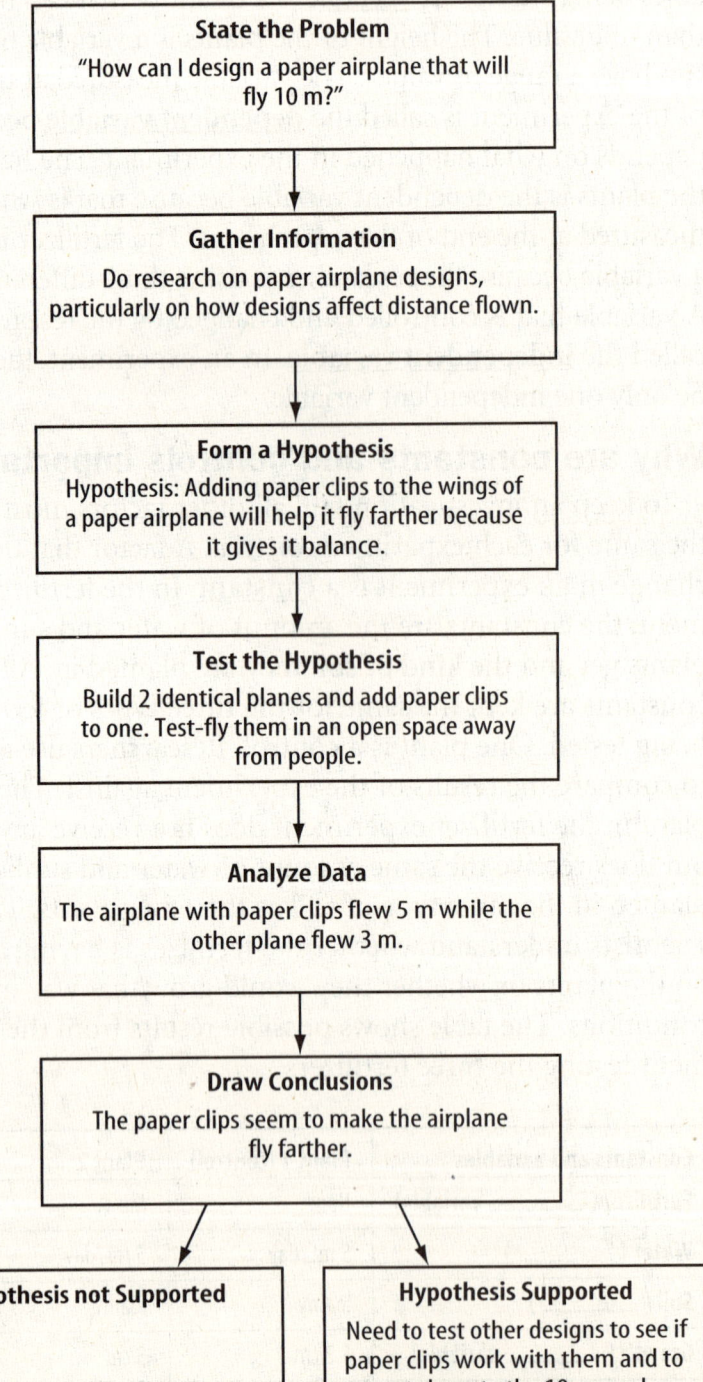

State the Problem
"How can I design a paper airplane that will fly 10 m?"

Gather Information
Do research on paper airplane designs, particularly on how designs affect distance flown.

Form a Hypothesis
Hypothesis: Adding paper clips to the wings of a paper airplane will help it fly farther because it gives it balance.

Test the Hypothesis
Build 2 identical planes and add paper clips to one. Test-fly them in an open space away from people.

Analyze Data
The airplane with paper clips flew 5 m while the other plane flew 3 m.

Draw Conclusions
The paper clips seem to make the airplane fly farther.

Hypothesis not Supported

Hypothesis Supported
Need to test other designs to see if paper clips work with them and to get closer to the 10-m goal.

4 CHAPTER 1 The Nature of Science

State the Problem Many scientific investigations begin when someone sees something happen in nature and wonders why or how it happened. The problem is stated as a "how" or "why" question. Sometimes a question may be asked because something didn't happen, or scientists are looking for something that will work better. For example, you may want to design a paper airplane that will fly 10 m. You can state the problem as "How can I design a paper airplane that will fly 10 m?"

Research and Gather Information It is important to study a problem before any testing is done. Sometimes, someone has already solved a similar problem. For the paper airplane problem, you would do research on paper airplane designs. You would try to find information on how different designs have affected the distances paper airplanes have flown.

Form a Hypothesis A <u>hypothesis</u> is an explanation for a question or problem based on what you know and what you observe. A scientist who forms a hypothesis must be certain that it can be tested. In your research you may have found that paper airplanes that are well-balanced fly greater distances. Your hypothesis might be that adding paper clips to the wings of a paper airplane will make it fly farther because the paper clips gave the airplane balance.

Test the Hypothesis Some hypotheses are tested by making observations. Sometimes building a model is the best way to test a hypothesis. Scientists often use an experiment to test a hypothesis. The <u>experiment</u> looks at how one thing affects another under controlled conditions. You could build two identical airplanes and add paper clips to one of them to test your hypothesis. Then, test-fly both airplanes in an open space.

Analyze Data An important part of any experiment is recording observations and organizing information. All results and observations should be recorded during an experiment. Many important discoveries have been made from unexpected results. The results from your experiment may be that the airplane with paper clips flew 5 m while the other plane flew 3 m. The information or data should be organized into an easy-to-read table or graph. Later in this chapter, you will learn how to show your data.

Think it Over

5. Research List two places where you might find information on paper airplane designs.

_____

_____

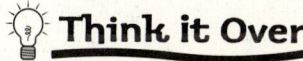

Think it Over

6. Classify Is the paper airplane without paper clips a variable, a constant, or a control in the experiment?

_____

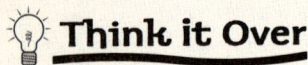

7. List two reasons why data must be organized.

_____
_____
_____

Understanding what the data and observations mean is also important. The data must be organized. Unorganized data may lead to a wrong conclusion. Scientists share their data with others through reports and meetings. Sometimes other people do not agree with the conclusions based on the data.

Draw Conclusions Scientists look at their data and decide if the data support the hypothesis. If the data is exactly the same after many experiments, the hypothesis is supported. If the data do not support the hypothesis, scientists may change the hypothesis or change the experiment.

Why must scientists be objective?

Scientists must be careful not to have a bias. A **bias** occurs in an experiment when a scientist expects something to happen and lets this influence how the results are viewed. Scientists try to reduce bias whenever they can. They reduce bias by doing the experiment many times and keeping very careful notes about what they observe. They also have other scientists repeat the same experiment to see if they get the same results.

Visualizing with Models

Sometimes scientists cannot see everything they are testing. They might be studying something too large or too small to see. It might even take too much time to see completely. In these cases, scientists use models. A **model** represents an idea, event, or object. The model helps people understand what is happening. Scientists have used models for a long time. Models make it easy to understand how and why things happen. Some models are very simple. For example, observing a small boat can help scientists understand how ships move through the oceans. ☑

8. Explain Why do scientists use models?

_____
_____
_____

What are high-tech models?

Scientific models don't have to be something you can touch. Many scientists use computers to build models. Computer models are used to solve difficult mathematical equations. NASA uses computers in experiments with space flights to solve equations that are too hard or would take too long to solve by hand.

Another type of model is a simulator. A simulator can create the same conditions found in real life. For example, a flight simulator is a model of an airplane. It can help a pilot pretend to be flying a plane. The pilot can test different ways to solve problems. The simulator reacts the same way a plane does when it flies, but there is no danger to the pilot or plane.

Scientific Theories and Laws

A scientific <u>theory</u> is a way of explaining things or events. These explanations are based on what has been learned from many observations and investigations. When these observations and investigations have been repeated many times and support the hypothesis, then the hypothesis becomes a theory. New information in the future may change the theory.

A <u>**scientific law**</u> is a statement about what happens in nature. It seems to be true all the time. A law explains what will happen under certain conditions, but it does not explain why or how it happens. Theories are used to explain how and why laws work. Gravity is an example of a scientific law. The law of gravity says that any one mass will attract another mass. To date, no experiments have been done that prove this law is not true.

The Limitations of Science

Science is used to explain many things about the world. However, science cannot explain everything. Questions about emotions or values are not questions science answers. A survey of peoples' opinions about these kinds of questions would not prove that these opinions are true for everyone. Scientists make predictions when they perform experiments. Then these predictions are tested and verified by using a scientific method.

Using Science—Technology

The words *science* and *technology* often are used in place of each other. However, the two words mean different things. <u>**Technology**</u> is the application of science. For example, science is used when a chemist develops a new material. When this new material is used on the space shuttle, technology is applied.

Sometimes technology comes before science. For example, when the steam engine was invented, no one knew exactly how it worked. Scientists studied it and learned about the steam engine. This led them to discover new ideas about the nature of heat.

Not all technology produces something good. Some people question the benefits of some technology, such as nuclear technology. Learning more about science can help society make decisions about these issues.

Think it Over

9. Think Critically Give an example of when technology is an application of science.

_____
_____
_____

FOLDABLES

B Compare Make the following Foldable to give examples showing how science and technology are different.

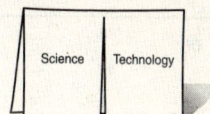

After You Read

Mini Glossary

bias: what is expected changes how the results are viewed

constant: a condition in an experiment that does not change

control: a standard in an experiment against which the results are compared

dependent variable: the condition in an experiment that results from the changes made to the independent variable

experiment: an investigation that tests a hypothesis by collecting information under controlled conditions

hypothesis: an explanation for a question or a problem that can be tested

independent variable: in an experiment, the condition that is tested

model: anything that represents an idea, event, or object to improve understanding

scientific law: a statement about what happens in nature that seems true all the time

scientific method: the steps a scientist follows when performing an investigation

technology: the application of science to help people

theory: an explanation that is supported by a large body of scientific evidence obtained from many different investigations and observations

variable: a quantity that can have more than a single value

1. Review the terms and their definitions in the Mini Glossary. Write a sentence using the terms *bias* and *experiment*.

   _____

   _____

2. Complete the chart below to organize the information you have learned in this section. Put the following steps for scientific methods in order.

   Analyze the data, Test the hypothesis, State the problem, Draw conclusions, Gather information, Form a hypothesis

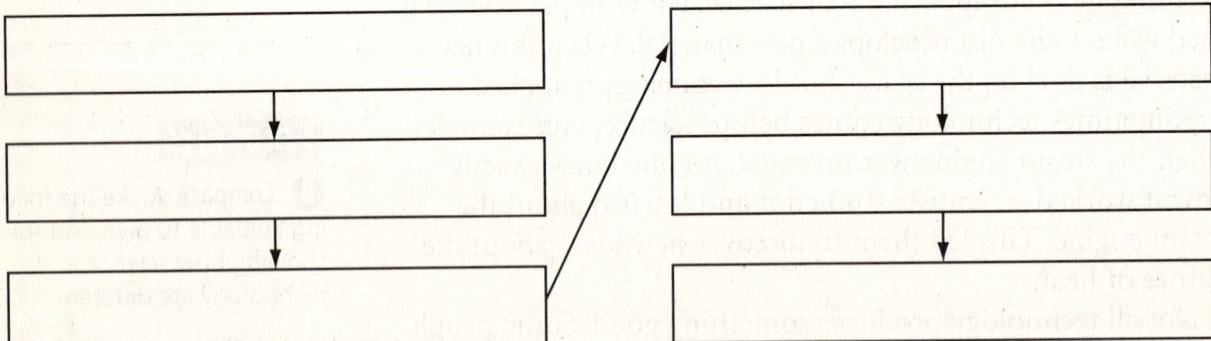

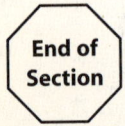

Science **Online** Visit gpscience.com to access your textbook, interactive games, and projects to help you learn more about the nature of science.

8   **CHAPTER 1**   The Nature of Science

chapter 1 The Nature of Science

section ❷ Standards of Measurement

● Before You Read

If someone asked you how wide your desk is, how would you measure it? Would you measure using inches, centimeters, feet, yards, or meters? Write why you selected this unit of measure.

_____

_____

_____

What You'll Learn
- SI units and symbols for length, volume, mass, density, time, and temperature
- how to convert related SI units

● Read to Learn

Units and Standards

A <u>standard</u> is an exact quantity that people agree to use to compare measurements. A standard is always exactly the same quantity when it is used anywhere in the world. If you and your friend both measure the width of your desk with your hands, will you get the same measure? You can't be sure, because you don't know if your hands are the same size.

Measurement Systems

A measurement is a way to describe the world using numbers. We use measurements to answer questions like how much, how long, or how far. For a measurement to make sense, it must include a number and a unit. Suppose you measure the width of your desk and say its width is 30. We would have to ask "Thirty what?" Your desk could measure 30 inches in width. Thirty is the number and inches is the unit.

In the United States, we commonly use units such as inches, feet, yards, miles, gallons, and pounds. These units are all part of the English system of measurement. Most other nations use the metric system. The metric system is based on multiples of ten.

Study Coach

Make an Outline Make an outline of the information in this section. Use headings as each part of the outline.

FOLDABLES

C Organize Information As you read this section, make the following Foldable to organize information about different types of measurements and units.

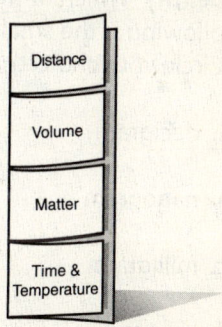

Reading Essentials   **9**

What is the International System of Units?

In 1960, an improvement was made to the metric system. This improvement is known as the International System of Units. This system is often abbreviated SI from the French *Le Systeme Internationale d'Unites*. The <u>SI</u> standards are accepted and used by scientists all over the world. Each type of SI measurement has a base unit. The base unit for length is the meter. The names and symbols for the seven base units are in the table below. All other SI units come from these seven base units.

Picture This

1. **Recognize** Circle the base units that you have seen before.

SI Base Units		
Quantity Measured	Unit	Symbol
Length	meter	m
Mass	kilogram	kg
Time	second	s
Electric current	ampere	A
Temperature	kelvin	K
Amount of substance	mole	mol
Intensity of light	candela	cd

What are SI prefixes?

The SI system is easy to use because it is based on multiples of ten. A prefix is added to the name of the base unit to indicate how many multiples of ten it should include. For example, the prefix *kilo-* means 1,000. That means that one kilometer is equal to 1,000 meters. This also means that one kilogram equals 1,000 grams. The most commonly used prefixes are shown in the table below.

Picture This

2. **Identify** Which of the following is the smallest? (Circle your choice.)

   a. decigram

   b. nanogram

   c. milligram

   d. kilogram

Common SI Prefixes		
Prefix	Symbol	Multiplying Factor
kilo-	k	1,000
deci-	d	0.1
centi-	c	0.01
milli-	m	0.001
micro-	μ	0.000 001
nano-	n	0.000 000 001

How do you convert between SI units?

Sometimes quantities are measured using different units. A teacher has 1.3 L of water for a class experiment. She needs 125 mL to conduct the experiment. To determine if she has enough water, she must first find out how many milliliters of water she has.

Conversion Factors A conversion factor is used to change measurements from one unit to another. A conversion factor is a ratio that equals one. For a ratio to equal one, the numerator and denominator must have the same value. The numerator of a conversion factor should be the new unit. The denominator should be the old unit. For example, if you are converting liters to milliliters, use the following conversion factor.

$$\frac{\text{new unit}}{\text{old unit}} = \frac{1000 \text{ mL}}{1 \text{ L}}$$

To find out how much water she has in centiliters, the teacher multiplies the amount of water she has by the conversion factor.

$$1.3 \text{ L} \times \frac{1000 \text{ mL}}{1 \text{ L}}$$

$$1.3 \cancel{\text{L}} \times \frac{1000 \text{ mL}}{1 \cancel{\text{L}}}$$

$$1.3 \times 1000 \text{ mL} = 1{,}300 \text{ mL}$$

The teacher has 1,300 mL of water. That is enough for her experiment!

Measuring Distance

In science, the word *length* is used to describe the distance between two points. The SI base unit of length is the meter, m. A baseball bat is about a meter long. Metric rulers and metersticks are commonly used to measure length. A meterstick is a little bit longer than a yardstick.

Applying Math

3. Convert Units A length of rope measures 3,000 millimeters. How long is it in meters?

$$3{,}000 \text{ mm} \times \frac{1 \text{ m}}{1{,}000 \text{ mm}}$$

$$\frac{3{,}000}{1} \times \frac{1 \text{ m}}{1{,}000}$$

$$\frac{3{,}000 \text{ m}}{1{,}000} = \boxed{}$$

4. Convert Units A small lizard weighs about 14,000 grams. How many kilograms does it weigh?

_____

_____

_____

How do you choose a unit of length?

When measuring distance, it is important to choose the proper unit. The unit you choose will depend on the object being measured. For example, you would measure the length of a pencil in centimeters. The length of your classroom would be measured in meters. The distance from school to your house would be measured in kilometers. By choosing the best unit, you can avoid very large or very small numbers. It is easier to say something is 21 km rather than saying it is 21,000 m.

Measuring Volume

<u>Volume</u> is the amount of space an object fills. The volume of a rectangular solid, such as a brick, is found by multiplying its length, width, and height ($V = l \times w \times h$). If the sides of the brick were measured in centimeters, cm, the volume would be expressed in cubic centimeters, cm^3. When you multiply all three measurements, you multiply "cm" three times, once with each measurement. The result is cm^3. If you were trying to find out how much space there is in a moving van, you would measure the van using meters. Its volume would be expressed in cubic meters, m^3. Let's find the volume of this van.

First find the length, width, and height of the van.

$$\text{Length} = 4 \text{ m}$$
$$\text{Width} = 2 \text{ m}$$
$$\text{Height} = 3 \text{ m}$$

Substitute these values into the formula for finding volume.

$$V = l \times w \times h$$
$$= 4 \text{ m} \times 2 \text{ m} \times 3 \text{ m}$$
$$= (4 \times 2 \times 3)(m \times m \times m)$$
$$= 24 \text{ m}^3$$

The volume of the moving van is 24 m^3.

Applying Math

5. Define In the calculations for finding the volume of the van, (m × m × m) is rewritten as m^3. The 3 in m^3 is called an exponent. What does an exponent represent?

Applying Math

6. Calculate What is the volume of a brick that has a length of 20 cm, a width of 6 cm, and a height of 5 cm? Show your work.

How do you measure the volume of a liquid?

Measuring the volume of a liquid in a container is different than measuring a solid object because the liquid does not have sides. To find the volume of a liquid, you first must know how much liquid the container can hold. This is called the capacity of the container. The most common units for expressing the volume of liquids are liters, L, and milliliters, mL. A milliliter is equal in volume to 1 cm³. So, the volume of 1 L equals 1,000 cm³. Look at food cans and bottles to see how these measurements are used.

Measuring Matter

Mass is the measure of how much matter is in an object. A golf ball and a table tennis ball are about the same size. If you pick up both, you notice a difference. The golf ball has more matter, or mass, than the table tennis ball.

What is density?

Another property of matter is density. The **density** of an object is the amount of mass in one cubic unit of volume of the object. You can find density by dividing an object's mass by its volume. The formula for density is:

$$D = \frac{m}{V}$$

In the formula, D represents density, m represents an object's mass, and V represents the object's volume.

Suppose an object weighs 10 g and has a volume of 2 cm³. Find the density of the object.

$$\frac{10 \text{ g}}{2 \text{ cm}^3} = 5 \text{ g/cm}^3$$

If two objects are the same size and one object has a greater mass, it also has a greater density. This is because there is more mass in one cubic unit of volume. The golf ball and the table tennis ball have about the same volume. However, the golf ball has a greater mass. This means that the golf ball also has a greater density.

Measuring Time

Sometimes scientists need to keep track of how long it takes something to happen. The unit of time in the SI system is second, s. Seconds are usually measured with a clock or a stopwatch.

✓ Reading Check

7. Measure in SI What are the most common units for expressing the volume of liquids?

_____

_____

Applying Math

8. Calculate Suppose an object weighs 15 g and has a volume of 5 cm³. What is the density of the object?

_____

💡 Think it Over

9. Compare Which has a greater density, a bowling ball or a volleyball?

_____

Measuring Temperature

Sometimes scientists need to measure how much something heats up or cools down. Temperature is a measure of how hot or cold something is. Later, you will learn the scientific meaning of temperature.

What is Fahrenheit?

The temperature measurement you are probably most familiar with is the Fahrenheit (F) scale. The Fahrenheit scale is based on the temperature of the human body, 98.6°. On this scale, water freezes at 32°F and boils at 212°F.

What is Celsius?

Scientists use the Celsius (C) scale to measure temperature. On this scale, water freezes at 0°C and boils at 100°C. The scale is divided into 100 equal divisions, or degrees, between the freezing point and the boiling point of water.

What is Kelvin?

The SI unit for measuring temperature is kelvin (K). On the Kelvin scale, 0 K is called absolute zero. This is the coldest possible temperature. Absolute zero is equal to −273°C, which is 273° below the freezing point of water. The divisions on the Kelvin and Celsius scales are the same size. This makes it easy to convert between the two scales. Water freezes at 0°C. To convert to Kelvin, you add 273 to the Celsius temperature. So, on the Kelvin scale, water freezes at 273 K. Water boils at 100°C. So, on the Kelvin scale, water boils at 373 K.

Reading Check

10. **Describe** What do you add to a temperature in Celsius to convert it to a temperature in kelvin?

_____
_____
_____

Picture This

11. **Label** each thermometer in the diagram with its correct temperature scale.

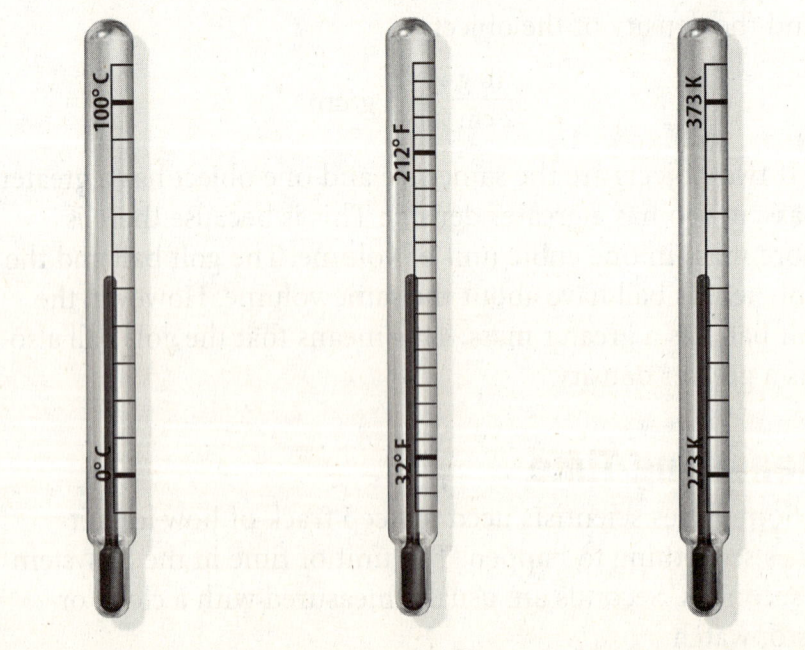

After You Read
Mini Glossary

density: the mass per unit volume of a material
mass: the measurement of the quantity of matter in an object
SI: the International System of Units
standard: an exact quantity that people agree to use to compare measurements
volume: the amount of space occupied by an object

1. Review the terms and their definitions in the Mini Glossary. Write, in your own words, a sentence that explains how mass affects an object's density.

   _____
   _____

2. Complete the chart below to organize the information from this section. For each unit include the unit's name, what it measures, and its symbol.

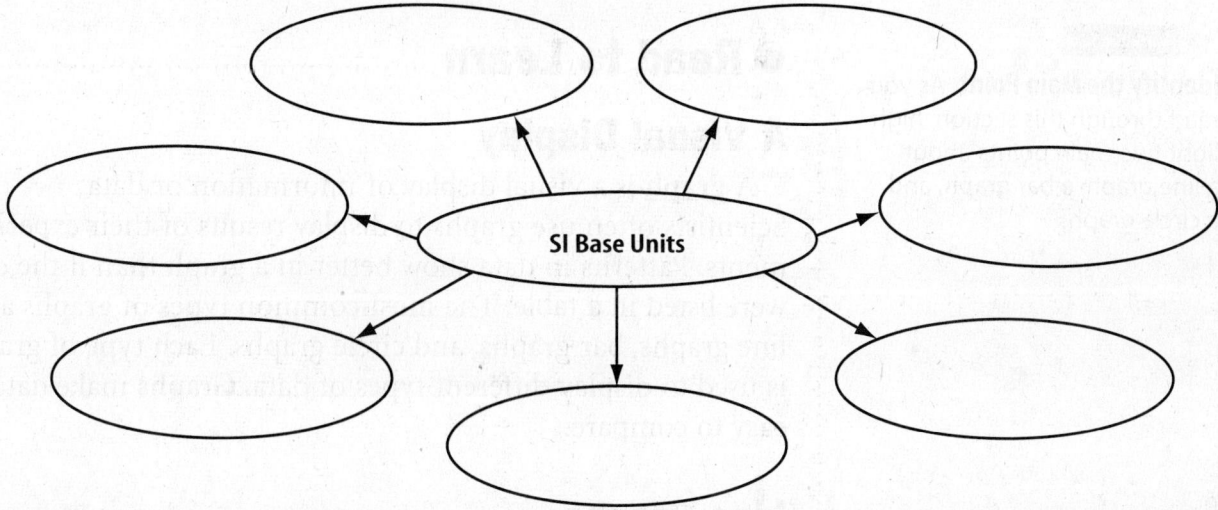

3. Think about what you have learned. Write a way to help you remember the meaning of volume.

   _____
   _____
   _____

 Visit gpscience.com to access your textbook, interactive games, and projects to help you learn more about the nature of science.

chapter 1 The Nature of Science

section ❸ Communicating with Graphs

What You'll Learn
- how graphs are used
- how to distinguish between dependent and independent variables

● Before You Read

Think about graphs that you have seen. Why do you think graphs are a useful way to display information?

_____

_____

_____

Mark the Text

Identify the Main Point As you read through this section, highlight two main points about a line graph, a bar graph, and a circle graph.

● Read to Learn

A Visual Display

A <u>graph</u> is a visual display of information or data. Scientists often use graphs to display results of their experiments. Patterns in data show better in a graph than if the data were listed in a table. The most common types of graphs are line graphs, bar graphs, and circle graphs. Each type of graph is used to display different types of data. Graphs make data easy to compare.

Line Graphs

Line graphs show change over time. A line graph can show more than one event on the same graph. For example, a builder had a choice of three brands of thermostats to install in school classrooms. He tested each brand to find the most efficient one. The builder set each thermostat at 20°C. He checked the classroom temperature every 5 minutes for 25 minutes and recorded the data.

The graph on the next page shows the results of the test. The break in the vertical axis between 0 and 15 means that the numbers 1–14 have been left out. This is done so there is more room to spread the scale and the graph is easier to read. The horizontal line represents the 20°C setting and is the control in this experiment. The three lines (A, B, and C) show how quickly each thermostat reached the setting of 20°C.

FOLDABLES

D Compare and Contrast Fold one piece of notebook paper into four equal sections. Use each section to draw an example of the different types of graphs.

Line Graphs	Bar Graphs
Circle Graphs	Variables

16 CHAPTER 1 The Nature of Science

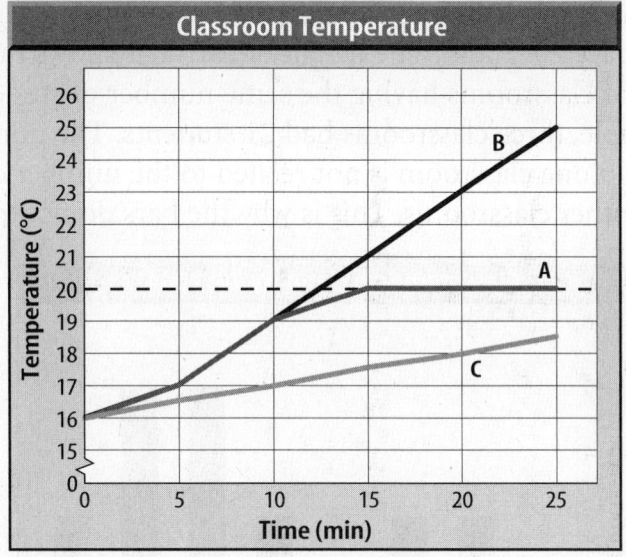

Applying Math

1. **Interpret Data** Which thermostat reached its temperature setting the quickest? Circle the answer.

   a. A

   b. B

   c. C

How do you construct line graphs?

There are several steps that you must follow to make a line graph. Always use the horizontal axis, or *x*-axis, for the independent variable. The vertical axis, or *y*-axis, is always used for the dependent variable. You should select a scale that makes your graph sensible and readable. For example, the scale on the *x*-axis would not make sense or be very readable if the units were by hours instead of minutes. All of lines A, B, and C that you see on this graph would be in the first tenth of the section ending with 5.

The data on a line graph are related. To draw the lines, you first plot points that show the relationship of the variables. For example, in line B, after 5 min the temperature was 17°C. You plot a point that aligns with 5 on the *x*-axis and 17 on the *y*-axis. Then plot the points representing the other data. Then you draw a line to connect the points.

You must also make sure that all the data use the same units. If some measurements were made in the Celsius scale and others were made in the Fahrenheit scale, the units must be converted to the same unit of measurement.

Bar Graphs

Bar graphs are useful for comparing data collected by counting. Each bar shows a number counted at a particular time. As on a line graph, the independent variable is plotted on the *x*-axis, and the dependent variable is plotted on the *y*-axis. The data are not related, as it is in a line graph, so the bars do not touch. ☑

Reading Check

2. **Describe** What kind of data are bar graphs useful for?

_____

_____

Reading Essentials **17**

The bar graph below shows classroom size on one particular day (January 20, 2004). The height of each bar shows the number of classrooms having the same number of students. For example, three classrooms had 21 students. The number of students in one classroom is not related to the number of students in other classrooms. This is why the bars do not touch.

Applying Math

3. **Interpret Data** How many classrooms had 26 students?

_____

Circle Graphs

A circle graph, or pie graph, is used to show how a certain quantity is broken down into parts. The circle represents the whole and the segments, or slices, are the parts of the whole. The segments are usually represented as percentages of the whole.

This circle graph below shows how much of each different kind of heating fuel is used in different buildings. According to the circle graph, gas is the heating fuel used the most.

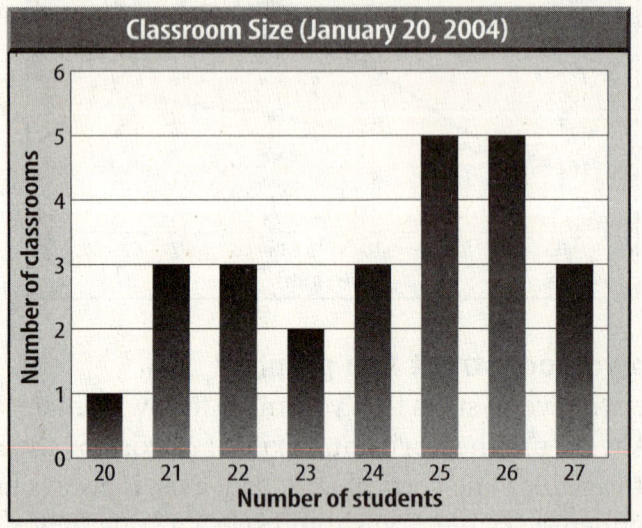

When you use a graph you should make sure that the conclusions you draw are based on accurate information. You should also select scales that help you make your graph easy to read. In a circle graph, label the sections so that anyone looking at your graph knows what data are shown. In bar and line graphs, always label both the *x*-axis and the *y*-axis.

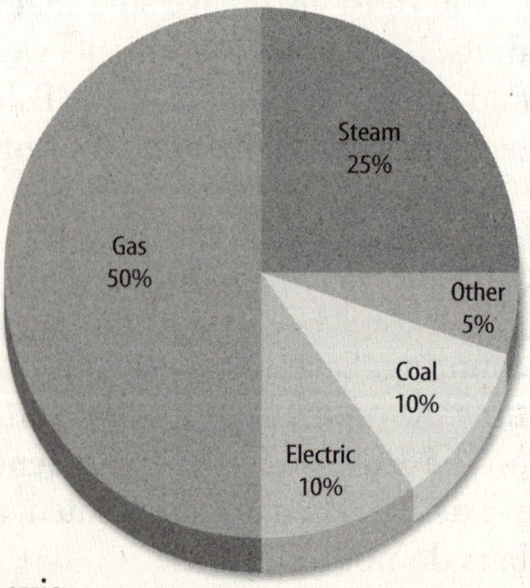

Applying Math

4. **Use Models** What do the percentages in a circle graph add up to? Why is this important?

_____

_____

18 CHAPTER 1 The Nature of Science

After You Read
Mini Glossary

graph: a visual display of information or data

1. Review the term *graph* and its definition in the Mini Glossary. Write a sentence that tells how a graph helps you understand data.

   _____

2. Complete the outline below. Tell what type of data each type of graph displays and then draw a small sample of each type of graph.

   Three Types of Graphs
   I. Line Graph
      A. data type:
      B. sample:

   II. Bar Graph
      A. data type:
      B. sample:

   III. Circle Graph
      A. data type:
      B. sample:

3. Think about what you have learned. Give an example of each type of data best suited for each type of graph.

   Line Graph: _____

   Bar Graph: _____

   Circle Graph: _____

 Visit gpscience.com to access your textbook, interactive games, and projects to help you learn more about the nature of science.

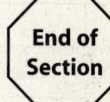

Motion

section ❶ Describing Motion

What You'll Learn
- the difference between displacement and distance
- the difference between speed and velocity
- how to graph motion

Mark the Text

Identify the Main Point Highlight the main point in each paragraph in this section. Highlight in a different color a detail or example that helps explain a point.

FOLDABLES

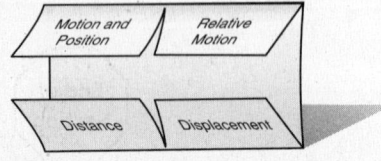

Ⓐ Find Main Ideas As you read this section, make the following Foldable to help you identify the main ideas about motion.

● Before You Read

Have you ever been on a roller coaster? You can feel the steep drops and quick turns in your body. Write how it feels to travel up a steep hill slowly and then to go down the other side quickly.

_____

_____

● Read to Learn

Motion

Distance and time are important in describing a race. The winner covers the distance in the shortest amount of time. It takes more time to run a 10-km race than to run a 5-km race because 10-km is a longer distance.

How are motion and position related?

You don't need to see something move to know that it has moved. Suppose you see a mail truck stopped next to a mailbox. When you look again later, the truck is farther down the street by a tree. You didn't see the truck move, but you know that motion has taken place because of the truck's new position.

A reference point is needed to tell where something is. Motion occurs when an object goes from one reference point to another. The mailbox was the first reference point for the mail truck. The tree was the second reference point.

What is relative motion?

Not all motion is as obvious as the mail truck's motion. When you are sitting still in a desk, you appear not to be moving. However, you are moving. You are not moving in relation to your desk or school building. You are moving in relation to the Sun because you are sitting on Earth.

Relative motion means that one thing moves in relation to another thing. Earth is moving in space in relation to the Sun. The Sun is the reference point for Earth's motion.

How are distance and displacement different?

Distance is how far something has moved. It is an important part of describing motion. In the 50-m dash, a runner travels 50 m between the start line and the finish line. The distance of the race is 50 m.

The SI unit of length or distance is the meter (m). Long distances are measured in kilometers (km). One kilometer is equal to 1,000 meters. Short distances are measured in centimeters (cm). One meter is equal to 100 centimeters.

Not all motion is in a straight line. In the figure, the runner jogged 50 m to the north. Then she turned around and jogged 30 m to the south. The total distance she jogged is 80 m. She is 20 m from the starting point. **Displacement** is the distance and direction of an object's position relative to the starting point. The runner's displacement is 20 m north.

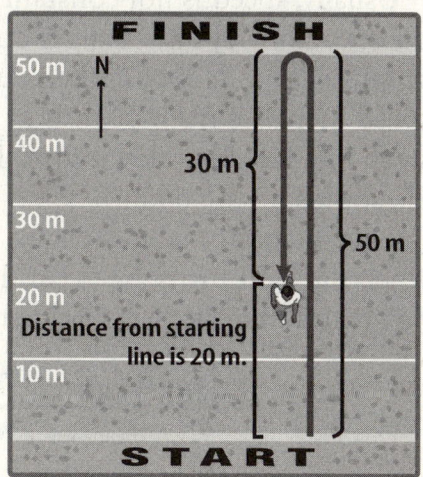

Displacement = 20 m north of starting line
Distance traveled = 50 m + 30 m = 80 m

Picture This
1. **Describe** another way a runner could have a displacement of 20 m north.

Speed

So far, motion has been described by the distance something has moved and by displacement from the starting point. You might also want to tell how fast something is moving. To do this, you need to know how far it travels in a certain amount of time. **Speed** is the distance an object travels per unit of time.

How is speed calculated?

The SI unit of distance is the meter (m). The SI unit of time is the second (s). So, in SI, speed is measured in meters per second (m/s). Sometimes it is easier to express speed in other units so that the numbers will not be very large or very small. Something that moves very quickly, such as a rocket, can be measured in kilometers per second (km/s). Very low speeds, such as geological plate movements, can be measured in centimeters per year (cm/y).

To calculate the speed of an object, divide the distance it traveled by the time it took to travel the distance. Here is a formula for calculating speed.

$$\text{speed} = \frac{\text{distance}}{\text{time}}, \text{ or } s = \frac{d}{t}$$

Applying Math
2. **Calculate** A train traveling at a constant speed covers a distance of 960 meters in 30 s. What is the train's speed? Show your work.

Reading Essentials **21**

What is motion with constant speed?

A speedometer measures the speed of a car. Suppose you look at the speedometer when you are riding on a freeway. The car's speed hardly changes. If the car is not speeding up or slowing down, it is moving at a constant speed. If you are traveling at a constant speed, you can measure your speed over any distance from millimeters to light years.

What is changing speed?

Usually, speed is not constant. The graph below shows how the speed of a cyclist changes during a 5-km ride. Follow the graph as the ride is described. As the cyclist starts off, his speed increases from 0 km/h to 20 km/h. Then he comes to a steep hill. He slows down to 10 km/h as he pedals up the hill. He speeds up to 30 km/h going down the other side of the hill. At the bottom, he stops for a red light. He speeds up when the light turns green. At the end of the ride, he slows down and then stops. The ride took 15 min.

Applying Math

3. Describe what is happening when the line on the graph is horizontal.

_____

_____

_____

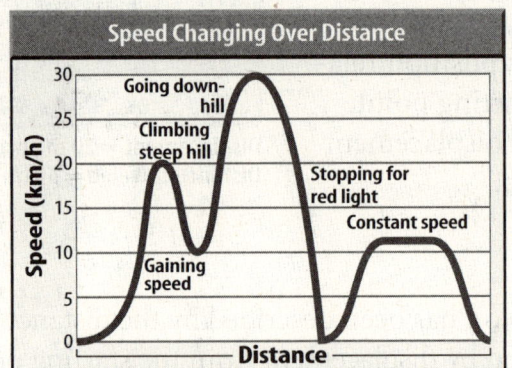

What is average speed?

Look at the graph of speed for the bicycle trip. Sometimes the bicycle was moving fast, sometimes it was moving slowly, and sometimes it was stopped. How could you describe the speed of the whole bike ride? Would you use the fastest speed or the slowest?

Average speed describes the speed of motion when speed is changing. **Average speed** is the total distance traveled divided by the total time of travel. It is calculated using the relationships among speed, distance, and time.

The total distance the cyclist traveled was 5 km. The total time was 15 minutes, or $\frac{1}{4}$ h. You can write $\frac{1}{4}$ h as 0.25 h. The average speed for the bicycle trip can be found using a mathematical equation.

Applying Math

4. Find the average speed in kilometers per hour of a race car that travels 260 km in 2 h.

$s = \frac{260}{2}$

$s =$ _____ km/h

$$\text{average speed} = \frac{\text{total distance}}{\text{total time}} = \frac{5 \text{ km}}{0.25 \text{ h}} = 20 \text{ km/h}$$

22 CHAPTER 2 Motion

What is instantaneous speed?

The speed shown on a car's speedometer is the speed at one point in time, or one instant. **Instantaneous speed** is the speed at one point in time. If an object is moving with constant speed, the instantaneous speed doesn't change. The speed is the same at every point in time. However, when a car speeds up or slows down, its instantaneous speed is changing. The speed is different at every point in time.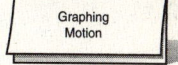

Graphing Motion

A distance-time graph shows the motion of an object over time. The graph below shows the motion of three swimmers during a 30-min workout. The line for Mary is straight. This means that she swam at a constant speed. Her speed was 80 m/min for the whole 30 min.

The line for Kathy is also straight. Kathy also swam at a constant speed. Her speed was 60 m/min for the whole 30 min. Notice that the line for Mary is steeper than the line for Kathy. This is because Mary swam faster than Kathy. Mary was swimming faster, so she swam farther than Kathy in the same amount of time.

The steepness of the line is called the slope. The slope of a line on the graph is the speed. A steeper slope means a greater speed. Mary was swimming faster than Kathy, so the slope of Mary's line is steeper than the slope of Kathy's line.

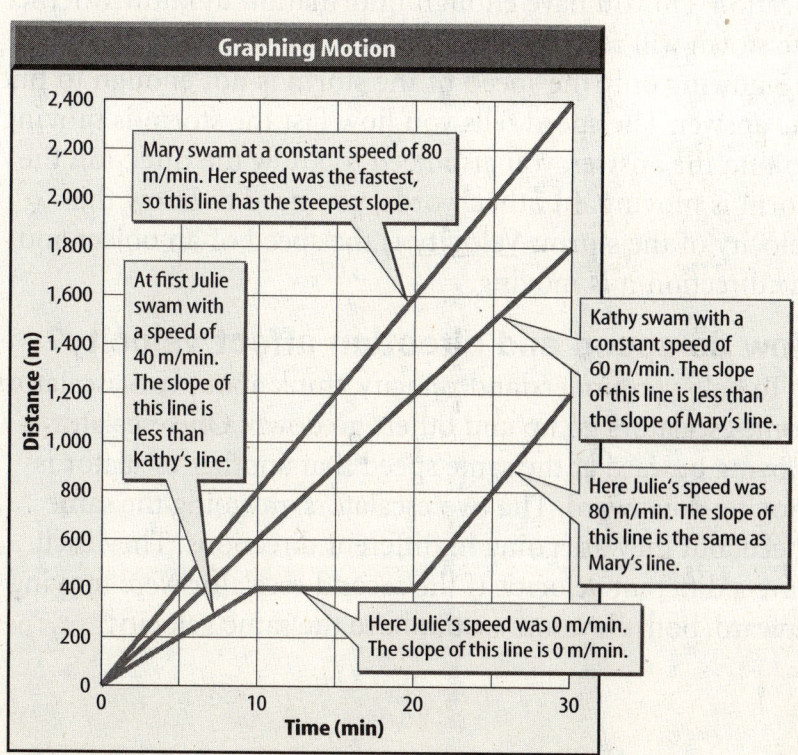

Reading Check

5. Describe a car's instantaneous speed as it stops for a red light.

___

FOLDABLES

B Making Graphs Make the following Foldable using graph paper. Draw and explain a motion graph on the inside.

Applying Math

6. Explain why the slopes are different for each swimmer.

___

Reading Essentials 23

How is changing speed graphed?

Julie's line is not straight. Julie did not swim at a constant speed for the entire 30-min practice. Julie swam at a constant speed for the first 10 min. Her rate was 40 m/min. Then she rested for 10 min. For this part of the practice, the line is horizontal. Her speed was 0 m/min, and the line has zero slope. She swam 800 m during the last 10 min. During this part of the practice, she swam as fast as Mary, so that part of her line has the same slope as Mary's line.

How do you draw a distance-time graph?

A distance-time graph plots data for distance and time. The distance traveled is plotted on the vertical axis. Time is plotted on the horizontal axis. Each axis has a scale, or a series of numbers, that covers the range of the data.

Mary swam the farthest during the practice, 2,400 m. So, the vertical scale must go to 2,400. The practice was 30 min long, so the horizontal axis must go to 30. Each axis is divided into equal parts. When each axis is finished, the data points are plotted on the graph. Then the data points for each swimmer are connected with a line.

Velocity

Suppose you hear that there is a storm nearby. The storm is traveling at a speed of 20 km/h and is 100 km east of your location. Do you have enough information to know whether the storm will reach you?

Knowing only the speed of the storm is not enough to find the answer. The speed tells you how fast the storm is moving. To find the answer, you also need to know the direction the storm is moving. In other words, you need to know the velocity of the storm. **Velocity** is the speed of an object and the direction it is moving.

How do speed and direction affect velocity?

To help you understand velocity, think about two escalators. Some escalators go up and others go down. One escalator is moving upward at the same speed that another escalator is moving downward. The two escalators are going the same speed, but they are going in different directions. They each have a different velocity. If the second escalator were moving upward, both elevators would have the same velocity.

Think it Over

7. Recall The vertical axis is called the *y*-axis. What is the horizontal axis called?

8. Describe how the velocity of an escalator that is going up is different from the velocity of one that is going down at the same speed.

Velocity depends on both speed and direction. Because of this, an object moving at a constant speed will have a changing velocity if it changes direction. A race car on an oval track has a constant speed. But, as the race car goes around the track, the direction in which the car is moving changes. This means that the velocity of the car is changing. An object has constant velocity if neither the speed nor direction it is moving changes. The light from a laser beam travels at a constant velocity.

Motion of Earth's Crust

Some motion is so slow that it is hard to see. The surface of Earth doesn't seem to change from year to year. But if you look at geological evidence of Earth over 250 million years, you will see that large changes have occurred.

According to the theory of plate tectonics, the continents are moving constantly over Earth's surface. The movement is shown in the figure below. The changes are so slow that we do not notice them.

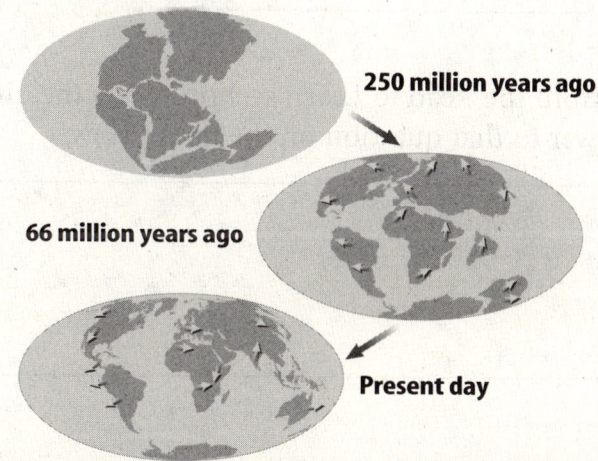

Picture This
9. **Describe** the movement of the continents over the past 250 million years.

___
___
___

How do continents move?

Earth is made of layers. The top layer is the crust. The layer below the crust is the upper mantle. Together, the crust and the top part of the upper mantle are the lithosphere. The lithosphere is broken into huge sections, plates.

Below the lithosphere, the layers are like putty. The plates slide slowly on these soft layers. The moving plates cause geological changes on Earth. These changes include the formation of mountain ranges, earthquakes, and volcanic eruptions.

The plates move very slowly. The speed of the plates is measured in centimeters per year. In California, two plates are sliding past each other along the San Andreas Fault. The average speed of the two plates is about 1 cm per year.

Think it Over
10. **Explain** Why is the speed of the movement of Earth's plates measured in centimeters per year instead of in meters per second?

___
___
___

Reading Essentials 25

● After You Read
Mini Glossary

average speed: the total distance traveled in a unit of time

displacement: the distance and direction that something moved from a starting point

distance: a measure of how far an object has moved

instantaneous speed: the speed of an object at a one point in time

speed: the distance an object travels in an amount of time

velocity: a measure of the speed of an object and the direction it is traveling

1. Review the terms and their definitions in the Mini Glossary. What is the difference between average speed and instantaneous speed?

___

___

___

___

2. Choose one of the question headings from the Read to Learn section. Write the question in the box below. Then write your answer to that question on the lines below.

Question:

Answer:

___

___

___

3. Use the data below to make a distance-time graph. Be sure to include labels and scales.

Bike Ride

Time (min)	Distance (m)
15	150
30	300
45	450
60	600

4. As you read, you highlighted important points in the text. How did highlighting the text help you learn the content of the section?

_____

_____

_____

 Visit **gpscience.com** to access your textbook, interactive games, and projects to help you learn more about motion.

Reading Essentials 27

chapter 2 Motion

section 2 Acceleration

What You'll Learn
- how acceleration, time, and velocity are related
- how positive and negative acceleration affect motion
- how to calculate acceleration

Study Coach

Outlining As you read the section, make an outline of the important information in each paragraph.

FOLDABLES

C Construct a Venn Diagram Make the following trifold Foldable to compare and contrast the characteristics of acceleration, speed, and velocity.

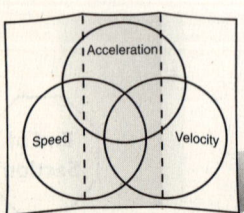

● Before You Read

Describe what happens to the speed of a bicycle as it goes uphill and downhill.

_____

_____

_____

● Read to Learn

Acceleration, Speed, and Velocity

A car sitting at a stoplight is not moving. When the light turns green, the driver presses the gas pedal and the car starts moving. The car moves faster and faster. Speed is the rate of change of position. **Acceleration** is the rate of change of velocity. When the velocity of an object changes, the object is accelerating.

Remember that velocity is a measure that includes both speed and direction. Because of this, a change in velocity can be either a change in how fast something is moving or a change in the direction it is moving. Acceleration means that an object changes it speed, its direction, or both.

How are speeding up and slowing down described?

When you think of something accelerating, you probably think of it as speeding up. But an object that is slowing down is also accelerating. Remember that acceleration is a change in speed. A car that is slowing down is decreasing its speed. It is also accelerating, because its speed is changing.

Imagine a car being driven down a road. If the speed is increasing, the car has positive acceleration. When the car slows down, the speed decreases. The decreasing speed is called negative acceleration. In both cases, the car is accelerating, but one acceleration is positive and one is negative.

28 CHAPTER 2 Motion

Acceleration has direction, just like velocity. In the figure below, both cars are accelerating because their speeds are changing. When a car's acceleration and velocity are in the same direction, the speed increases and the acceleration is positive. Car A has positive acceleration. When a car is slowing down, the acceleration and velocity are in opposite directions. The acceleration is negative. Car B has negative acceleration.

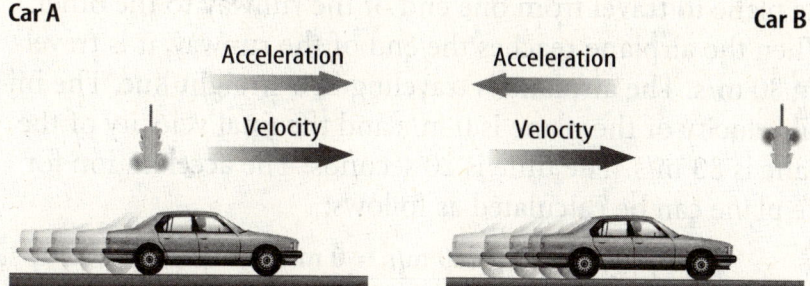

Picture This
1. **Describe** the acceleration of the cars in each figure.

   _____
   _____

Does changing direction affect acceleration?

A change in velocity is either a change in an object's speed or its direction. When a moving object changes direction, its velocity changes and it is accelerating. The speed of a horse moving around on a carousel remains constant, but it is constantly changing direction. So, the horse is accelerating.

Calculating Acceleration

Acceleration is the rate of change in velocity. To calculate acceleration, you first find the change in velocity. To find change in velocity subtract the beginning velocity of an object from the velocity at the end of its movement. Beginning velocity is called the initial velocity, or v_i. Velocity at the end is called the final velocity, or v_f.

$$\text{change in velocity} = \text{final velocity} - \text{initial velocity}$$
$$= v_f - v_i$$

Reading Check
2. **Use Variables** Write what v_f and v_i mean.

   _____
   _____
   _____

If motion is in a straight line, the change in speed can be used to calculate the change in velocity. The change in speed is the final speed minus the initial speed.

To find acceleration, divide the change in velocity by the length of time during which the velocity changed.

$$\text{acceleration } (a) = \frac{\text{change in velocity}}{\text{time } (t)} \quad \text{or} \quad a = \frac{(v_f - v_i)}{t}$$

The SI unit for velocity is meters per second (m/s). To find acceleration, velocity is divided by the time in seconds (s). So, the unit for acceleration is m/s^2.

Applying Math
3. **Calculate** Suppose a bird takes off from a tree and flies in a straight line. It reaches a speed of 10 m/s. What is the change in the bird's velocity?

   _____
   _____

Applying Math

4. Explain Why is the acceleration of an object moving at a constant velocity always 0?

_____
_____
_____

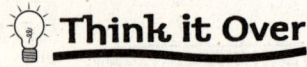

5. Think Critically A car that is slowing down is still moving forward. Why is this considered negative acceleration?

_____
_____
_____

How is positive acceleration calculated?

How is the acceleration of an object that is speeding up different from that of an object that is slowing down? The acceleration of an object that is speeding up is always positive. The acceleration of an object that is slowing down is always negative.

Suppose an airplane is sitting at the end of a runway. The plane takes off and moves down the runway. It takes 20 s for the plane to travel from one end of the runway to the other. When the airplane reaches the end of the runway, it is traveling 80 m/s. The airplane is traveling in a straight line. The initial velocity of the plane is 0 m/s and the final velocity of the plane is 80 m/s. The time is 20 seconds. The acceleration for the plane can be calculated as follows:

$$a = \frac{(v_f - v_i)}{t} = \frac{(80 \text{ m/s} - 0 \text{ m/s})}{20 \text{ s}} = 4 \text{ m/s}^2$$

The airplane is speeding up as it goes down the runway. The final speed is greater than the initial speed. The acceleration is positive.

How is negative acceleration calculated?

Now imagine a skateboarder moving in a straight line. The skateboarder is moving at a speed of 3 m/s. It takes the person 2 s to come to a stop. The initial velocity is 3 m/s and the final velocity is 0 m/s. The total time is 2 seconds. The calculation for the skateboarder's acceleration is as follows:

$$a = \frac{(v_f - v_i)}{t} = \frac{(0 \text{ m/s} - 3 \text{ m/s})}{2 \text{ s}} = -1.5 \text{ m/s}^2$$

The skateboarder is slowing down. The final speed is less than the initial speed. The acceleration has a negative value.

Amusement Park Acceleration

Roller coasters are exciting rides. People who design roller coasters use the laws of physics. The steep drops and loops of steel roller coasters give the rider large accelerations. When riders move down a steep hill, gravity will cause them to accelerate toward the ground. When riders go around a sharp turn, they are also accelerated. This acceleration makes them feel as if a force were pushing them toward the side of the car.

After You Read

Mini Glossary

acceleration: the rate of change in velocity

1. Review the term and its definition in the Mini Glossary. Explain why a change in velocity affects acceleration.

   _____

   _____

   _____

2. Complete the chart to organize information about how acceleration is calculated.

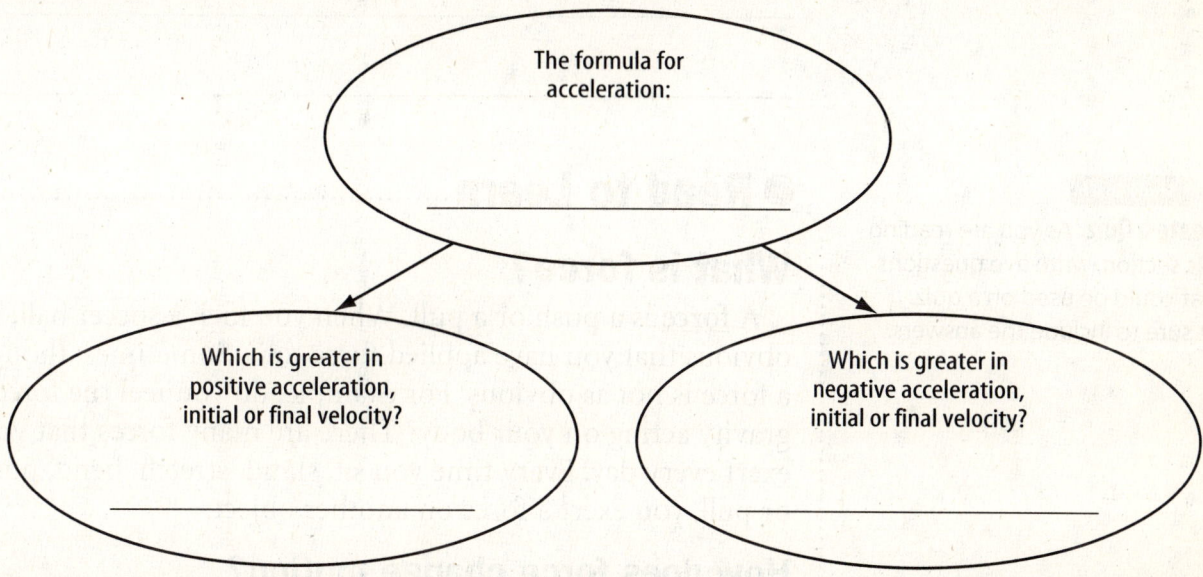

3. As you read the section, you made an outline describing the points covered in each paragraph. How did you decide what to write as the major points in your outline?

   _____

   _____

   _____

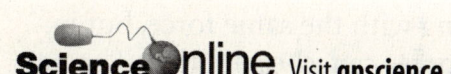 Visit gpscience.com to access your textbook, interactive games, and projects to help you learn more about acceleration.

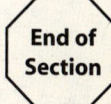

Reading Essentials **31**

Motion

section ❸ Motion and Forces

What You'll Learn
- how force and motion are related
- Newton's first law of motion
- what forces and motion are present during a car crash

Study Coach

Create a Quiz As you are reading this section, write five questions that could be used on a quiz. Be sure to include the answers.

● Before You Read

When you hit a baseball with a bat, you apply a force that moves another object. Think of three more examples from sports in which force is used to move something. Write your examples on the lines below.

_____

_____

_____

● Read to Learn

What is force?

A **force** is a push or a pull. When you kick a soccer ball, it is obvious that you have applied force to it. Sometimes, though, a force is not as obvious. For example, do you feel the force of gravity acting on your body? There are many forces that you exert every day. Every time you sit, stand, stretch, bend, push, or pull, you exert a force on another object.

How does force change motion?

Think about a tennis player hitting a ball. What happens to the motion of the ball when the racket hits it? The force of the racket hitting the ball makes the ball stop. Then the force makes the ball move in a different direction.

What are balanced forces?

Not all forces change velocity. Suppose two students are pushing on opposite sides of a box. As shown in the figure, if both students are pushing with an equal force, the box does not move. When two or more forces act on an object at the same time, the forces combine. This is called a **net force**. When two students are pushing with the same force, but in opposite directions, the two forces cancel each other. The net force on the box is zero. Forces on an object that are equal in size and opposite in direction are called **balanced forces**.

32 CHAPTER 2 Motion

+ → ← = 0
Net force = 0

Picture This

1. Describe Why are the forces in the figure said to be balanced?

_____
_____
_____

What is the result of unbalanced forces?

Not all forces in opposite directions cancel each other. Think about two students pushing on the opposite sides of a box. What happens if one student pushes harder than the other as in the first figure? The box will move in the direction of the larger force. The student who is pushing harder will move the box in the direction of the force. The net force that moves the box is the difference between the two forces. They are unbalanced forces.

Suppose both students push on the same side of the box as in the second figure. The students are both exerting force in the same direction. The forces are combined, or added together, because they are exerted on the box in the same direction. The net force is equal to both forces added together.

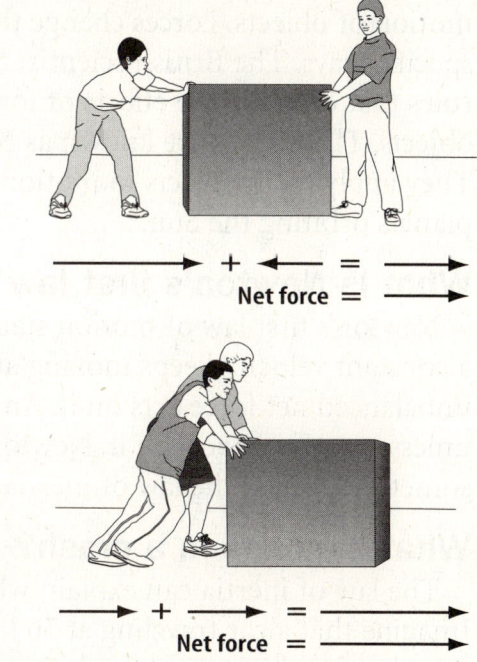

+ ⟶ ← = ⟶
Net force = ⟶

+ ⟶ ⟶ = ⟶
Net force = ⟶

Picture This

2. Draw in the space below an example showing the result of unbalanced forces on opposite sides of an object.

Inertia and Mass

Inertia is the tendency of an object to resist any change in its motion. If an object is moving, it will keep moving in a straight line at a constant speed until a force changes its direction or speed. Inertia causes an object to resist changes in direction and speed. A dirt bike will move in a straight line with a constant speed unless a force acts on it. A force can turn the wheel and change the direction. Another force, friction, can slow the speed of the bike. An object that is not moving also has inertia. It will remain motionless until an unbalanced force causes it to move.

D Find Main Idea As you read, use quarter or half sheets of paper to help you identify the main ideas about inertia and Newton's first law of motion.

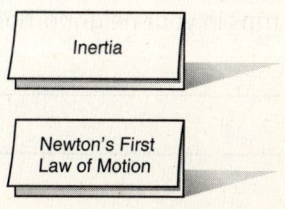

✓ **Reading Check**

3. Compare the inertia of a car to the inertia of a bicycle.

_____

_____

_____

✓ **Reading Check**

4. Restate Newton's first law of motion in your own words.

_____

_____

_____

💡 **Think it Over**

5. Think Critically Why do you think you should wear your safety belt, even on short trips in your neighborhood?

_____

_____

_____

Why can different objects have different inertia?

The inertia of an object is related to its mass. The greater the mass is of an object, the greater its inertia is. Remember, mass is the amount of matter in an object. A bowling ball has much greater mass than a table-tennis ball does. Therefore, a bowling ball has greater inertia than a table-tennis ball. Imagine hitting a bowling ball with a table-tennis paddle. The bowling ball would not move very much. Imagine hitting a table-tennis ball with the same amount of force. The table-tennis ball would move quite easily. ✓

What are Newton's laws of motion?

You have seen many examples of how forces change the motion of objects. Forces change the motion of objects in specific ways. The British scientist Sir Isaac Newton stated rules that describe the effects of forces on the motion of objects. These rules are known as Newton's laws of motion. They apply to all objects in motion, from billiard balls to planets orbiting the Sun.

What is Newton's first law of motion?

Newton's first law of motion states that an object moving at a constant velocity keeps moving at that velocity unless an unbalanced net force acts on it. An object at rest stays at rest unless a net force acts on it. Newton's first law of motion is sometimes called the law of inertia. ✓

What happens in a crash?

The law of inertia can explain what happens in a car crash. Imagine that a car traveling at 50 km/h crashes head-on into a stationary object. It crumples and comes to a stop within 0.1 s. Passengers who are not wearing safety belts continue to move forward at the same speed the car was traveling. Within about 0.02 s after the car stops, the unbelted passengers slam into the dashboard, steering wheel, windshield, or the backs of the front seats. They are still moving at the original speed of 50 km/h. This is about the same speed they would reach falling from a three-story building.

How do safety belts help?

A person who is wearing a safety belt will be attached to the car. The person will slow down as the car slows down. Safety belts also prevent people from being thrown out of cars. About half of the people who die in car crashes would survive if they wore safety belts. Thousands of others would suffer fewer serious injuries.

After You Read
Mini Glossary

balanced force: equal but opposite forces acting on an object
force: a push or pull on an object
inertia: the tendency of an object to resist a change in motion
net force: a combination of forces acting on an object

1. Review the terms and their definitions in the Mini Glossary. Choose two terms that are related and write a sentence using both terms.

   _____
   _____
   _____

2. Complete the chart below with information from this section.

Effects of a Car Crash

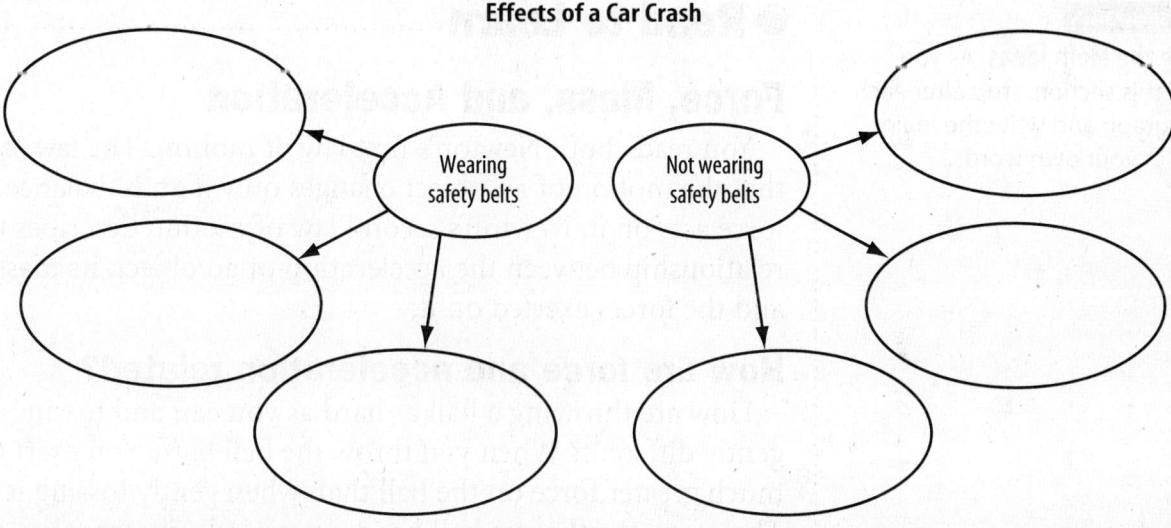

3. Think about what you have learned in this section. You wrote five quiz questions as you read the section. How did writing these questions help you learn the content of this section?

   _____
   _____
   _____

 Visit **gpscience.com** to access your textbook, interactive games, and projects to help you learn more about motion and forces.

Reading Essentials 35

Forces

section ❶ Weathering

What You'll Learn
- how force, mass, and acceleration are related
- the three different types of friction
- how air resistance affects falling objects

Study Coach

State the Main Ideas As you read this section, stop after each paragraph and write the main idea in your own words.

FOLDABLES

A Compare Use the Foldable to help you understand how the relationship of force and acceleration compares to the relationship of mass and acceleration.

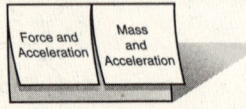

Before You Read

Have you ever seen a paper cup fall from a high shelf to the floor? This is a long distance for anything to fall. Yet, the paper cup is seldom damaged. Why do you think this is?

_____

_____

Read to Learn

Force, Mass, and Acceleration

You read about Newton's first law of motion. The law says that the motion of an object changes only if an unbalanced force acts on it. Newton's second law of motion describes the relationship between the acceleration of an object, its mass, and the forces exerted on it.

How are force and acceleration related?

How are throwing a ball as hard as you can and tossing it gently different? When you throw the ball hard, you exert a much greater force on the ball than when gently tossing it. Therefore, the thrown ball has a greater velocity when it leaves your hand than it does when you toss it. The ball you throw hard has a greater change in velocity. And, the change in velocity happens over a shorter period of time.

Remember that acceleration is the change in velocity divided by the time it takes for the change to happen. So, the ball you throw harder has greater acceleration because it has a greater change in velocity over a shorter period of time.

Are mass and acceleration related?

Imagine that you throw a softball and a baseball as hard as you can. Will they have the same speed? No, they will not. They will have different speeds because a softball has more mass than a baseball.

36 CHAPTER 3 Forces

The softball has less velocity after it leaves your hand, even though you applied the same amount of force. If it takes the same amount of time to throw both balls, then the softball would have less acceleration. This means that the acceleration of an object depends on its mass, as well as the force placed on it. As you can see, force, mass, and acceleration are all related.

Newton's Second Law

<u>Newton's second law of motion</u> states that the net force acting on an object causes the object to accelerate in the direction of the net force. To find the acceleration of an object, you need to know its mass and the net force that is acting on it. The acceleration of an object can be found using the following equation. ☑

$$\text{acceleration (meters/second}^2) = \frac{\text{net force (in newtons)}}{\text{mass (in kilograms)}}$$

$$a = \frac{F_{net}}{m}$$

Suppose you are pushing a friend on a sled. The mass of your friend and the sled together is 70 kg. The net force on the sled is 35 N. Find the acceleration of the sled.

$$a = \frac{35 \text{ N}}{70 \text{ kg}} = 0.5 \text{ m/s}^2$$

The acceleration of the sled is 0.5 m/s^2.

How can the second law help find net force?

If you know an object's mass and acceleration, you also can use Newton's second law to find the net force. Multiply both sides of the above equation by m to get the following equation.

$$F_{net} = ma$$

Use this equation to find net force. For example, think about a tennis serve. The tennis ball touches the racket for only a few thousandths of a second. The ball's velocity changes over a very short time. Suppose the ball leaves the racket with a speed of 100 km/h and has an acceleration of $5{,}000 \text{ m/s}^2$. The ball's mass is 0.06 kg. Here is how to find the force placed on the ball by the racket.

$$\begin{aligned} F_{net} &= ma \\ &= (0.06 \text{ kg})(5{,}000 \text{ m/s}^2) \\ &= 300 \text{ kg m/s}^2 \\ &= 300 \text{ N} \end{aligned}$$

Reading Check

1. **Define** What is Newton's second law of motion?

_____

_____

_____

Applying Math

2. **Demonstrate** Show how multiplying both sides of the acceleration equation, $a = \frac{F_{net}}{m}$, by m results in the net forct equation $F_{net} = ma$.

_____

_____

_____

Friction

Suppose you give a skateboard a push with your hand. Newton's first law of motion states that if the net force acting on an object is zero, the object will continue to move in a straight line with a constant speed. Does the skateboard keep moving with constant speed after it leaves your hand?

You know what happens. The skateboard slows down and soon stops. Remember that when an object slows down, its velocity is changing. If its velocity is changing, then it is accelerating. And if an object is accelerating, then a net force must be acting on it. So, what force is acting on the skateboard?

The force that brings the skateboard to a stop is friction. **Friction** is the force that opposes the sliding motion of two surfaces that are touching each other. The amount of friction between two surfaces depends on two things. The first is the kinds of surfaces that are touching. The second is the amount of force pressing the surfaces together.

What causes friction?

Something that seems very smooth, like polished metal, may actually be rough. You can see the dips and bumps on the surface when you look at it under a microscope. If two surfaces are pushed tightly together, welding, or sticking, happens where the bumps touch each other. These places are called microwelds. *Micro-* means "very small." Friction is caused by the microwelds that form where the surfaces are in contact.

What makes things stick together?

The greater the force pushing two surfaces together, the stronger the microwelds will be. More of the surface bumps will be touching each other. You can see microwelds in the figure below. To move one surface over another, a force must be applied to break these microwelds.

Think it Over

3. Infer Which would have more friction, a car traveling down a street or a toy remote-control car traveling on the same street? Why?

_____
_____
_____

Picture This

4. Observe Do surfaces touch more when there is less force or greater force?

_____
_____
_____

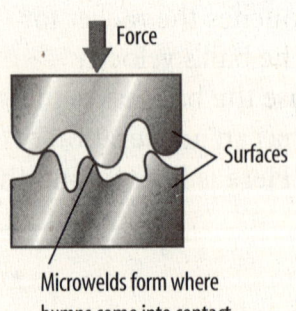

Microwelds form where bumps come into contact.

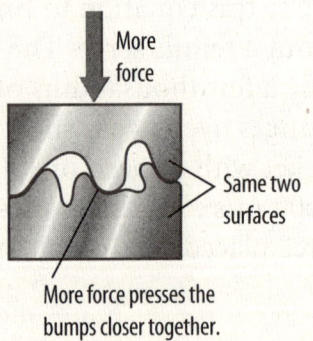

More force presses the bumps closer together.

What is static friction?

Suppose you fill a cardboard box with books, but it is too heavy to lift. You try to push it, but you cannot get it to move. This means there must be a force that is working against you. The force is friction caused by microwelds between the bottom of the box and the floor. This is called static friction. **Static friction** is the force that keeps two surfaces at rest from sliding across each other. In this case, your push is not large enough to break the microwelds between the box and floor. So, the box will not start moving. ✓

What is sliding friction?

Suppose you ask your friend to help move the box. As you and your friend push together, the box starts to move. The two of you have applied enough force to break the microwelds between the box and the floor. If you stop pushing, the box will stop moving. Another force, sliding friction, is opposing the motion of the box. **Sliding friction** is the force that works against the motion of two surfaces that are sliding across each other. Sliding friction is caused by the microwelds breaking and then forming again as the box slides along the floor.

Have you ever seen a car's wheels spinning in the mud or snow? There is sliding friction between the spinning wheels and the surface. Sand or gravel on the surface increases the sliding friction, so the wheels will stop slipping.

What is rolling friction?

A wheel digs into the surface that it is rolling over. This causes both the wheel and the surface to be deformed. There is static friction in the deformed area where the wheel and surface are in contact. This makes a frictional force called rolling friction. **Rolling friction** is the frictional force between a rolling object and the surface it rolls on.

Air Resistance

Objects falling toward Earth are being pulled downward by the force of gravity. **Air resistance** is a force that opposes the movement of objects through the air. Air resistance is similar to friction. You can feel air resistance on your face when you ride your bike very fast.

Like friction, air resistance acts in the direction opposite to the object's motion. In the case of a falling object, air resistance pushes up as gravity pulls down. If there were no air resistance, only gravity would affect falling objects. All objects would fall at the same rate.

Reading Check

5. **Complete** the sentence: The force that keeps two surfaces at rest from sliding over each other is _____.

Think it Over

6. **Compare and Contrast** How do static friction and sliding friction differ?

_____
_____
_____

Air resistance causes different objects to fall with different accelerations and different speeds. The amount of air resistance depends on an object's size and shape. Imagine dropping two identical plastic bags. One is crumpled into a ball and the other is spread out, resembling a parachute. When the bags are dropped, the crumpled bag falls faster than the spread-out bag. The downward force of gravity on both bags is the same. But, the upward force of air resistance on the crumpled bag is less. So, the net downward force on the crumpled bag is greater.

What is terminal velocity?

Imagine an object falling toward Earth. As the object falls, gravity causes it to accelerate. This causes the upward force of air resistance to increase. At some point, the upward force of air resistance becomes equal to the downward force of gravity. This means that the net force is zero. So, from this point on, the object will fall at a constant speed. This constant speed is called terminal velocity. Terminal velocity is the highest speed a falling object can reach. The terminal velocity of an object depends on its size, shape, and mass.

Look at the figure below. The air resistance force on an open parachute is much greater than the air resistance on the sky diver with a closed parachute. With the parachute open, air resistance increases. This makes the terminal velocity of the sky diver become small enough that the sky diver can land safely.

Picture This

7. **Compare** Measure the length of each arrow beside the sky diver. Compare the length of the gravity arrow to the lengths of the air resistance and net force arrows.

_____

_____

_____

Gravity = air resistance + net force

After You Read

Mini Glossary

air resistance: a force that opposes the movement of objects through air

friction: the force that opposes the sliding motion of two surfaces that are touching each other

Newton's second law of motion: the net force acting on an object causes the object to accelerate in the direction of the net force

sliding friction: the force that opposes the motion of two surfaces sliding past each other

static friction: the force that keeps two surfaces at rest from sliding over each other

1. Review the definition of air resistance in the Mini Glossary. Give an example of air resistance.

2. Complete the chart below to organize the different kinds of friction.

Describe static friction.	Give an example of static friction.
Describe sliding friction.	Give an example of sliding friction.
Describe rolling friction.	Give an example of rolling friction.

3. As you read this section, you wrote the main idea of each paragraph in your own words. How did you decide what the main idea of each paragraph was?

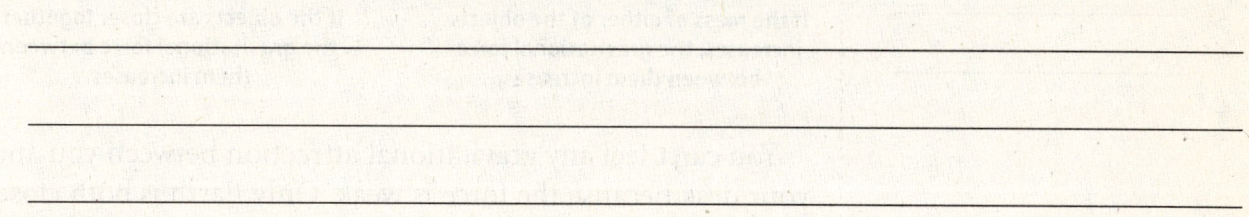 Visit **gpscience.com** to access your textbook, interactive games, and projects to help you learn more about Newton's second law of motion.

End of Section

chapter 3 Forces

section 2 Gravity

What You'll Learn
- how gravity and weight are related
- the difference between mass and weight

Before You Read

Think about what you know about gravity. Write what you know about gravity on the lines below.

_____

_____

Mark the Text

Identify the Main Point Highlight the main point in each paragraph. Use a different color to highlight details or examples that help explain the main point.

Read to Learn

What is gravity?

You have learned that objects falling toward Earth are being pulled downward by the force of gravity. There is also a gravitational attraction between you and your desk, you and your science book, and even between you and the planet Jupiter. Everything that has mass is attracted by the force of gravity. **Gravity** is an attractive force between two objects. Gravity increases as the mass of either object increases, or as the objects move closer to each other.

If the mass of either of the objects increases, the gravitational force between them increases.

If the objects are closer together, the gravitational force between them increases.

Picture This

1. **Interpret Illustrations** What two things can cause the gravitational force between objects to increase?

_____

_____

You can't feel any gravitational attraction between you and your desk because the force is weak. Only Earth is both close enough and has a large enough mass that you can feel its gravitational attraction. The Sun has much more mass than Earth. But it is too far away to exert a gravitational force that you could notice. Your desk is close, but it doesn't have enough mass to exert an attraction you can feel.

42 CHAPTER 3 Forces

The Law of Universal Gravitation

The law of universal gravitation lets us find the force of gravity between any two objects if their masses and the distance between them are known. Here is an equation for the law:

$$\text{gravitational force} = (\text{constant}) \times \frac{(\text{mass 1}) \times (\text{mass 2})}{\text{distance}^2}$$

$$F = G\frac{m_1 m_2}{d_2}$$

In this equation, G is the constant called the universal gravitational constant. A constant is an amount that never changes. The variable d is the distance between the two masses, m_1 and m_2.

Why is gravity called a long-range force?

The law of universal gravitation states that the gravitational force between two masses decreases as the distance between them increases. Suppose the distance between two objects increases from 1 m to 2 m. Then the gravitational force between them becomes one fourth as large. If the distance increases from 1 m to 10 m, the gravitational force between the objects is one hundredth as large. No matter how far apart two objects are, the gravitational force between them never completely goes to zero. Gravity is called a long-range force. No matter how long the distance is between two objects, gravity never disappears.

How did gravity help astronomers find other planets?

The motion of every planet in our solar system is affected by the gravitational forces of the other planets. In the 1840s, the farthest known planet was Uranus. Its motion could not completely be explained by the gravitational attraction of the other known planets. Was another planet affecting its motion? Using the law of universal gravitation and Newton's laws of motion, astronomers discovered another planet, Neptune.

Earth's Gravitational Acceleration

If you dropped a marble and a bowling ball at the same time, which one would hit the ground first? Suppose the effects of air resistance were small enough to be ignored. When all forces except gravity can be ignored, a falling object is said to be in free fall. If there were no air resistance, all objects near Earth's surface would fall with the same acceleration. The marble and the bowling ball would hit the ground at the same time.

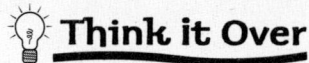 **Think it Over**

2. Determine What do the 1 and 2 in $m_1 m_2$ mean?

_____

_____

_____

FOLDABLES

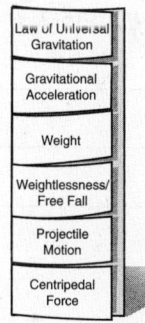 **Organize Information** Make the following 6-tab Foldable to help you organize information about gravity.

 **Reading Check**

3. Communicate On the lines below, write why some astronomers thought there might be a planet beyond Uranus.

_____

_____

_____

The acceleration of an object in free fall is about 9.8 m/s². This acceleration is sometimes called the acceleration of gravity. It is given the symbol g. The force of Earth's gravity on a falling object is the object's mass times the acceleration of gravity. This can be expressed by the following equation.

force of gravity (N) = **mass** (kg) × **acceleration of gravity** (m/s²)
$$F = mg$$

You can use this equation to find the gravitational force on a sky diver with a mass of 60 kg.

$$F = mg = (60 \text{ kg})(9.8 \text{ m/s}^2) = 588 \text{ N}$$

The gravitational force on the sky diver is 588 N.

How is weight calculated?

Earth always exerts a gravitational force on objects. The gravitational force exerted on an object is its **weight**. Weight is found using the following equation. The letter g represents the acceleration of gravity.

weight (N) = **mass** (kg) × **acceleration of gravity** (m/s²)
$$W = mg$$

Is there a difference between weight and mass?

Weight and mass are not the same. Weight is a force. Mass is a measure of the amount of matter in an object. But weight and mass are related. The greater an object's mass, the stronger the gravitational force between the object and Earth. So, weight increases as mass increases. An object has the same mass on Earth and as on the moon. However, the gravitational force of Earth is greater than the moon's. So an object weighs more on Earth than on the moon.

Weightlessness and Free Fall

Suppose you are standing on a scale in an elevator that is not moving. The scale would record your normal weight. But, what would happen if you were standing on the scale and the elevator were falling rapidly? If you and the scale were in free fall, then you would no longer push down on the scale. The dial would say you have zero weight, even though your weight has not changed.

You may have seen pictures of astronauts floating inside a space shuttle. They are experiencing weightlessness. The astronauts are not really weightless. However, they do weigh less because they are farther from Earth's gravitational pull.

Applying Math

4. **Calculate** Find the weight of a 50-kg person on Earth. Remember, g = 9.8m/s². Show your work.

Reading Check

5. **Explain** Why do astronauts in the space shuttle weigh less?

An orbiting space shuttle is in free fall. It is falling around Earth, not straight downward. Everything in the space shuttle is falling at the same rate, much like the way you and the scale were falling in the elevator. Objects in the shuttle seem to be floating because they are all falling with the same acceleration.

Projectile Motion

You probably have noticed that thrown objects do not always move in a straight line. Their path curves downward. Anything that is thrown or shot through the air is called a projectile. Earth's gravity causes projectiles to follow a downward, curved path.

What happens when an object has both horizontal and vertical motion?

When you throw a ball, the force exerted on the ball by your hand pushes the ball forward. This force gives the ball horizontal motion. When you let go of the ball, gravity pulls it downward, giving it vertical motion. The ball has constant horizontal velocity, but increasing vertical velocity. Gravity exerts an unbalanced force on the ball. It changes the direction of the ball's path from only forward, to forward and downward. The result of these two motions is that the ball moves in a downward curve.

Are horizontal and vertical distance always the same?

Suppose an automatic ball machine launches a ball in a horizontal direction from 1 m above the ground. Would it take longer to reach the ground than if you dropped the ball from the same height? Surprisingly, it would not. Look at the figure below. A dropped ball and one thrown horizontally from the same height will hit the ground at the same time. Both balls travel the same vertical distance in the same amount of time. However, the ball thrown horizontally travels a greater total distance than the ball that is dropped.

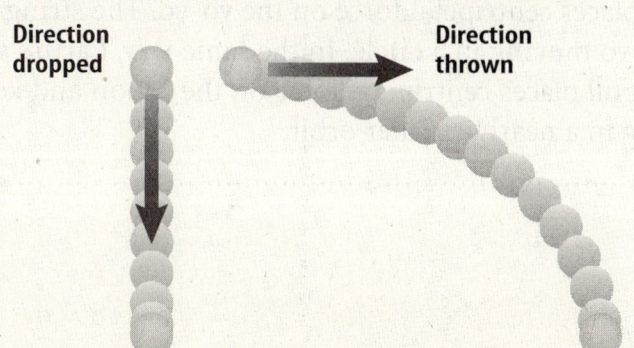

✓ Reading Check

6. **Describe** What does gravity do to the path of an object that is thrown horizontally?

_____

_____

Picture This

7. **Interpret a Figure** Which ball has fallen the greater vertical distance in one second, the ball that was dropped or the ball that was thrown?

_____

_____

Centripetal Force

Look at the path of the ball as it travels through the curved tube in the figure below. When the ball enters a curve, even if its speed does not change, it is accelerating. This is because its direction is changing. When the ball goes through a curve, the change in its direction is toward the center of the curve. Acceleration toward the center of a curved path is called **centripetal acceleration**.

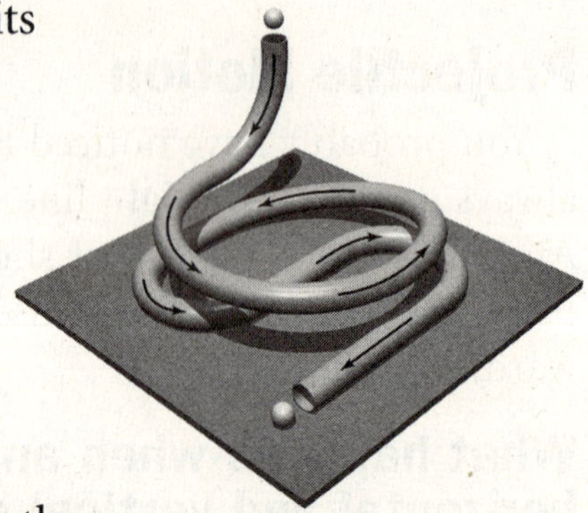

The ball has centripetal acceleration. So, according to the second law of motion, the direction of the net force on the ball must be toward the center of the curved path. The net force exerted toward the center of a curved path is called a **centripetal force**. An object that moves in a circle is doing so because a centripetal force is acting on it in a direction toward the center. The centripetal force is the force exerted by the walls of the tube on the ball.

How does centripetal force depend on traction?

When a car rounds a curve, a centripetal force must be acting on the car to keep it moving in a curved path. If it does not, the car will slide off the road. This centripetal force is the frictional force, or traction, between the tires and the road. The road may be slippery because of rain or ice. As tires get older, they get smoother and their traction decreases. If either or both of these situations occur, the car may slide in a straight line and not follow the curve. This is because the centripetal force, traction, is not strong enough to keep the car moving around the curve.

Can gravity be a centripetal force?

Imagine swinging a yo-yo on a string above your head. The string places centripetal force on the yo-yo. The string keeps the yo-yo moving in a circle. In the same way, Earth's gravitational pull places centripetal force on the Moon and keeps it moving in a nearly circular orbit.

Picture This

8. **Trace** the path of the ball at right with a pen. Try to move the pen at a constant speed.

Think it Over

9. **Describe** a real-world situation that involves centripetal force.

_____

_____

● **After You Read**

Mini Glossary

centripetal acceleration: acceleration toward the center of a curved path

centripetal force: a force that moves an object in the direction of the center of a curved path

gravity: a force that pulls two objects together

weight: the gravitational force placed on an object

1. Review the terms and their definitions in the Mini Glossary. Explain the difference between the mass of an object and its weight.

   _____
   _____
   _____
   _____

2. Choose one of the question headings in the Read to Learn section. Write the question in the space below. Then write your answer to the question.

Question:

   Answer: _____
   _____
   _____

3. Think about what you have learned in this section. How did identifying the main point and supporting details of each paragraph help you learn the new material?

   _____
   _____
   _____

Science Online Visit **gpscience.com** to access your textbook, interactive games, and projects to help you learn more about gravity.

End of Section

Reading Essentials **47**

Forces

section ❸ The Third Law of Motion

What You'll Learn
- Newton's third law of motion
- how to find momentum

Study Coach

Create a Quiz As you read this section, write five questions that could be included on a quiz. Be sure to include the answer.

FOLDABLES

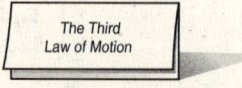

● **Finding Main Ideas** While reading this section, make a Foldable to help you identify the main ideas in Newton's third law of motion.

● **Before You Read**

Think about a time you reacted to someone else's action. For example, a time when you smiled because someone complimented you. Another one of Newton's laws of motion says that to every action there is an equal and opposite reaction. Write about an action/reaction situation you have seen.

_____

_____

● **Read to Learn**

Newton's Third Law

If you push against a wall while wearing in-line skates, you will roll backwards. The action of pushing against the wall produced a reaction—moving backward. This is an example of Newton's third law of motion.

Newton's third law of motion describes action and reaction. It states that when one object applies a force on a second object, the second object applies a force on the first object that is equal in strength and opposite in direction. In other words "to every action force there is an equal and opposite reaction force."

How do action and reaction happen?

When a force is applied in nature, a reaction force occurs at the same time. When you jump on a trampoline, you exert a downward force on the trampoline. At the same time, the trampoline exerts an equal upward force on you.

How do action and reaction make you move?

The action and reaction forces are equal. So how do you move by walking if each time you push on the ground, Earth pushes back with an equal force? The forces are acting on objects that have different masses. Earth has more mass than you do. Even though the forces are equal, their net force is not equal. Unequal net forces determine the direction you move.

48 CHAPTER 3 Forces

What is rocket propulsion?

Suppose you are standing on skates holding a softball. You exert a force on the softball when you throw it. Newton's third law says the softball exerts a reaction force on you. This force pushes you in the direction opposite the softball's motion.

Rockets use this same principle to move. In a rocket engine, burning fuel produces hot gases. The rocket engine applies a force on the gases and causes them to escape out of the back of the rocket. By Newton's third law, the gases apply a reaction force on the rocket and push it in the opposite direction.

Momentum

A moving object has a property called momentum. Momentum is related to how much force is needed to change an object's motion. The **momentum** of an object is the product of its mass and its velocity. Momentum can be found using the following equation. The symbol p represents momentum. The unit for momentum is kg·m/s.

$$\text{momentum (kg·m/s)} = \text{mass (kg)} \times \text{velocity (m/s)}$$
$$p = mv$$

Two cars can have the same velocity. But the bigger car has more momentum, because it has more mass. An archer's arrow can have a large momentum because of its high velocity, even though its mass is small. A walking elephant may have a low velocity, but because of its large mass, it has a large momentum.

Suppose a sprinter with a mass of 80 kg has a speed of 10 m/s. What is the sprinter's momentum? Substitute the known values into the momentum equation.

$$p = mv$$
$$= (80 \text{ kg})(10 \text{ m/s})$$
$$= 800 \text{ kg·m/s}$$

The sprinter's momentum is 800 kg·m/s.

How are force and momentum related?

Recall that acceleration is the difference between final and initial velocity, divided by the time. Also recall that the net force on an object is its mass times its acceleration. When you combine these two relationships, you get the following equation.

$$F = \frac{(mv_f - mv_i)}{t}$$

FOLDABLES

D Finding Main Ideas Use two quarter-sheets of notepaper to organize notes on momentum and the law of conservation of momentum.

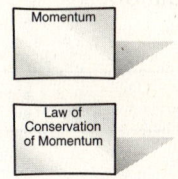

Think it Over

1. **Compare** Which has more momentum, a car traveling at 12 km/h or a bicycle traveling at the same speed? Explain why.

   _____
   _____
   _____

Applying Math

2. **Apply** What is the momentum of a bicycle with a mass of 18 kg traveling at 20 m/s?

   _____
   _____
   _____

Reading Check

3. Define What are mv_f and mv_i in the net force equation?

_____
_____
_____

Think it Over

4. Recognize Cause and Effect In a game of pool, why will the balls eventually stop after a collision?

_____
_____
_____

In the equation, mv_f is final momentum and mv_i is initial momentum. This equation shows that net force exerted on an object is its change in momentum divided by the time over which the change occurs.

When you catch a ball, your hand applies a force on the ball that stops it. The force your hand exerts on the ball and the force the ball exerts on your hand are equal. The force depends on the mass and initial velocity of the ball and how long it takes to come to a stop. The ball's final velocity is zero.

What is the law of conservation of momentum?

Momentum can be passed from one object to another. When a cue ball hits a group of balls that are motionless, the cue ball slows down and the other balls move. The momentum that the group of balls gained equals the momentum that the cue ball lost. But the total momentum of all the balls before and after the collision is the same. Total momentum has not been lost, nor has new momentum been created. This is an example of the law of conservation of momentum. If a group of objects applies forces on each other, their total momentum does not change.

What happens when objects collide?

In a game of pool, suppose one ball is moving in one direction, and another ball moving the same direction strikes it from behind. The ball that is struck will continue to move in the same direction, but more quickly. The striking ball has given it more momentum in the same direction.

What if two balls of equal mass are moving toward each other with the same speed? They would have the same momentum, but in opposite directions. If the balls collided, each would reverse direction, and move with the same speed as before the collision.

● After You Read
Mini Glossary

momentum: the product of a moving object's mass and velocity

Newton's third law of motion: to every action force there is an equal and opposite reaction force

1. Review the terms and their definitions in the Mini Glossary. Describe a real-world example of Newton's third law of motion.

   _____

   _____

   _____

2. Match the terms with the correct statements. Put the letter of the statement in Column 2 on the line in front of the term it matches in Column 1.

   Column 1

   _____ 1. rocket propulsion

   _____ 2. momentum

   _____ 3. conservation of momentum

   _____ 4. Newton's third law of motion

   Column 2

   a. To every action force there is an equal and opposite reaction.

   b. Momentum cannot be created or destroyed.

   c. the product of a moving object's mass and velocity

   d. An engine applies a force on hot gases and the gases apply a force in the opposite direction.

3. You created quiz questions to help you learn the material in this section. How can you use these questions to help you prepare for a test on the whole chapter?

   _____

   _____

   _____

Science Online Visit **gpscience.com** to access your textbook, interactive games, and projects to help you learn more about Newton's third law of motion.

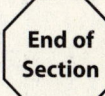

End of Section

Energy

section 1 The Nature of Energy

What You'll Learn
- the different forms of energy
- how energy can be stored

Study Coach

State the Main Ideas As you read this section, stop after each paragraph and write the main idea of what you read in your own words.

Before You Read

Energy comes in many forms. List as many types of energy as you can think of on the lines below.

_____

_____

Read to Learn

What is energy?

Changes are taking place all around you all the time. For example, lightbulbs are heating the air around them, and the wind may be blowing leaves in the trees. Even you are changing as you breathe, blink, or move around in your desk.

All changes involve energy. Imagine a baseball flying through the air. It hits a window and breaks the glass into small pieces. The moving baseball causes the solid pane of glass to change into the small pieces. The moving baseball has energy.

Does change require energy?

Energy is the ability to cause change. The moving baseball caused the glass to change, so the baseball had energy. Anything that causes change must have energy. For example, you use energy when you comb your hair. You also use energy when you move between classes, open a book, or write with a pen. You even use energy when you yawn or sleep. ✓

Reading Check

1. **Define** What is energy?

_____

_____

What are some different forms of energy?

Turn on an electric light, and a dark room becomes brighter. Turn on your CD player, and sound comes through the headphones. In both of these cases, energy moves from one place to another. These changes are different from each other. They are also different from the change caused by the baseball hitting the window. Energy has many different forms. Some forms are electrical, chemical, radiant, and thermal.

| Electrical energy | Chemical energy | Radiant energy |

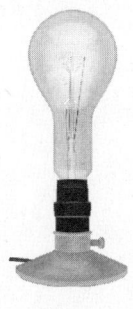  

The figure shows some forms of energy and some objects associated with these forms of energy. The lightbulb uses electrical energy to light a room. Chemical energy is stored in the food you eat and in the fuel in a car. Radiant energy from the Sun travels to Earth and warms the planet. Energy plays a role in every activity.

How is energy like money?

Suppose you have $100. You could have one hundred-dollar bill, two fifty-dollar bills, 100 one-dollar bills, or 10,000 pennies. You could start with the $100 in one form and change it into another form. But, no matter what form it is in, it is still $100. This is also true of energy. It is the same no matter what form it is in. Energy from the Sun that warms you and the energy from the food you eat are just different forms of the same thing—energy! ✓

Kinetic Energy

When you think of energy, you might think of moving objects. An object in motion, like the baseball, does have energy. **Kinetic energy** is the energy a moving object has because of its motion. The kinetic energy of a moving object depends on the object's mass and speed. You can find the kinetic energy of an object using the following equation.

kinetic energy (joules) = $\frac{1}{2}$**mass** (kg) × [**speed** (m/s)]^2

$$KE = \frac{1}{2}mv^2$$

Energy is measured using the SI unit called the **joule**. The letter J stands for joule. If you drop a baseball from about 0.5 m, it will have a kinetic energy of about 1 joule, or 1 J.

Picture This
2. **Illustrate** In the space below, draw another representation of one of the types of energy shown in the figure.

Reading Check
3. **Explain** how energy and money are alike.

_____
_____
_____

FOLDABLES

A **Find Main Ideas** Fold a piece of paper into 12 sections and label. Fill in the main ideas about kinetic and potential energy.

The Nature of Energy	Define	Examples of...	Calculate
Kinetic Energy			
Potential Energy			

Reading Essentials **53**

Applying Math

4. **Explain** Look at the problem about the jogger. Why does (3 m/s)² become (9 m²/s²)?

_____

_____

_____

Think it Over

5. **Apply** Describe a situation where potential energy becomes kinetic energy.

_____

_____

_____

FOLDABLES

B Compare and Contrast Make the following Foldable to compare and contrast the properties of different types of potential energy.

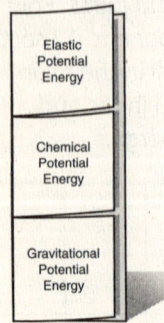

A jogger whose mass is 60 kg is moving at a speed of 3 m/s. Use the equation to find the jogger's kinetic energy.

$$KE = \frac{1}{2}mv^2$$
$$= \frac{1}{2}(60 \text{ kg})(3 \text{ m/s})^2$$
$$= \frac{1}{2}(60 \text{ kg})(9 \text{ m}^2/\text{s}^2)$$
$$= \frac{1}{2}(540 \text{ kg} \cdot \text{m}^2/\text{s}^2)$$
$$= 270 \text{ kg} \cdot \text{m}^2/\text{s}^2$$

The kinetic energy of the jogger is 270 J.

Potential Energy

An object with energy does not have to be moving. Objects that are at rest have stored energy. For example, an apple that is hanging from a tree has stored energy. If the apple stays in the tree, it will keep its stored energy because of its position above the ground. If the apple falls to the ground, a change happens. Because the apple can cause change, it has energy. Stored energy due to position is called **potential energy**. The stored energy of position, potential energy, will change to energy of motion, kinetic energy, when the apple falls.

What is elastic potential energy?

Energy can be stored in other ways, too. Suppose you take a rubber band and stretch it. If you let the rubber band go, it will fly through the air. Where did this kinetic energy come from? The stretched rubber band had something called elastic potential energy. **Elastic potential energy** is energy stored by an object that can stretch or shrink, like a rubber band or a spring. When you let the rubber band fly through the air, its elastic potential energy becomes kinetic energy.

What is chemical potential energy?

Where does your body get the energy to make it move? The food that you eat each day has stored energy. To be more exact, food's energy is stored in chemical bonds between atoms. Natural gas stores energy in the same way. Energy stored in chemical bonds is called **chemical potential energy**. In natural gas, energy is stored in the bonds that hold the carbon and hydrogen atoms together. This energy is released when the gas is burned.

What is gravitational potential energy?

Anything that can fall has stored energy called gravitational potential energy. **Gravitational potential energy** (GPE) is energy that is stored by objects that are above Earth's surface. The apple in the tree has GPE. The GPE of an object depends on two things—the object's mass and its height above the ground. Gravitational potential energy can be found using the following equation.

gravitational potential energy (J) = mass (kg) × acceleration of gravity (m/s²) × height (m)

$$GPE = mgh$$

On Earth, the acceleration of gravity is 9.8 m/s² and has the symbol g. Like all forms of energy, gravitational potential energy is measured in joules.

Suppose a ceiling fan has a mass of 7 kg and is 4 m above the ground. What is the gravitational potential energy of the ceiling fan?

$$GPE = mgh$$
$$= (7 \text{ kg})(9.8 \text{ m/s}^2)(4 \text{ m})$$
$$= 274 \text{ kg m}^2/\text{s}^2$$

The ceiling fan has a GPE of 274 kg m²/s², or 274 J.

How is potential energy stored?

So far, you have studied three types of potential energy. The table below lists the ways each is stored.

Potential Energy	Way It Is Stored
Elastic	stored by an object that can stretch or shrink
Chemical	stored in chemical bonds
Gravitational	stored by objects that are above Earth's surface

How does GPE change?

By looking at the equation for gravitational potential energy, you can see that two things can change an object's gravitational potential energy. The acceleration of gravity, g, is always 9.8 m/s². It is the constant. So, the two factors in the equation that can change are mass, m, and height, h. They are the variables.

So if you change the mass or height of an object, its gravitational potential energy will also change.

Reading Check

6. **List** two things that determine an object's GPE.

Applying Math

7. **Interpret** In the formula $GPE = mgh$, which symbols represent the constants and which symbols represent the variables?

Look at the vase near the bottom of the bookcase. If you fill the vase with water, you increase its GPE by increasing its mass. If you move the vase to a higher shelf, you also increase its GPE by increasing its height. The gravitational potential energy of an object can increase if you change its mass or move the object higher above the ground.

If two objects are at the same height, then the object with the greater mass will have more GPE. If two objects have the same mass, the one that is higher above the ground will have the greater GPE.

Picture This

8. Reasoning Which probably has a greater GPE, a feather on a high shelf or a large book on the next shelf down? Explain your reasoning.

_____
_____
_____

What does GPE change into?

What would happen if the vase on the top shelf fell? As the vase falls, it starts moving. It now has both GPE and kinetic energy. As the vase gets closer to the ground, its GPE decreases. At the same time, its kinetic energy increases. The GPE changes into kinetic energy.

Look at the two vases in the figure. If the vase on the top shelf falls, it will start with more GPE and end with more kinetic energy when it hits the ground. This is why a vase that falls from a high shelf is more likely to break than a vase that falls from a lower shelf.

Reading Check

9. Determine As an object falls, what does its gravitational potential energy change to? (Circle your choice.)

a. chemical energy

b. kinetic energy

c. thermal energy

d. radiant

After You Read
Mini Glossary

chemical potential energy: energy stored in chemical bonds
elastic potential energy: energy stored by things that stretch or shrink
gravitational potential energy: energy stored by things that are above Earth
joule: the standard unit for measuring energy
kinetic energy: energy in the form of motion
potential energy: energy stored in a motionless object

1. Review the terms and their definitions in the Mini Glossary. Explain the difference between kinetic energy and potential energy.

   _____

   _____

2. Complete the chart below. Fill in the first column with the three kinds of potential energy. Fill in the second column with an example of something that stores each type of potential energy.

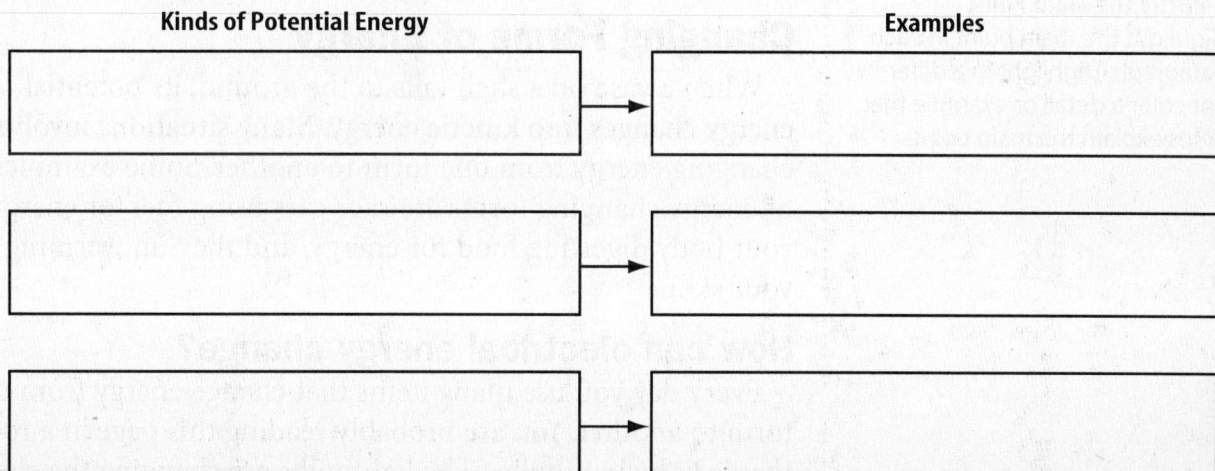

Kinds of Potential Energy	Examples

3. Think about what you have learned. How did you decide what was the main idea of each paragraph?

   _____

   _____

 Visit **gpscience.com** to access your textbook, interactive games, and projects to help you learn more about the nature of energy.

chapter 4 Energy

section 2 Conservation of Energy

What You'll Learn
- how energy can be changed from one form to another
- how energy is conserved

● **Before You Read**

The motion of objects appear to change all the time. Imagine a person swinging on a swing. Explain the person's motion while swinging.

_____

_____

_____

Mark the Text

Identify the Main Point Highlight the main point in each paragraph. Highlight in a different color a detail or example that helps explain the main point.

● **Read to Learn**

Changing Forms of Energy

When a vase on a shelf falls to the ground, its potential energy changes into kinetic energy. Many situations involve changing energy from one form to another. Some examples of energy changing forms are race cars using fuel for energy, your body digesting food for energy, and the Sun warming your skin.

How can electrical energy change?

Every day you use many items that change energy from one form to another. You are probably reading this page in a room that is lit by lightbulbs. The lightbulbs are changing the electrical energy that they receive into light energy.

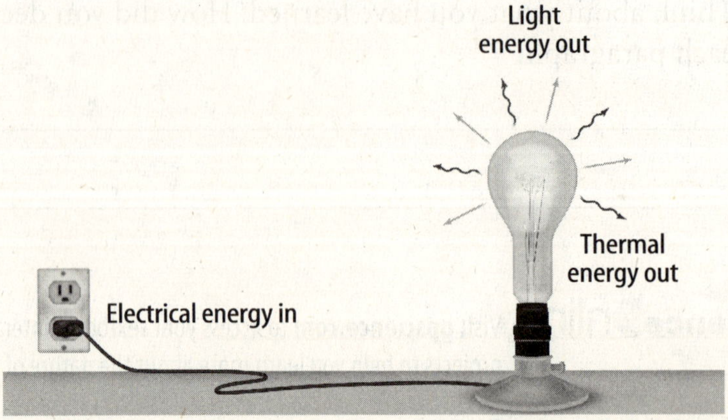

Picture This

1. **Relate** Name other items that change electrical energy into heat.

   _____

   _____

Light energy out

Thermal energy out

Electrical energy in

58 CHAPTER 4 Energy

The heat you can feel around the bulb lets you know that some of the electrical energy also is changing to thermal energy, as shown in the figure on the previous page.

How can chemical energy change?

Chemical energy can be changed into kinetic energy. A fuel, such as gasoline, stores energy in the form of chemical potential energy. Cars and buses usually run on gasoline. As shown in the figure, an electrical spark causes a small amount of fuel in the engine to burn. This changes the chemical potential energy stored in the gasoline molecules into thermal energy. The thermal energy heats up gases and they expand. The expanding gases cause parts of the car to move. The moving parts have kinetic energy. Chemical energy has been changed into thermal energy and then into kinetic energy.

✓ **Reading Check**

2. **List** three forms of energy that are involved in the running of a car engine.

_____

_____

_____

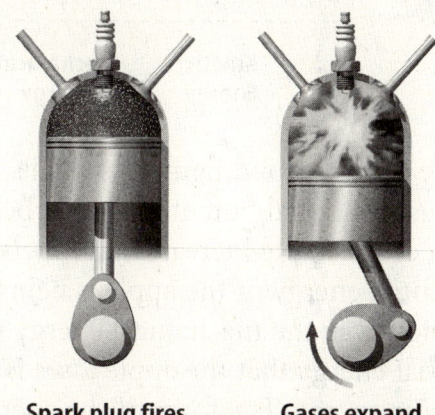

Spark plug fires   **Gases expand**

Picture This

3. **Identify** Circle the moving parts in the figure. What type of energy is this?

_____

Do all energy changes result in motion?

Not all changes in energy result in motion that can be seen. Nor do they result in sound, heat, or light. For example, every green plant changes light energy from the Sun into chemical energy that is stored in the plant. When you eat an ear of corn, the chemical potential energy in the corn is changed to other forms of energy by your body.

Changes Between Kinetic and Potential Energy

Recall that a rubber band has elastic potential energy. When a stretched rubber band is let go, the potential energy is changed into kinetic energy. **Mechanical energy** is the total amount of potential and kinetic energy in a system. The mechanical energy of the rubber band is the total of its potential energy and its kinetic energy at any one time. Mechanical energy comes from where an object is and the movement of the object.

FOLDABLES

C **Collect Information** Make note cards from two half-sheets of paper as shown. Write on each note card what you learn about mechanical energy and the law of conservation of energy.

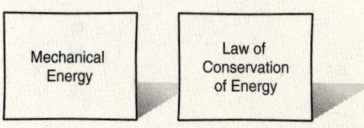

Reading Essentials **59**

Think it Over

4. Explain Does the mechanical energy of an object falling from a shelf change? Explain.

_____
_____
_____

Picture This

5. Interpret What do the arrows in the figures mean?

_____
_____
_____

Does mechanical energy of an object change?

What happens to the mechanical energy of an object as its potential energy is changed into kinetic energy? Look at the apple in the tree below. It has gravitational potential energy because Earth is pulling down on it. The apple does not have kinetic energy while it hangs in the tree. The apple's gravitational potential energy and its mechanical energy are the same.

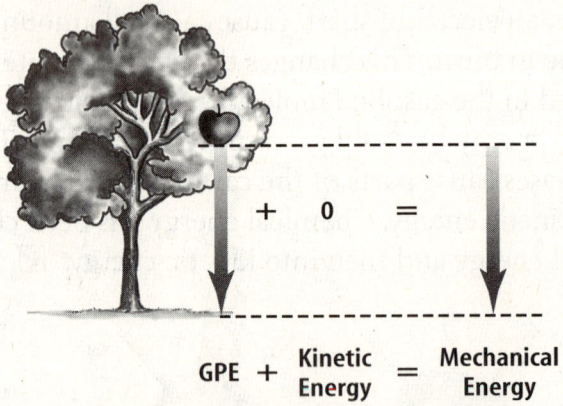

Look at the apple in the second figure. As it falls, the apple loses height, so its gravitational potential energy becomes less. As the velocity of the apple increases, its kinetic energy increases. The potential energy of the apple is being changed to kinetic energy. However, the mechanical energy will not change. The potential energy that the apple loses is being gained back as kinetic energy. The form of the energy changes, but the total amount of the energy remains the same.

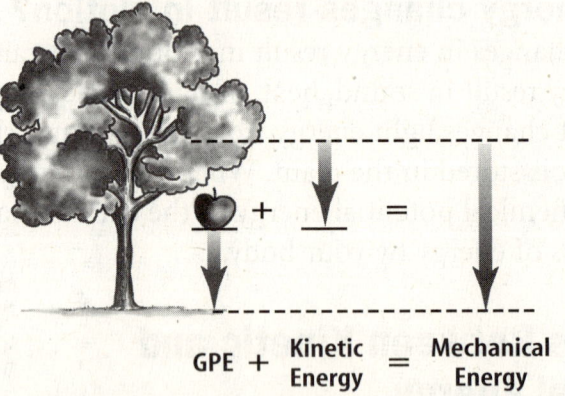

This can be explained using the mechanical energy equation. As one value of the mechanical energy equation decreases, the other value increases by the same amount. The sum of the two stays the same. Therefore, the mechanical energy of an object stays the same.

How does energy change in projectile motion?

Energy changes also occur during projectile motion. During projectile motion, an object moves through the air in a curved path. Look at the figure below. As the ball leaves the bat, it has mostly kinetic energy. As the ball rises, its gravitational potential energy becomes greater, but its kinetic energy becomes less due to decreasing velocity. As the ball falls, its gravitational potential energy becomes less, but its kinetic energy becomes greater due to increasing velocity. However, the total mechanical energy of the ball does not change as the ball moves through the air.

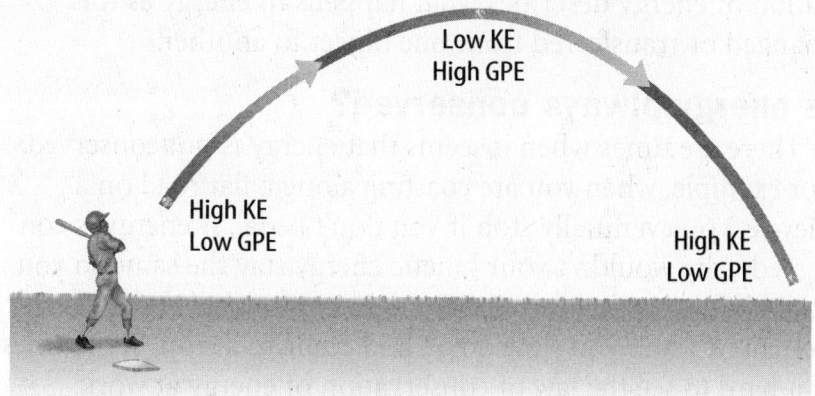

Picture This
6. Observe What does the symbol KE represent in the figure?

What happens to energy during a swing?

When you ride on a swing, part of the fun is the way you feel just as you drop from the highest part of the swing's path. Energy is constantly changing during this ride on a swing. The push that gets you moving is kinetic energy. As the swing rises, you lose speed. This means the kinetic energy is changing into gravitational potential energy. At the top of the path, the GPE is its greatest and the kinetic energy is at its lowest. As the swing starts its downward path and its speed increases, the GPE changes into kinetic energy. At the bottom of the swing's path, the kinetic energy is greatest and the GPE is at its lowest. As you swing back and forth, kinetic and potential energy are constantly changing back and forth.

Think it Over
7. Analyze Why is kinetic energy at its lowest at the top of the swing?

The Law of Conservation of Energy

As a batted ball speeds up or slows down, its kinetic and potential energy are always changing. But the amount of mechanical energy always stays the same. The kinetic and potential energy continually change form back and forth and no energy is destroyed.

✔ **Reading Check**

8. **Explain** What does the law of conservation of energy state?

_____

_____

_____

💡 **Think it Over**

9. **Describe** a time when you may not see the law of conservation of energy.

_____

_____

_____

_____

This is true for all forms of energy. Energy can change from one form to another, but the total amount of energy never changes. Another way to say this is that energy is conserved. The <u>law of conservation of energy</u> states that energy cannot be created or destroyed. This means that the total amount of energy in the universe is always the same. It just changes from one form to another. ✔

You might have heard the phrase *energy conservation* before. Conserving energy means reducing the need for energy so we use fewer energy resources such as coal and oil. This is not the same as the law of conservation of energy. The law of conservation of energy describes what happens to energy as it is changed or transferred from one object to another.

Is energy always conserved?

There are times when it seems that energy is not conserved. For example, when you are coasting along a flat road on a bicycle, you eventually stop if you don't pedal. If energy is conserved, why wouldn't your kinetic energy stay the same so you could keep coasting forever? It might appear that energy is destroyed when you slow down and come to a stop. Sometimes is it hard to see the law of conservation of energy at work.

How does friction affect energy?

Suppose you are swinging on a swing. If you stop pumping and no one is pushing you, you will soon stop swinging. It would seem that the mechanical (kinetic and potential) energy of the swing is lost. Wouldn't this go against the law of conservation of energy?

If the energy of the swing decreases, then the energy of some other object must increase by the same amount. What object has an energy increase? With every motion, the swing's ropes or chains rub on their hooks, and air pushes on the rider. Friction and air resistance cause the temperature of the hooks and air to increase a little. The mechanical energy is not destroyed. It is changed into thermal energy. So, the total amount of energy stays the same—it just changes forms.

Where does the Sun get its energy?

Have you ever wondered how the Sun gives off enough energy to light and warm Earth? The Sun and other stars have a special way of changing their energy. It is called nuclear fusion. Nuclear fusion is the reaction that takes place when nuclei join together. Nuclear fusion uses the law of conservation of energy when it changes the potential energy of a small amount of mass into a huge amount of energy.

What is nuclear fission?

Another way to change a small amount of mass into a huge amount of energy is through nuclear fission. Both nuclear fusion and nuclear fission involve the nuclei of matter. In nuclear fusion, the nuclei are fused, or joined. The nuclei are not fused in nuclear fission. They are broken apart. In either process, fusion or fission, mass is changed into energy. In both processes, the total amount of energy is conserved if the energy content of the masses used are included. Nuclear fission is how nuclear power plants produce energy. The figure below shows nuclear fission. In both nuclear fusion and nuclear fission, mass is changed into energy. In nuclear fission, the mass of the large nucleus on the left is greater than the combined mass of the other two nuclei and the neutrons. But once again, the total amount of mass and energy does not change during these reactions. ✓

Reading Check

10. Compare and contrast nuclear fusion and nuclear fission.

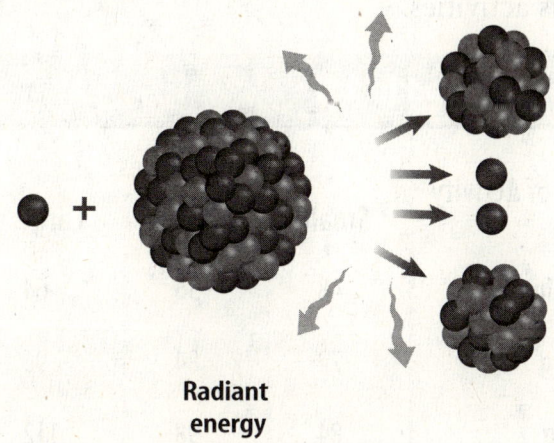

Radiant energy

Picture This

11. Determine How can you determine that this figure models fission, not fusion?

Energy and the Human Body

With your right hand, reach up and feel your left shoulder. With that simple move, stored potential energy in your body is changed to kinetic energy when you move your arm. Does your shoulder feel warm? Some of the stored chemical potential energy also is being used to keep your body at about the same temperature. Do you feel warmer if someone wraps their arms around you? Some of the body's potential energy is changed into heat that the body gives off to it's surroundings.

How does your body store and use energy?

Your body requires energy to stand still. The law of conservation of energy applies to the chemical and physical changes that are going on in your body. Your body stores potential energy in the form of fat and other chemical compounds. The potential energy is the fuel for processes such as the beating of your heart, digesting of food, and moving muscles.

What are Calories?

The potential energy stored in your body comes from the food you eat. Your body breaks down the food you eat into molecules that can be used as fuel. The chemical potential energy in these molecules supplies the cells of your body with the energy they need to function. Your body also can use the chemical potential energy stored in fat for its energy needs.

The food Calorie (C) is a unit nutritionists use to measure how much energy you get from different foods. One Calorie is equal to about 4,184 J. These Calories produce the energy needed by your body. Look at the labels on food packages. They provide information about the Calories contained in a serving, as well as the amount of protein, fat, and carbohydrates. To maintain a healthy weight, you must have a proper balance between the food you eat and the energy your body uses. The table below shows the amount of energy used in doing various activities.

Applying Math

12. **Displaying Data** On the graph below, make a bar graph comparing the number of calories used by a medium-framed person for the following activities: standing, walking, playing tennis, and bicycling.

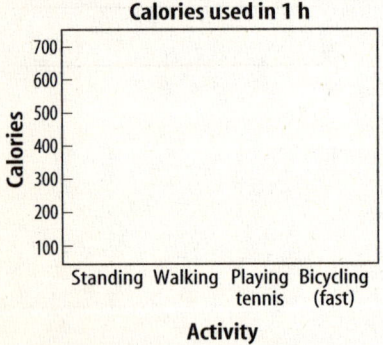

Calories Used in 1 h			
Type of Activity	**Body Frames**		
	Small	Medium	Large
Sleeping	48	56	64
Sitting	72	84	96
Eating	84	98	112
Standing	96	112	123
Walking	180	210	240
Playing tennis	380	420	460
Bicycling (fast)	500	600	700
Running	700	850	1,000

After You Read

Mini Glossary

law of conservation of energy: energy may change from one form to another, but the total amount of energy never changes.

mechanical energy: the total amount of potential and kinetic energy in a system

1. Review the terms and their definitions in the Mini Glossary. Describe a real-world example in which the amount of potential and kinetic energy change, but the total amount of mechanical energy stays the same.

   _____

   _____

   _____

2. In this section, you learned how potential energy can be changed in the human body. Complete the chart below by naming three different processes the potential energy in your body is used for.

   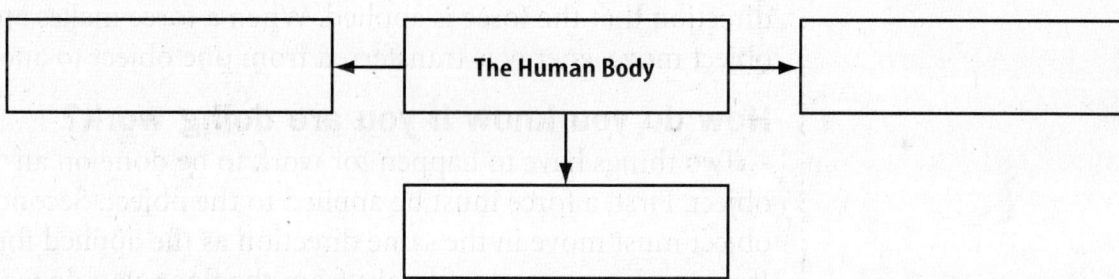

3. Think about what you have learned. How did highlighting the main points and details or examples help you learn the new material?

   _____

   _____

   _____

 Visit **gpscience.com** to access your textbook, interactive games, and projects to help you learn more about the conservation of energy.

chapter 5
Work and Machines

section ❶ Work

What You'll Learn
- what work is
- how work and energy are related
- how to calculate work and power

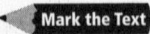
Selective Underlining As you read, underline the key idea in each paragraph.

Reading Check

1. **Identify** What two things have to happen for work to be done?

_____

_____

● Before You Read

Name two things in your life that require work. Name two things that require power.

_____

● Read to Learn

Work

Suppose you read a chapter in your history book for homework. You are sure that you have done a lot of work. But, a scientist will tell you that this is not work. Scientists have a more specific meaning for the word *work*. **Work** is done when a force causes an object to move in the same direction that the force is applied. When a force makes an object move, energy is transferred from one object to another.

How do you know if you are doing work?

Two things have to happen for work to be done on an object. First, a force must be applied to the object. Second, the object must move in the same direction as the applied force. If you pick up a stack of books from the floor, you do work on the books. The books move upward in the direction of the force you apply. However, you are not doing work when you stand still and hold a stack of books. You are applying an upward force on the books, but they are not moving. ☑

What does direction have to do with work?

When you picked up the stack of books, your arms applied a force upward. Your arms did work on the books. Suppose you start walking while still holding the books. You are still applying an upward force on the books, but they are moving forward, not upward. The direction of motion is not in the same direction as the force applied by your arms. Your arms are not doing any work on the books.

Work and Energy

How are work and energy related? Recall that energy is the ability to cause change. Energy is also the ability to do work. One object can transfer energy to a second object by doing work on the second object. When work is done, energy is transferred from one object to another. If you carry a box up some stairs, as shown in the figure, you do work on the box. Your work transfers chemical potential energy from your moving muscles to the box's energy in two forms. The kinetic energy of the box increased because of the movement, and the gravitational potential energy increased because it is higher. Can you feel energy leave your muscles as you carry a heavy box up stairs?

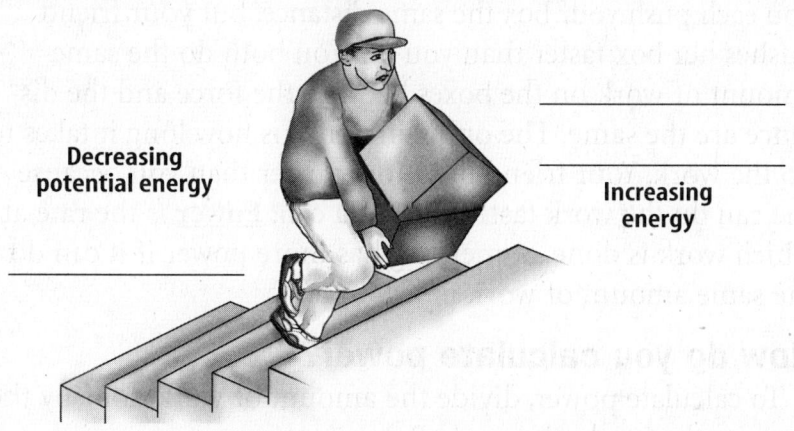

Decreasing potential energy

Increasing energy

Calculating Work

Scientists need to know how to calculate the amount of work that is done. Work is done when a force makes an object move. More work is done when the force is increased or the object is moved a greater distance. You multiply *force* times *distance* to calculate work. Like energy, work is measured in joules.

Work (in joules) = **force** (in newtons) × **distance** (in meters)

$$W = Fd$$

Suppose you want to calculate the amount of work a painter does when he lifts a can of paint. The can weighs 40 newtons, and he lifts it a distance of 2 meters. To lift a can weighing 40 newtons, the painter must apply a force of 40 newtons.

$$W = Fd$$
$$= (40 \text{ N})(2 \text{ m})$$
$$= 80 \text{ joules}$$

The amount of work the painter does is 80 joules.

FOLDABLES

A Organize Make the following Foldable to define, explain, and calculate ideas about work and power.

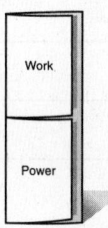

Picture This

2. **Label** Fill in the blanks by the figure with the potential energies that are decreasing and increasing as the box is carried up the stairs.

Applying Math

3. **Calculate** If a bag of groceries weighs 25 newtons and you carry it for 20 feet, how much work have you done?

Reading Essentials **67**

Think it Over

4. Apply A girl throws a bowling ball down the alley at the pins. When is she doing work on the ball?

_____

_____

_____

Think it Over

5. Apply How are work and power related?

_____

_____

_____

When is work done?

When a pitcher throws a ball to a catcher, she applies force to the ball only when it is in her hand. Suppose her hand moves the ball 1 meter before she releases it. The ball moves 10 meters after it leaves her hand. She does work on the ball only when it is in her hand. So, she does work on the ball for a distance of 1 meter only. The distance in the work equation is only the distance an object moves while force is being applied to the object.

Power

Suppose you and a friend want to see who can push a box of books up a ramp faster. Each box of books weighs the same. You each push your box the same distance, but your friend pushes her box faster than you do. You both do the same amount of work on the boxes because the force and the distance are the same. The only difference is how long it takes to do the work. Your friend has more power than you because she can do the work faster than you can. **Power** is the rate at which work is done. Something has more power if it can do the same amount of work in less time.

How do you calculate power?

To calculate power, divide the amount of work done by the time it takes to do the work. Below is the power equation.

$$\text{Power (in watts)} = \frac{\text{work (in joules)}}{\text{time (in seconds)}}$$

$$P = \frac{W}{t}$$

The SI unit for power is the watt (W). One watt equals one joule of work done in one second. The watt is a very small unit. Power usually is given in kilowatts. One kilowatt equals 1,000 watts.

Find the power of a machine that can do 5,000 joules of work in 20 seconds. Use the power equation.

$$P = \frac{W}{t}$$

$$P = \frac{5000 \text{ joules}}{20 \text{ seconds}}$$

$$P = 250 \text{ watts}$$

The power of the machine is 250 watts.

How is power calculated when energy is transferred?

You have learned that when work is done, energy is transferred from one object to another object. You also know that power is the rate at which work is done. Power is also the rate at which energy is transferred. To calculate the power when energy is transferred, divide the amount of energy transferred by the time it took to transfer it. Below is the equation for calculating power when energy is transferred.

$$\text{Power (in watts)} = \frac{\text{energy transferred (in joules)}}{\text{time (in seconds)}}$$

$$P = \frac{E}{t}$$

How is energy transferred when no work is done?

Sometimes energy is transferred when no work is done. A lightbulb can be used in a household lamp. The figure shows that when you plug in a lamp, electrical energy is transferred from the circuit in the wall to the filament in the lightbulb. The filament changes the electrical energy into heat and light. To calculate power, you divide the amount of electrical energy that is transferred to the lightbulb each second. This lightbulb changes electrical energy into light and heat at a rate of 100 joules per second. To calculate the power of this lightbulb, use the equation above.

$$P = \frac{E}{t}$$

$$P = \frac{100 \text{ joules}}{1 \text{ second}}$$

$$P = 100 \text{ watts}$$

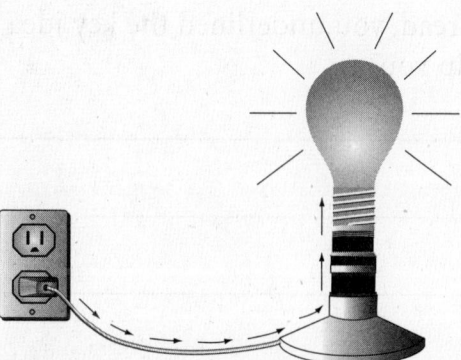

The power of the lightbulb is 100 watts.

Applying Math

6. **Calculate** Suppose a lightbulb changes electrical energy into light and heat at a rate of $\frac{100 \text{ joules}}{2 \text{ seconds}}$. How many watts of power will the lightbulb have?

Picture This

7. **Trace** Use a highlighter to trace the transfer of energy from the plug to the lightbulb.

After You Read
Mini Glossary

power: the rate at which work is done

work: the transfer of energy from one object to another

1. Review the definitions for the terms *work* and *power*. Use a real-world situation to explain the relationship between work and power.

   _____
   _____
   _____

2. Complete the chart below to organize the equations you have learned about using energy.

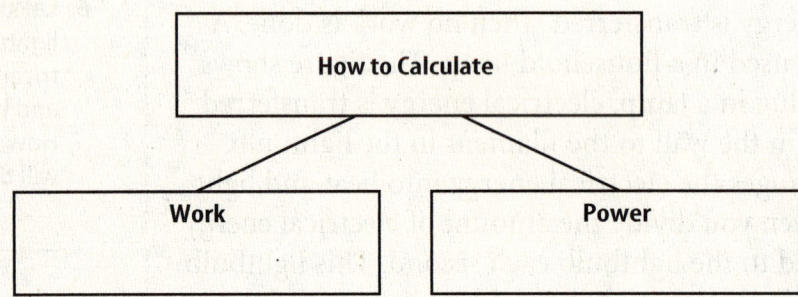

3. Look back at the "Before You Read" section to see the examples of what you listed for things that require work and power. Now that you have learned the definitions that scientists use for these words, do you think a scientist would agree with your list? Tell why or why not.

   _____
   _____

4. Think about what you have learned. As you read, you underlined the key idea in each paragraph. How did underlining the text help you?

   _____
   _____
   _____

 Visit gpscience.com to access your textbook, interactive games, and projects to help you learn more about work and machines.

Chapter 5: Work and Machines

section ❷ Using Machines

Before You Read

List a machine that you used in the previous week. How did this machine help you?

_____

_____

What You'll Learn
- how machines make work easier
- how to calculate mechanical advantage
- how to calculate the efficiency of a machine

Read to Learn

What is a machine?

A <u>machine</u> is a device that makes work easier. Some machines, such as engines, have many moving parts. Some machines are simple. Some simple machines that make work easier are knives, scissors, and doorknobs.

Making Work Easier

There are three ways that machines make work easier. One way increases the force on an object. For example, a screwdriver increases the force on a screw. The second way increases the distance over which a force is applied as in a leaf rake. Machines also make work easier by changing the direction of an applied force. A pulley changes a downward force to an upward force.

How can force be increased using a machine?

You have learned that *work = force × distance*. Remember, either the force or the distance increases. When distance increases, the force decreases because the amount of work stays the same. When you use a car jack, you apply a downward force to the handle. The upward force exerted by the jack is greater than the downward force you applied to the handle. The distance that you push the handle downward is greater than the distance the car is pushed upward. The jack increases the applied force but it doesn't increase the amount of work done. The car jack is an example of a machine that increases an applied force.

Study Coach

Main Idea—Detail Notes Take notes as you read. Make two columns. List main ideas in the left column. List details that support the main ideas in the right column.

Reading Check

1. **Determine** When a car jack increases force, what is decreased?

_____

💡 Think it Over

2. Describe a job made easier by increasing distance.

_____

_____

_____

Picture This

3. Recognize Highlight the parts of the figure that tell the three ways machines make doing work easier.

How does increasing distance decrease force?

Suppose you want to pick up some leaves that cover your yard. You could use a leaf rake to help make your work easier. The leaf rake allows you to increase the distance over which you apply force. You move your hands a small distance at the top of the rake handle. The bottom of the rake moves a greater distance as it moves the leaves. Increasing the distance at the bottom of the rake makes your work easier. The work stays the same and the distance is increased. Therefore, the force needed to rake the leaves is less.

Why do you want a machine that will change direction?

Some machines change the direction of the force you apply. Sometimes it is easier to apply force in a certain direction. To raise a flag up a flagpole, it is easier to pull down on the rope than to climb to the top of the pole and pull up on the rope. Some machines change the direction of a downward force to a sideways force. When you use an ax to split wood, you use a downward, or vertical, force toward the wood. The blade changes the downward force into a sideways, or horizontal, force that splits the wood apart. The figure explains the three ways machines make doing work easier.

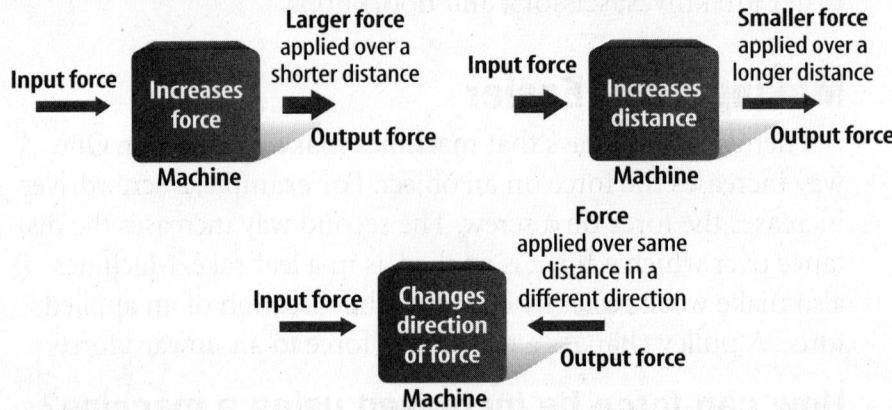

The Work Done by Machines

When you use a machine, you are trying to move an object that resists moving. For example, you can use a crowbar to pry the lid off a wooden crate. You slip the end of the crowbar under the edge of the crate lid and push down on the handle. When you move the handle downward, you do work on the crowbar. The crowbar does work on the lid to lift it up. You are working against the friction between the nails in the lid and the crate. If you used a crowbar to move a large rock, you would be working against gravity.

What are input forces and output forces?

Even though machines make work easier, they do not decrease the amount of work to be done. Instead, a machine changes the way in which you do the work. Two forces are involved when a machine is used to do work. A force is applied to the machine, and the machine applies a force to the object. Think of the work involved in prying open the wooden crate. You did work on the crowbar. The force you apply to the machine is called the **input force**. So, you used an input force to push down on the crowbar. F_{in} stands for the input force. The crowbar did work on the lid. The force applied by the machine is called the **output force**. The crowbar used an output force on the lid to push it up. F_{out} stands for the output force.

In the figure below, a claw hammer is used to pry a nail out of wood. An input force is applied to the hammer. The hammer applies an output force to pull the nail out.

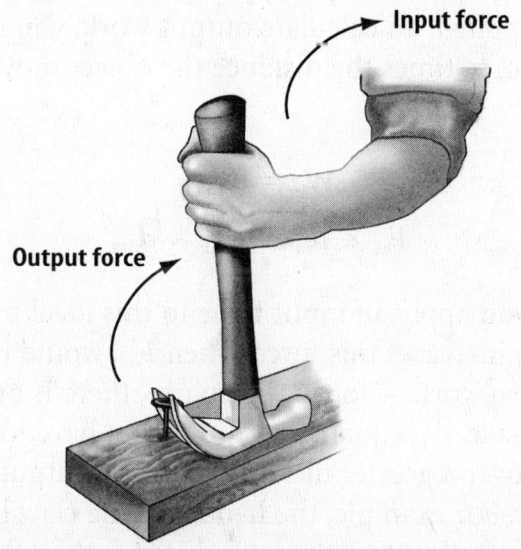

Input force

Output force

Picture This

4. **Label** the input force with F_{in} and the output force with F_{out}.

What are input work and output work?

When you use a machine, there are two types of work. The first type of work is done by you on the machine. The second type of work is done by the machine. You do work when you apply force to the crowbar. This is called the input work. W_{in} stands for the input work. The work done by the machine is called the output work. W_{out} stands for the output work.

Think it Over

5. **Compare** What is the difference between F_{in} and W_{in}?

_____
_____
_____
_____

Reading Check

6. **Explain** With a real machine, why is W_{out} always smaller than W_{in}?

_____

_____

_____

How do machines use conservation of energy?

You learned that energy is conserved. It cannot be created or destroyed. It always remains the same. When you do work on a machine, you transfer your energy to the machine. When the machine does work on an object, its energy is transferred to the object. The amount of energy the machine transfers cannot be greater than the amount of energy you transfer to it. So, W_{out} is never greater than W_{in}.

However, a machine does not transfer all of its energy to the object. Some of the energy changes to heat because of friction. The heat energy cannot be used to do work. So, W_{out} is always smaller than W_{in}.

What is an ideal machine?

Suppose you could build a perfect machine. One in which friction does not change any of the energy to heat. Then the input work would equal the output work. To calculate input work, you multiply the input force times the distance over which it is applied. To calculate output work, you multiply the output force times the distance the object moves. So, for a machine:

$$W_{in} = W_{out}$$
$$F_{in} \times d_{in} = F_{out} \times d_{out}$$

Suppose you apply an input force to this ideal machine, and the machine increases this force. Then F_{out} would be greater than F_{in}. Since work = force × distance, there is only one way you could make W_{in} equal W_{out}. You would have to apply the input force over a greater distance than the output force is applied over. For example, the handle of the claw hammer removing the nail moves a greater distance than the nail.

Mechanical Advantage

Some machines make work easier by making the output force greater than the input force. The number of times the applied force is increased by a machine is called the mechanical advantage (MA) of the machine. The **mechanical advantage** is the ratio of the output force to the input force.

$$\text{mechanical advantage} = \frac{\text{output force (in newtons)}}{\text{input force (in newtons)}}$$

$$MA = \frac{F_{out}}{F_{in}}$$

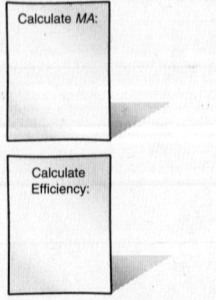

FOLDABLES

B Calculate Make the following note cards from quarter sheets of notebook paper and take note on how to perform the calculations in this section.

For example, suppose that using a pulley system you only need 300 N to lift a piano that weighs 1,500 N. The force you exert on the pulley system in 300 N. The force the pulley system exerts on the piano is 1,500 N. The *MA* of the pulley system is 1,500 N divided by 300 N, or 5.

What is the ideal mechanical advantage?

The mechanical advantage of a machine without friction is called the ideal mechanical advantage, or *IMA*. You can calculate the *IMA* by dividing the input distance by the output distance.

Efficiency

As you have learned, some of the energy put into a real machine is changed into heat by friction. So, the output work of a machine is always less than the work put into it. **Efficiency** is the comparison of the amount of work put into a machine to the amount of work the machine puts out. A high-efficiency machine produces less heat from friction. More of the input work is changed to useful output work.

How do you calculate efficiency?

To calculate the efficiency of a machine, divide the output work by the input work. Efficiency is written as a percentage.

$$\text{efficiency} = \frac{W_{out}}{W_{in}} \times 100\%$$

To calculate the efficiency of a machine with W_{in} of 50 joules and W_{out} of 40 joules, you divide 40 by 50, then multiply by 100%.

$$40 \div 50 = 0.8, \text{ or } 80\%$$

The efficiency of the machines is 80%.

An ideal machine does not produce friction. So, the output work equals the input work. The efficiency of an ideal machine is 100 percent. A real machine does produce friction. This causes the output work to be less than the input work. Therefore, the efficiency of a real machine is always less than 100 percent.

How can machines be made more efficient?

A machine can be made more efficient by reducing friction. Adding oil or grease to the surfaces that rub together fills the gaps between the surfaces. This allows the surfaces to slide across each other more easily.

Applying Math

7. Draw Conclusions What is the *MA* of a machine that has an F_{out} equal to its F_{in}? Explain.

Applying Math

8. Explain Using the efficiency equation, explain why the efficiency of a machine is always less than 100%.

After You Read

Mini Glossary

efficiency: comparison of the amount of work put into a machine to the amount of work put out by the machine

input force: the force applied to a machine

machine: a device used to make work easier

mechanical advantage: the ratio of a machine's output force to its input force

output force: the force applied by a machine

1. Review the definitions of the glossary words. Write one sentence using the terms *input force* and *output force*.

   _____

   _____

   _____

2. Complete the chart below to organize information about the three ways machines make work easier.

   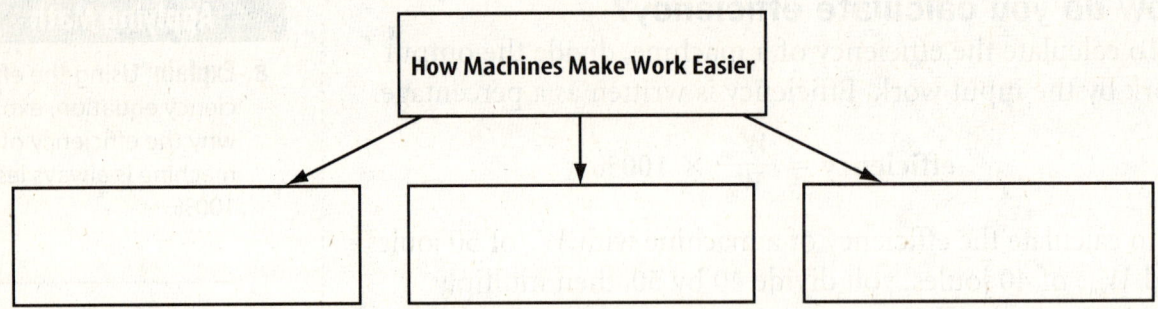

3. How did writing notes about the main ideas and details in two columns help you learn?

   _____

   _____

   _____

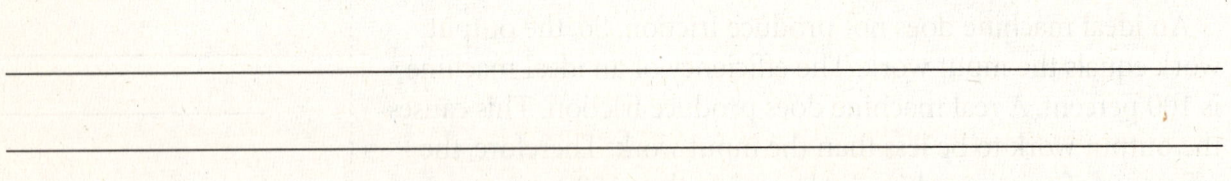

Visit **gpscience.com** to access your textbook, interactive games, and projects to help you learn more about work and machines.

76 CHAPTER 5 Work and Machines

Chapter 5: Work and Machines

section 3 Simple Machines

● Before You Read

Think of a job you did using a machine. How did the machine help you do the work?

_____

_____

_____

What You'll Learn
- six types of simple machines
- how simple machines make work easier
- how to calculate the ideal mechanical advantage of simple machines

● Read to Learn

Types of Simple Machines

You use a simple machine when you cut your food with a knife. A screwdriver is also a simple machine. A **simple machine** is a machine that does work with only one movement of the machine. There are six types of simple machines: lever, pulley, wheel and axle, inclined plane, screw, and wedge. The pulley and the wheel and axle are different forms of the lever. The screw and wedge are different forms of the inclined plane.

Levers

A wheelbarrow, a rake, and a baseball bat are all examples of levers. A **lever** is a bar that pivots, or turns around, a fixed point. The fixed point is called the fulcrum.

A lever has an input and an output arm. The input arm is the distance from the fulcrum to the point where the input force is applied. The output arm is the distance from the fulcrum to the point where the lever exerts the output force. The output force can be greater or less than the input force. As the output force changes, the distance from one end of the bar to the fulcrum changes. There are three classes of levers.

What are the three classes of levers?

The class of a lever is based on the location of the fulcrum, the input force, and the output force.

Study Coach

KWL Chart Draw three vertical lines to make three columns on your paper. Label the three sections: *Know, Want to Know,* and *Learned.* List what you already know about simple machines in the first column. List what you want to know in the second column. As you read this section, list the new things you have learned.

FOLDABLES

C Vocabulary Make the following vocabulary Foldable to define the six types of simple machines.

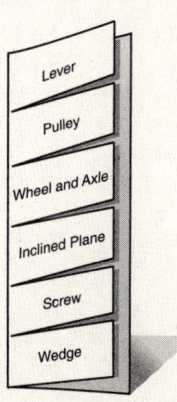

Picture This

1. Draw and Label Draw a picture of a real-life first-class lever. Label the fulcrum, the input force, and the output force.

Think it Over

2. Determine A hockey stick is a third-class lever similar to a baseball bat. You hold a hockey stick near the top of the stick. Where is the output force applied?

a. at your right hand

b. at the top of the stick

c. at the bottom of the stick

d. at the fulcrum

First-Class Lever The first figure shows a first-class lever. The fulcrum is located between the input and output forces. The first-class lever always changes the direction of the force. If there is a downward input force, then the output force moves upward. A crowbar, a pair of scissors, and a seesaw are all examples of first-class levers.

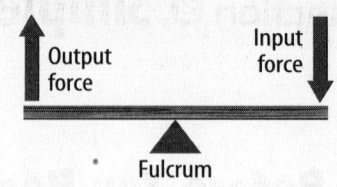

Second-Class Lever The middle figure shows a second-class lever. The output force is between the input force and the fulcrum. Both the input force and the output force move in the same direction. A wheelbarrow is an example of a second-class lever. When you pick up the handles, you apply an input force. The wheel is the fulcrum.

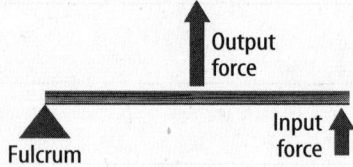

Third-Class Lever The last figure shows a third-class lever. The output force is farther away from the fulcrum than the input force. Therefore, the output force is always less than the input force in a third-class lever. A baseball bat is a third-class lever. A right-handed batter applies the input force with the right hand. The left hand is the fulcrum. The output force is exerted by the bat above the right hand. What is the advantage of a third-class lever? It increases the distance over which the output force is applied.

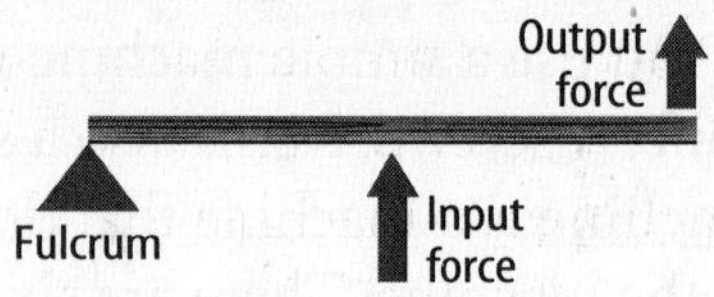

How is the ideal mechanical advantage of a lever calculated?

To calculate the ideal mechanical advantage (*IMA*) of any machine, divide the input distance by the output distance. For a lever, the input distance is the length of the input arm. The output distance is the length of the output arm.

$$\text{ideal mechanical advantage} = \frac{\text{length of the input arm}}{\text{length of the output arm}}$$

$$IMA = \frac{L_{in}}{L_{out}}$$

Pulleys

To raise a sail upward, a sailor pulls down on a rope. The rope uses a simple machine to change the direction of the input force. The machine is called a pulley. A **pulley** is a grooved wheel with a rope, chain, or cable wrapped around it. There are two types of pulleys, fixed and movable. There are also systems of fixed and movable pulleys. All three are shown in the figure at the bottom of the page.

What is a fixed pulley?

A fixed pulley is a modified first-class lever. See Figure A below. It changes the direction of the input force. Fixed pulleys, such as the one on a sail or a flagpole are attached to a structure above your head that will not move. When the cord is pulled down, an object goes up.

An elevator uses a fixed pulley. The cable goes over the fixed pulley at the top of the elevator shaft. When the cable is pulled downward, it lifts the elevator upward. A fixed pulley does not change the amount of force exerted. It does not change the distance over which the force is exerted. It changes the direction of the input force. The input and output force remain the same.

What is a movable pulley?

A movable pulley has one end of the rope fixed and the wheel is free to move. See Figure B below. The movable pulley doesn't change the direction of a force. It decreases the amount of input force needed to lift the object.

Suppose a 4-N weight is hung from a movable pulley. The rope is attached to the ceiling. The ceiling acts like someone helping you lift the object. The part of the rope attached to the ceiling will support half of the weight. You only need to exert enough force to lift the other half of the weight. Both you and the ceiling exert 2 N of force. Since you exert 2 N, but the total output force lifting the weight is 4 N, the mechanical advantage of a movable pulley is 2.

Think it Over

3. Apply What is the mechanical advantage of a fixed pulley? Explain.

_____
_____
_____

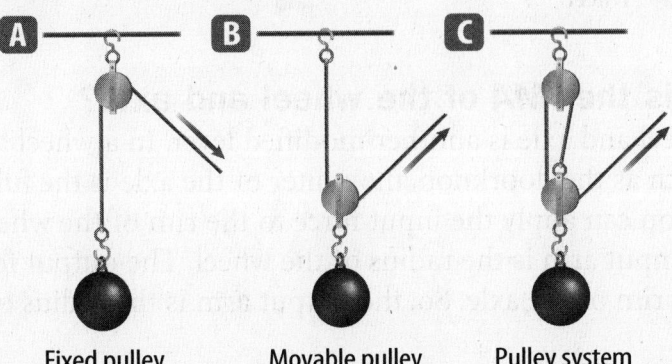

Fixed pulley   Movable pulley   Pulley system

Picture This

4. Label On Figure C, label the fixed pulley and the movable pulley.

Think it Over

5. Evaluate Which type of pulley would be best to lift a very heavy weight? Why?

_____
_____
_____
_____

Picture This

6. Identify Which figure, A or B, shows a wheel and axle as used in a doorknob?

_____

What is a block and tackle?

A block and tackle is a system of fixed and movable pulleys used together. Figure C has one fixed and one movable pulley. This system changes the direction of the force. It also decreases the input force needed to lift the object. The more sections of the rope a system uses to pull up an object, the greater the output force is. If you have three sections of rope pulling on the object, as shown in the art on the previous page, the output would be three times the input force. The *IMA* of a pulley system is equal to the number of section of rope pulling up on the object.

Wheel and Axle

Could you turn a doorknob if it were the shape of a pencil? It would be possible, but difficult. A doorknob makes it easier to open a door because of a machine called a wheel and axle. A **wheel and axle** is a simple machine that has an axle attached to the center of a larger wheel. The wheel and axle are both circular objects of different sizes that are attached and turn together. The larger object is the wheel and the smaller object is the axle.

In some devices, such as a doorknob, the input force is used to turn the wheel. The axle exerts the output force. The axle is smaller than the wheel, so the output force is greater than the input force. This is shown in Figure A. In other devices, such as a ferris wheel, the input force is used to turn the axle. The wheel exerts the output force. The wheel is larger than the axle, so the output force is less than the input force. This is shown in Figure B.

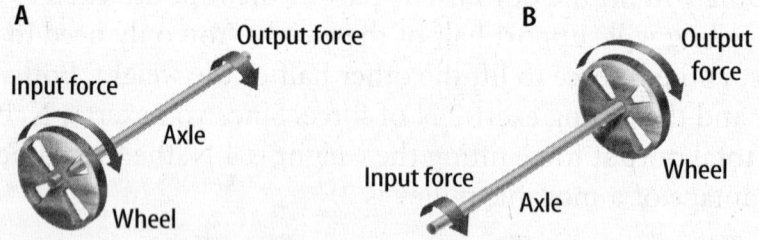

What is the *IMA* of the wheel and axle?

A wheel and axle is another modified lever. In a wheel and axle, such as the doorknob, the center of the axle is the fulcrum. You can apply the input force to the rim of the wheel. So, the input arm is the radius of the wheel. The output force is at the rim of the axle. So, the output arm is the radius of the axle.

To calculate the *IMA* of a wheel and axle, use this equation:

$$\text{ideal mechanical advantage} = \frac{\text{radius of the wheel (meters)}}{\text{radius of axle (meters)}}$$

$$IMA = \frac{r_w}{r_a}$$

The ideal mechanical advantage of a wheel and axle can be increased by increasing the radius of the wheel.

How do gears work?

A gear is a wheel and axle with the wheel having teeth around its rim. When the teeth of two gears come together, one gear makes the other gear turn. When two gears of different sizes come together, they turn at different speeds. Each time the larger gear turns once, it causes the smaller gear to turn more than one time.

If the input force is applied to the larger gear, the output force of the smaller gear is less than the input force. Gears may also change the direction of the force. When the larger gear in the figure turns counterclockwise, the smaller gear turns clockwise.

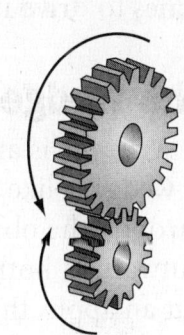

Inclined Planes

An <u>inclined plane</u> is a sloping surface that reduces the amount of force it takes to do work. For example, a ramp is an inclined plane. It is easier to put a heavy object on the ramp and push it up than to lift it up. The inclined plane makes the work easier.

How does an inclined plane make work easier?

You do the same work by lifting a box straight up or pushing it up an inclined plane. As an inclined plane becomes longer, the force needed to move the object becomes less. When you push a box up an inclined plane, the input force is applied over a longer distance. Therefore, it takes less input force to push the box up the ramp than to lift it straight up. To calculate the *IMA* of an inclined plane, use this equation:

$$\text{ideal mechanical advantage} = \frac{\text{length of slope (meters)}}{\text{height of slope (meters)}}$$

$$IMA = \frac{l}{h}$$

To increase the *IMA* of an inclined plane, make the plane longer.

Picture This

7. Apply What direction would the smaller gear turn if you turned the larger gear clockwise?

_____

Applying Math

8. Calculate Find the ideal mechanical advantage of a ramp used to load bags on an airplane. The length of the ramp is 15 meters, and the height of its slope is 3 meters.

_____

The Screw

A <u>screw</u> is an inclined plane wrapped in a spiral around a post. The inclined plane forms the threads on the screw. You apply the input force by turning the screw. The threads change the input force to an output force that pulls the screw into the material. Friction between the threads and the material holds the screw tightly in place. Some examples of screws are jar lids, a corkscrew, a drill bit, and the bottom of a lightbulb.

The ideal mechanical advantage of a screw depends on the length of the plane. The longer the plane is, the closer the spacing of the threads is. Therefore, the *IMA* is greater if the threads are close together. You have to turn the screw more times to drive it into the object, but you use less force.

The Wedge

A <u>wedge</u> is an inclined plane with one or two sloping sides. A wedge is like the screw because the inclined plane moves through the object. A knife is a wedge. The sharp edge slopes outward at both sides to form an inclined plane. When you cut an apple, the downward input force of the knife is changed to a horizontal output force pushing the apple pieces apart. The *IMA* of a wedge increases as it gets longer and thinner.

Compound Machines

Some machines that you use everyday are made up of several simple machines. A **compound machine** is two or more simple machines that work together. The can opener shown below is an example of a compound machine. To open a can, you first squeeze the handles together. The handles act as a lever. They increase the force applied on a wedge. The wedge pierces the can. Then you turn the wheel and axle to open the can.

Reading Check

9. **Describe** How do you apply an input force to a screw?

Picture This

10. **List** What three types of machines make up a can opener?

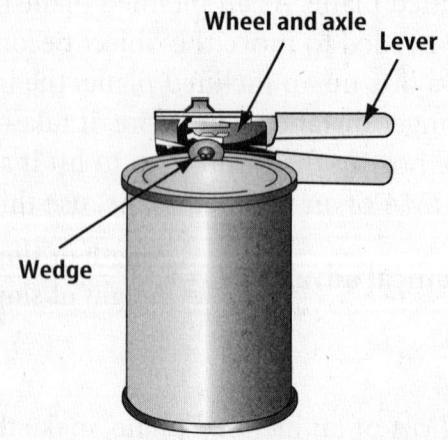

After You Read
Mini Glossary

compound machine: a machine made from two or more simple machines that work together

inclined plane: a straight, slanted surface that reduces how much force it takes to do work

lever: a bar that is able to pivot or turn around a fixed point

pulley: a wheel with a groove that has a rope, chain, or cable running along the groove.

screw: an inclined plane wrapped in a spiral around a post

simple machine: is a machine that does work with only one movement of the machine

wedge: an inclined plane with one or two sloping sides

wheel and axle: a simple machine that has an axle attached to the center of a larger wheel

1. Review the meanings of the vocabulary words. Two of the terms are modified inclined planes. Describe how one of them might be used in a real-life situation.

   _____
   _____

2. Use the chart to fill in the six types of simple machines you learned about in this section.

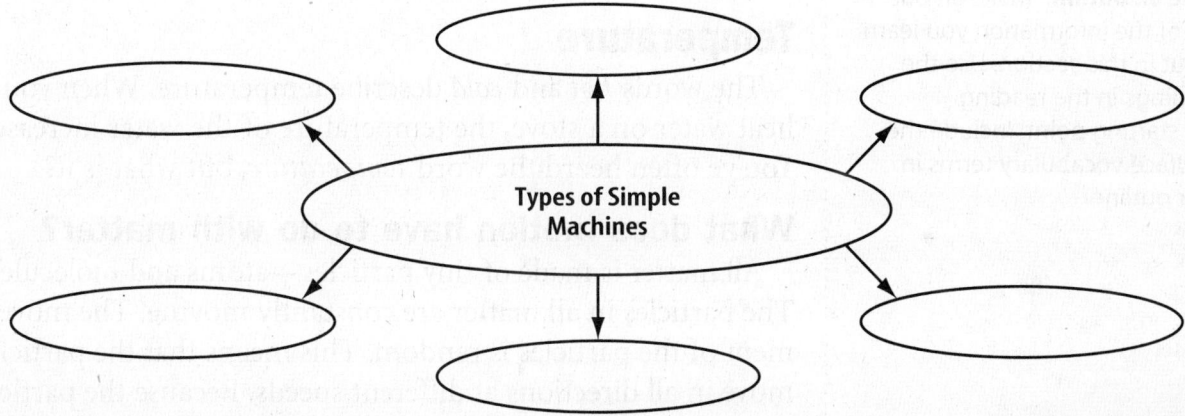

3. Look back at the *Know, Want to Know,* and *Learned* chart you made at the beginning of this section. How did making this chart help you remember what you learned?

   _____
   _____

 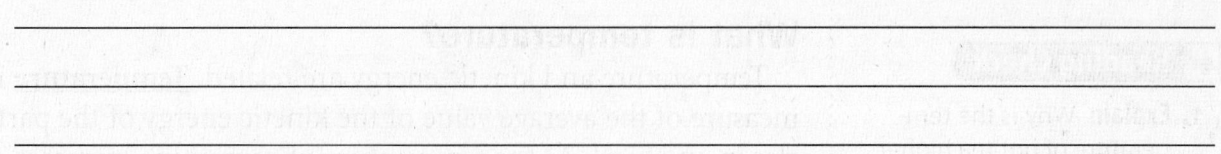 Visit **gpscience.com** to access your textbook, interactive games, and projects to help you learn more about work and machines.

chapter 6

Thermal Energy

section ❶ Temperature and Heat

What You'll Learn
- what temperature is
- how thermal energy depends on temperature
- how thermal energy and heat are related
- calculate the change in thermal energy

Before You Read

You wake up in the morning and get out of bed. Does the floor feel cold or warm on your bare feet? On the lines below, write a sentence that compares how it feels to step on a bare floor and on a rug on a cold morning.

___

Study Coach

Make an Outline Make an outline of the information you learn about in this section. Use the headings in the reading as a starting point. Include the boldface vocabulary terms in your outline.

Read to Learn

Temperature

The words *hot* and *cold* describe temperature. When you heat water on a stove, the temperature of the water increases. You've often heard the word *temperature*, but what is it?

What does motion have to do with matter?

All matter is made of tiny particles—atoms and molecules. The particles in all matter are constantly moving. The movement of the particles is random. This means that the particles move in all directions at different speeds. Because the particles are moving, they have kinetic energy. The faster they move, the more kinetic energy they have. When an object is hot, the particles move faster. As it cools, the particles move more slowly. Particles move faster in hot objects than in cooler objects.

What is temperature?

Temperature and kinetic energy are related. **Temperature** is a measure of the average value of the kinetic energy of the particles in an object. As the temperature of something increases, the average speed of its particles increases. The temperature of hot tea is higher than the temperature of iced tea because the particles in the hot tea are moving faster. Recall from Chapter 1 that in SI units, temperature is measured in kelvins (K). The Celsius scale is used more commonly than the Kelvin scale. One kelvin is the same size as one Celsius degree. ✓

✓ Reading Check

1. **Explain** Why is the temperature of hot tea higher than the temperature of iced tea?

___

84 **CHAPTER 6** Thermal Energy

Thermal Energy

When you take butter out of the refrigerator, it is cold and hard. If you let it sit at room temperature for a while, it gets warmer and softer. The air in the room is at a higher temperature than the butter. So, particles in the air have more kinetic energy than the butter particles. Air particles and butter particles bump into each other. The collisions transfer kinetic energy from the air to the butter. The butter particles begin moving faster and the temperature of the butter increases.

Potential Energy to Kinetic Energy Molecules also have potential energy that can be changed into kinetic energy. How can molecules have potential energy? Think about a ball. Earth exerts an attractive gravitational force on the ball. When you hold the ball above your head, the ball and Earth are separated. This gives the ball potential energy. The molecules in a substance also exert attractive forces on each other. They have potential energy when they are separated. ☑

As molecules move farther apart, their potential energy increases. As they move faster, their kinetic energy increases. <u>Thermal energy</u> is the sum of the kinetic and potential energy of all the molecules in an object. The figure shows that if either potential or kinetic energy increases, thermal energy increases.

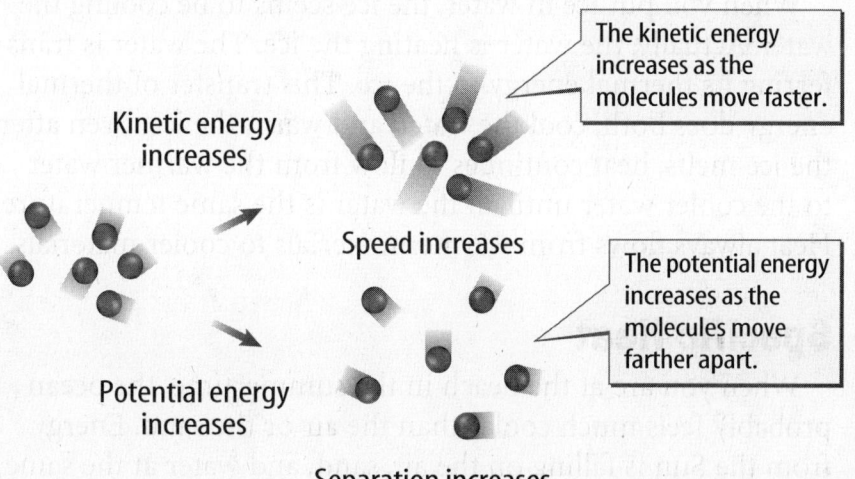

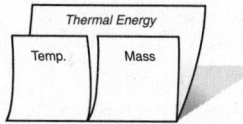

FOLDABLES

A Relate Make the following Foldable to tell how temperature and mass are related to thermal energy.

Reading Check

2. Identify What gives particles in a material potential energy?

_____
_____
_____
_____

How are thermal energy and temperature related?

When the temperature of an object increases, the average kinetic energy of its particles increases. When the average kinetic energy of its particles increases, the object's thermal energy increases. Therefore, the thermal energy of an object increases as its temperature increases.

Picture This

3. Use Scientific Illustrations In the space below, sketch the particles in the figure if separation increased even more.

How are thermal energy and mass related?

Suppose you have a glass of water and a beaker of water. They are at the same temperature. The beaker contains twice as much water as the glass. The average kinetic energy of the water molecules is the same in both containers, since they are the same temperature. But there are twice as many water molecules in the beaker as there are in the glass. So the total kinetic energy of the water molecules in the beaker is twice as large as that of the water molecules in the glass. The water in the beaker has twice as much thermal energy as the water in the glass has. If the temperature doesn't change but the mass of the object increases, the thermal energy in the object increases.

Heat

Have you ever noticed that your chair felt warm when you sat down? You could tell that someone had been sitting in it recently. The chair felt warm because thermal energy from the person's body flowed to the chair and increased its temperature.

Heat is thermal energy that flows from something at a higher temperature to something at a lower temperature. Recall that joules are the units that energy is measured in. Heat is a form of energy, so it is measured in joules.

When you put ice in water, the ice seems to be cooling the water. Actually, the water is heating the ice. The water is transferring its thermal energy to the ice. This transfer of thermal energy does both, cool the water, and warm the ice. Even after the ice melts, heat continues to flow from the warmer water to the cooler water until all the water is the same temperature. Heat always flows from warmer materials to cooler materials.

Specific Heat

When you are at the beach in the summertime, the ocean probably feels much cooler than the air or the sand. Energy from the Sun is falling on the air, sand, and water at the same rate. But the temperature of the water has changed less than the temperature of the air or sand has changed.

As a substance or material absorbs heat, how much its temperature changes depends on how much heat is added. But it also depends what the material is made of. For example, think about 1 kg of sand and 1 kg of water. It takes six times as much heat to raise the temperature of water 1°C than it takes to raise the temperature of sand 1°C. The ocean water at the beach would have to absorb six times as much heat as the sand to be at the same temperature.

Think it Over

4. Communicate Write in your own words the difference between temperature and heat.

Think it Over

5. Infer Read the Before You Read exercise on the first page of this section again. Use what you learned about specific heat. Why does a carpeted floor feel different to your bare feet than a bare floor feels on a cold morning?

The **specific heat** of a substance is the amount of heat needed to raise the temperature of 1 kg of that substance by 1°C. Specific heat is measured in joules per kilogram per degree Celsius [J/(kg°C)]. The table shows the specific heat of some familiar materials.

Specific Heat of Some Common Materials	
Substance	Specific Heat [J/(kg°C)]
Water	4,184
Wood	1,760
Carbon (graphite)	710
Glass	664
Iron	450

Applying Math

6. **Use a Table** How much heat is needed to raise the temperature of 1 kg of iron from 25°C to 30°C? Show your work.

How does water cool things?

Look at the table of specific heat above. Compared with the other common materials, water has the highest specific heat. The figure below shows why this is. Water molecules form strong bonds with each other. When heat is added, the strong bonds make it hard for the molecules to move. Some of the added heat has to break these bonds before the molecules can start moving faster. In metals, electrons can move freely. When heat is added, no strong bonds have to be broken before the electrons can start moving faster.

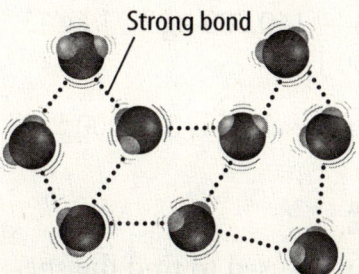

When heat is added to water, some of the added heat has to break some of these bonds before the molecules can start moving faster.

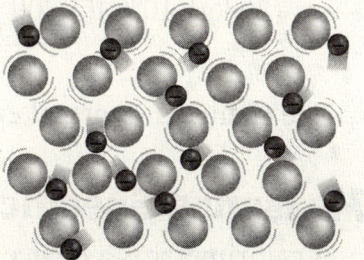

In metals, electrons can move freely. When heat is added, no strong bonds have to be broken before the electrons can start moving faster.

Picture This

7. **Interpret Scientific Illustrations** Explain how strong bonds give water a high specific heat.

A coolant is a substance that is used to absorb heat. Water can absorb heat without a large change in temperature, so it is useful as a coolant. Water is used in the cooling systems of automobile engines. As long as the water temperature is lower than the engine temperature, heat will flow from the engine to the water. Compared to other materials that could be used in an engine, the temperature of water will increase less.

How does thermal energy change?

The thermal energy of an object changes when heat flows into or out of the object. You can use the following equation to calculate the change in thermal energy.

change in thermal energy (J) =

mass (kg) × change in temperature (°C) × specific heat $\left(\frac{J}{kg°C}\right)$

$$Q = m(T_f - T_i)C$$

In the equation, Q stands for the change in thermal energy. C stands for the object's specific heat. T_f is the final temperature and T_i is the initial temperature.

Suppose a wooden chair with a mass of 20 kg is sitting in sunlight. As it sits, it warms from 15°C to 25°C. The specific heat of wood is 1,700 J/(kg°C). What is the change in the thermal energy of the chair?

The mass of the chair is 20 kg, so m = 20 kg.
The temperature warms from 15°C to 25°C.
So T_i = 15°C and T_f = 25°C.
The specific heat of wood is 1,700 J/(kg°C), so
C = 1,700 J/(kg°C).

$$Q = m(T_f - T_i)C$$
$$= (20 \text{ kg})(25°C - 15°C)(1,700 \text{ J/kg°C})$$
$$= (20 \text{ kg})(10°C)(1,700 \text{ J/kg°C}) = 140,000 \text{ kg °C J/kg°C}$$
$$= 340,000 \text{ J}$$

The change in thermal energy of the chair is 340,000 J.

Measuring Specific Heat

A calorimeter is a device that can be used to find the specific heat of a material. The specific heat of a material can be determined if the mass of the material, its change in temperature, and the amount of thermal energy absorbed or released are known. In a calorimeter, an object that has been heated transfers heat to a known mass of water. The transfer of heat continues until the object and the water are the same temperature. The energy absorbed by the water can be calculated by measuring the water's temperature change. The thermal energy released by the object equals the thermal energy absorbed by the water. ✓

Applying Math

8. Calculate The specific heat of copper is 385 J/kg°C. Find the change in thermal energy for a copper pipe with a mass of 8 kg when it is heated from 12°C to 21°C. Show your work.

✓ Reading Check

9. Explain What does a calorimeter measure?

After You Read

Mini Glossary

heat: thermal energy that flows from something at a higher temperature to something at a lower temperature

specific heat: the amount of heat needed to raise the temperature of 1 kg of a substance by 1°C

temperature: the average value of the kinetic energy of the particles in an object

thermal energy: the sum of the kinetic and potential energy of all the molecules in an object

1. Review the terms and their definitions in the Mini Glossary. What is the difference between temperature and thermal energy?

   _____
   _____
   _____

2. Complete the chart below to organize information from this section about temperature.

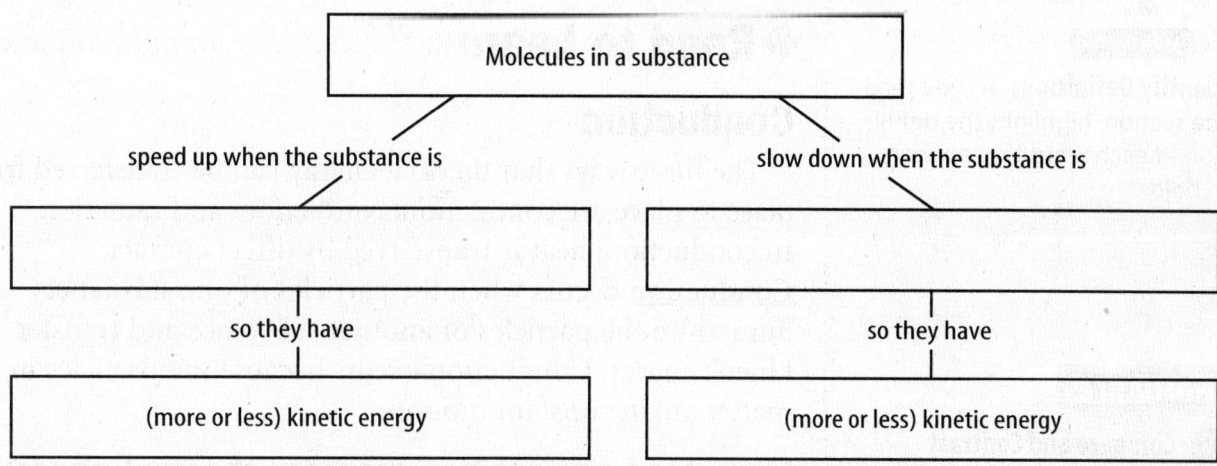

3. You made an outline as you read this section. How did that strategy help you learn the material?

   _____
   _____
   _____

 Visit gpscience.com to access your textbook, interactive games, and projects to help you learn more about temperature and heat.

chapter 6 Thermal Energy

section ❷ Transferring Thermal Energy

What You'll Learn
- the three ways heat is transferred
- the difference between insulators and conductors
- how insulators control the transfer of thermal energy

● Before You Read

There is an old joke that says a sweater is something you put on when your mother is cold. On the lines below, write about a time when you felt warm, but someone else felt cold.

_____

_____

_____

Mark the Text

Identify Definitions As you read the section, highlight the definition of each word that appears in bold.

FOLDABLES

B Compare and Contrast
Make the Foldable shown below to help you understand how conduction, convection, and radiation are similar and different.

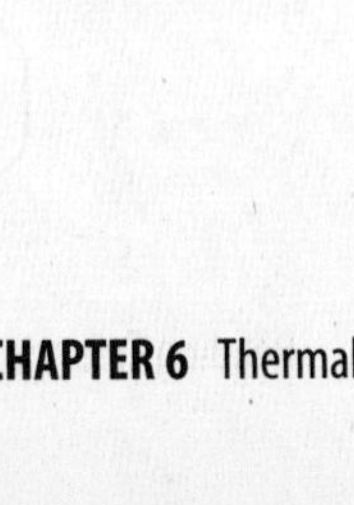

[Foldable diagram: Conduction / Convection / Radiation]

● Read to Learn

Conduction

The three ways that thermal energy can be transferred from place to place are conduction, convection, and radiation. In conduction, heat is transferred by direct contact. **Conduction** occurs when the particles of one substance bump into the particles of another substance and transfer kinetic energy. Conduction occurs because the particles in matter are in constant motion.

How does conduction transfer thermal energy?

The figure shows how thermal energy is transferred when one end of a metal spoon is heated by a flame. The kinetic energy of the particles near the flame increases. The heated particles bump into surrounding particles. In these collisions, thermal energy is transferred from particles with more kinetic energy to particles with less kinetic energy. The collisions continue until thermal energy is transferred to the other end of the spoon. When heat is transferred by conduction, thermal energy is transferred from place to place without transferring matter. Thermal energy is transferred by the collisions between particles, not by movement of matter.

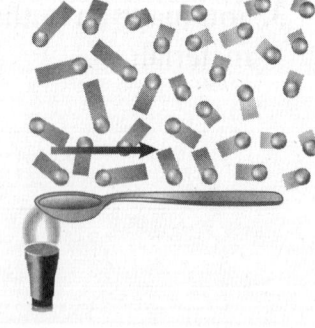

90 CHAPTER 6 Thermal Energy

What are heat conductors?

Heat can be transferred by conduction in all materials. But heat moves more quickly in some materials than in others. Heat moves faster by conduction in solids and liquids than in gases. In gases, particles are farther apart. Collisions with other particles happen less often in gases than in solids or liquids.

The best conductors of heat are metals. Remember that an atom has a nucleus surrounded by electrons. In metals, there are electrons that are not bound to individual atoms. They can move easily through the metal and bump into other electrons. This helps the electrons transfer thermal energy. Silver, copper, and aluminum are some of the best conductors of heat.

Convection

Thermal energy can also be transferred by a method called convection. **Convection** is the transfer of thermal energy in a fluid by the movement of warmer and cooler fluids from one place to another. A fluid is a substance that can flow. It can be a liquid or a gas. The movement of fluids from one place to another causes currents. These currents transfer heat.

How is heat transferred by convection currents?

When a pot of water is heated on a stove, heat is transferred by convection currents. First, the water molecules at the bottom of the pot gain thermal energy from the stove. The water becomes less dense as it is heated. The water at the top of the pot is still cool, so it is more dense than the warm water. Since the warm water is less dense, it rises. As it rises, it is replaced at the bottom of the pot by the cooler, more dense water. The cycle continues until all the water in the pot is at the same temperature. This rising-and-sinking action is a convection current. Also, as the warmer water rises, it heats some of the cooler water that it comes in contact with by conduction. In a convection current, both conduction and convection transfer thermal energy.

How do convection and conduction differ?

Both convection and conduction transfer heat. However, convection is different from conduction. Convection transfers thermal energy by moving particles from one place to another. Warm particles change places with cooler particles. In conduction, no particles move. Thermal energy is transferred by particles bumping into surrounding particles. But, in conduction, no particles actually move from place to place as they do in convection.

FOLDABLES

C Compare Make the following Foldables to compare conductors and insulators.

Reading Check

1. **Explain** why metals are good conductors.

_____
_____
_____

Think it Over

2. **Explain** A convection oven has a fan to move the air inside it. Is the air in a regular oven also affected by convection? Why or why not?

_____
_____
_____

Reading Essentials **91**

Picture This

3. **Make a Drawing** In the first box, draw an ice-covered mountain. Then draw arrows to show how radiation is reflected and absorbed by an ice-covered mountain range. In the second box, draw trees to represent a rain forest. Then draw arrows to show how radiation is reflected and absorbed by a dark rain forest.

Radiation

How does heat travel through space? There is almost no matter between the Sun and Earth. So heat does not travel by conduction or convection. Instead, heat is transferred from the Sun to Earth by radiation.

Radiation is the transfer of energy by electromagnetic waves. They carry energy through solids, liquids, and gases. But they can also carry energy through empty space. Energy transferred by radiation is often called radiant energy. When you sit by a fireplace, you feel warm because heat is transferred by radiation from the fire to your skin.

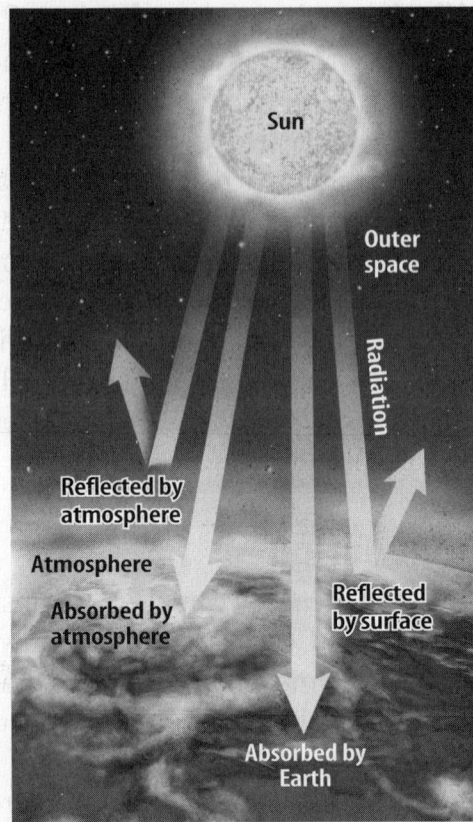

How does matter affect radiant energy?

When radiation strikes a material, three things can happen. The material absorbs some of the energy. It also reflects some of the energy. And some of the energy may be transmitted, or passed through, the material. The amount of energy absorbed, reflected, and transmitted depends on the type of material. Light-colored materials reflect more radiant energy, while dark-colored materials absorb more radiant energy. When material absorbs radiant energy, the thermal energy of the material increases. The figure above shows what happens to radiant energy from the Sun as it reaches Earth.

How is radiation different in solids, liquids, and gases?

Radiant energy can travel through the space between molecules in a solid, liquid, or gas. Molecules can absorb the radiation and emit some of the energy they absorbed. This energy then travels through the space between molecules. It is absorbed and emitted by other molecules. The transfer of energy by radiation works best with gases. The molecules in gases are much farther apart than in solids or liquids. This extra space allows the radiant energy to travel more easily through gases.

92 CHAPTER 6 Thermal Energy

Controlling Heat Flow

You probably do things every day to control the flow of heat. For example, you put on a jacket when you leave your house on a cool morning. And you might wear an oven mitt on your hand when you reach into the oven to pull out a hot pan. In both cases, you use different materials to help control the flow of heat. Your jacket reduces the flow of heat from your body to the surrounding air. The oven mitt reduces the flow of heat from the hot dish to your hand.

Animal Adaptations Almost all living things have special features that help them control the flow of heat. The antarctic fur seal has a coat that is about 10 cm thick. This coat helps keep the seal from losing heat. The emperor penguin has a thick layer of blubber and thick, closely spaced feathers. This helps the penguin reduce the loss of body heat. These adaptations help the animals survive in a climate where the temperature is often below freezing. The scaly skin of the desert spiny lizard has the opposite effect. The lizard's skin reflects the Sun's rays and keeps the lizard from becoming too hot. The leathery skin also prevents water loss. An animal's color also can also help it keep warm or cool. The black feathers on a penguin's back allow it to absorb radiant energy.

Insulators

When you cook, you want the pan to conduct heat easily from the stove to your food. But you don't want the heat to move easily to the handle of the pan. Most pans have handles that are made from insulators. An **insulator** is a material in which heat flows slowly. Materials that are insulators include wood, some plastics, fiberglass, and air. An insulator is the opposite of a conductor. Remember, conductors are materials through which heat flows quickly. Metals like silver, copper, and aluminum are excellent conductors. Materials that are good insulators are poor conductors of heat. That is why many pan handles are made of wood or plastic.

Air as an Insulator Gases, such as air, are usually better insulators than solids or liquids. Some types of insulators contain many pockets of trapped air. These air pockets conduct heat poorly and keep convection currents from forming. When you wear a fleece jacket, the fibers in the fleece trap air between the jacket and your body. This air slows down the flow of your body heat to the colder air outside the jacket. Your body heat warms the air trapped by the fleece and you are wrapped in a blanket of warm air.

Reading Check

4. Describe How do animals' colors help them control their temperature?

Think it Over

5. Draw Conclusions Some of the best cooking pans are made of copper. Why do you think this is?

How are buildings insulated?

Materials that are insulators help keep warm air from flowing out of buildings in cold weather. They also keep warm air from flowing into buildings in warm weather. These materials are called insulation. Building insulation is usually made of materials, such as fiberglass or foam, which contain pockets of trapped air. Insulation is placed between a building's outer walls and inner walls and between the ceiling and the attic. It reduces the flow of heat between the building and the surrounding air. ✓

Insulation also helps furnaces and air conditioners work more effectively. Since the heat doesn't escape or enter the building as easily, the furnaces and air conditioners do not have to work as much. This can save a great deal of energy. In the United States, about 55 percent of the energy used in homes is used for heating and cooling.

How does a thermos work?

Have you used a thermos bottle like the one in the figure? A thermos bottle lets you keep soup hot or iced tea cold. It does this by reducing the flow of heat into and out of the liquid in the bottle. The temperature of the liquid hardly changes over a number of hours. A thermos bottle has two glass walls. The air between the two walls is removed so there is a vacuum between the glass layers. The vacuum contains almost no matter. This vacuum prevents heat transfer by conduction or convection between the liquid and the air outside the thermos bottle.

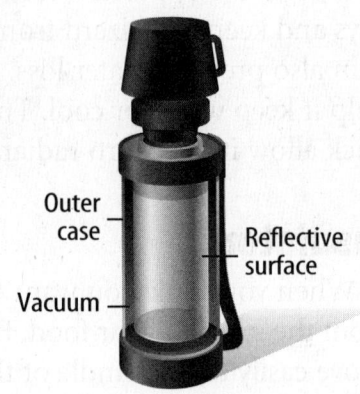

The inside and outside glass surface of a thermos bottle is coated with aluminum. The coating makes each surface highly reflective. Electromagnetic waves are reflected at each surface. The inner reflective surface prevents radiation from transferring heat out of the liquid. The outer reflective surface prevents radiation from transferring heat into the liquid.

Think about the things you do to stay warm or cool. Sitting in the shade reduces the heat transferred to you by radiation. Opening or closing windows reduces heat transferred by convection. Putting on a jacket reduces the heat transferred from your body by conduction. In what other ways do you control the flow of heat?

Reading Check

6. **Identify** What does building material insulation contain?

___

Picture This

7. **Explain** How does the shiny surface of a thermos bottle prevent heat transfer?

___

After You Read

Mini Glossary

conduction: the transfer of thermal energy by collisions between particles in matter

convection: when thermal energy is transferred in a fluid by the movement of warmer and cooler fluid from place to place

insulator: a material in which heat flows slowly

radiation: the transfer of energy by electromagnetic waves

1. Review the terms and their definitions in the Mini Glossary. Write a sentence on the lines below that shows your understanding of the term *insulator*.

   _____

   _____

2. Complete the table to organize information about how heat is transferred.

How Heat Is Transferred	Definition	Example
Conduction		
Convection		
Radiation		

3. Think about what you have learned. How did identifying definitions help you as you read the section?

   _____

   _____

   _____

Science Online Visit gpscience.com to access your textbook, interactive games, and projects to help you learn more about transferring thermal energy.

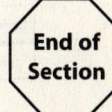

Chapter 6: Thermal Energy

section 3 Using Heat

What You'll Learn
- types of heating systems
- laws of thermodynamics
- how car engines work
- how refrigerators keep food cold

Study Coach

Create a Diagram As you read about internal combustion engines in this section, make a diagram of the steps of the four-stroke cycle. Make sure you label your diagram.

FOLDABLES

D Build Vocabulary Make the following Foldable to help you learn the vocabulary words from this section. You will need to use more than one Foldable.

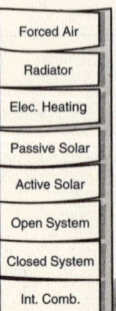

Before You Read

Cars and buses use heat engines. List three more kinds of devices you know about that use engines.

_____

Read to Learn

Heating Systems

Most buildings have some kind of heating system. All heating systems need a source of energy. In the simplest and oldest heating system, wood or coal is burned in a stove. The heat that is produced is transferred to the air by conduction, convection, and radiation. One disadvantage of this type of heating system is that the transfer of heat is slow.

What are some common heating systems?

Forced-Air Systems The forced-air system is the most common type of heating system today. In this system, fuel is burned in a furnace. This heats a volume of air. A fan blows the warm air through large pipes, called ducts. The ducts lead to openings, or vents, in each room. Cool air returns through other vents to the furnace where it is reheated.

Radiator Systems In the past, many homes and buildings were heated by radiators. In a radiator system, fuel is burned in a furnace and heats a tank of water. Pipes carry the hot water or steam to radiators in the building. The thermal energy in the water or steam is transferred to the air by conduction. The warm air then moves through the room by convection. After the water cools it flows back to the water tank and is reheated.

Electric Heating Systems An electric heating system has no central furnace. It has electrically heated coils in floors and walls. The coils heat the surrounding air by conduction. The heat is distributed through the room by convection. Electric heating systems are not as common as forced-air systems.

96 **CHAPTER 6** Thermal Energy

Solar Heating

The Sun emits an enormous amount of radiant energy that strikes Earth every day. The radiant energy from the Sun can be used to help heat homes and buildings.

What kinds of solar heating systems are there?

Passive Solar Heating In passive solar heating systems, materials inside a building absorb radiant energy from the Sun during the day. The materials heat up. At night, the building begins to cool. Thermal energy absorbed by these materials helps keep the building warm.

Active Solar Heating Active solar heating systems use solar collectors. **Solar collectors** absorb radiant energy from the Sun. The collectors usually are placed on the roof or on the south side of a building. Radiant energy from the Sun heats air or water in the solar collectors. An example of an active solar collector is shown in the figure above. The black metal plate absorbs radiant energy from the Sun. The absorbed energy heats water in pipes. A pump circulates the hot water to radiators in the building. The cooled water is pumped back to the collectors to be reheated.

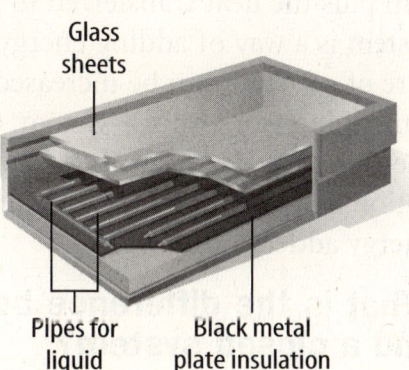

Thermodynamics

There is a way to increase the thermal energy of an object without adding heat. Have you ever rubbed your hands together to warm them? No heat is flowing to your hands, but they get warmer. Their thermal energy and temperature increase. You did work on your hands by rubbing them together. The work you did increased the thermal energy of your hands. Thermal energy, heat, and work are related. **Thermodynamics** is the study of the relationship among thermal energy, heat, and work.

How do heat and work increase thermal energy?

You can warm your hands by placing them near a fire. Heat is added to your hands by radiation. You can also rub your hands together as you hold them near a fire. The thermal energy of your hands increases even more. Both the heat from the fire and the work you do increase the thermal energy of your hands.

Think it Over

1. **Infer** How do you suppose passive solar heating systems got that name?

   _____
   _____
   _____

Picture This

2. **Conclude** Many parts of a solar collector are black. Why do you think they are black?

   _____
   _____
   _____

Reading Check

3. **Name** the two things that can increase thermal energy in a system.

   _____
   _____

In this example, your hands are a system. A system can be a group of objects, such as a galaxy or a car's engine. It can also be as simple as a ball. A system is anything you can draw a boundary around. The heat transferred to a system is the amount of heat that crosses the boundary. The work done on a system is the work done by something outside the system's boundary.

What is the first law of thermodynamics?

The <u>first law of thermodynamics</u> says that the increase in thermal energy of a system equals the work done on the system plus the heat transferred to the system. Doing work on a system is a way of adding energy to a system. So, the temperature of a system can be increased by adding heat to the system, doing work on the system, or both. The first law of thermodynamics is another way of stating the law of conservation of energy. The increase in energy of a system equals the energy added to the system.

What is the difference between an open system and a closed system?

A system is an open system if heat flows across the boundary or if work is done across the boundary. In other words, energy is added to the system. If no heat flows across the boundary and no outside work is done, then the system is closed. The first law of thermodynamics says that the thermal energy of a closed system doesn't change. Processes within the system may be changing one form of energy into another. But the total energy of the system stays the same. Energy cannot be created or destroyed. The total energy stays constant in a closed system.

What is the second law of thermodynamics?

When heat flows from a warm object to a cool object, the thermal energy of the warm object decreases. The thermal energy of the cool object increases. According to the first law of thermodynamics, the increase in thermal energy of the cool object equals the decrease in thermal energy of the warm object. Can heat flow from a cold object to a warm object? This process wouldn't break the first law of thermodynamics. The decrease in thermal energy of the cool object would only have to be equal to the increase in thermal energy of the warm object. However, the natural flow of heat from a cool object to a warm object never happens. It breaks another law.

Think it Over

4. Compare The law of conservation of momentum says that momentum can be transferred from one object to another, but it cannot be created or destroyed. How is this similar to the first law of thermodynamics?

_____

_____

_____

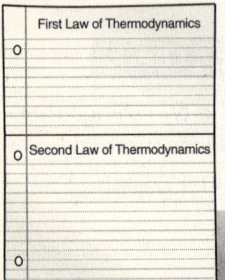

Find Main Ideas Make the following Foldable to take notes about the first and second laws of thermodynamics.

The **second law of thermodynamics** says that it is impossible for heat to flow from a cool object to a warmer object unless work is done. For example, if you hold an ice cube in your hand, no work is done. So, heat flows only from your hand to the ice, not the other way around.

Converting Heat to Work

When you rub your hands, friction converts the work you do on your hands to heat. Your hands become warmer. In this example, work was converted completely into heat. Is it possible to convert heat completely into work? This would not break or violate the first law of thermodynamics. But the second law of thermodynamics states it is impossible to build a device that converts heat completely into work.

A **heat engine**, such as a car, is a device that changes heat into work. The engine converts chemical energy in gasoline into heat. Then it changes some of the thermal energy into work by rotating the car's wheels. About 25 percent of the heat released by burning gasoline is converted into work. The rest is transferred to the engine's surroundings.

How does an internal combustion engine work?

The heat engine in a car is an internal combustion engine. In an **internal combustion engine**, the fuel burns inside the engine in chambers, or cylinders. Car engines usually have four or more cylinders. Each cylinder has a piston inside it. The pistons move up and down. Each up-and-down movement of the piston is called a stroke. The figure shows the four-stroke cycle of a car engine and explains how it works.

✔ **Reading Check**

5. Identify What does a car engine convert into heat?

_____

_____

Picture This

6. Interpret a Diagram Circle stroke where the spark plug ignites the fuel.

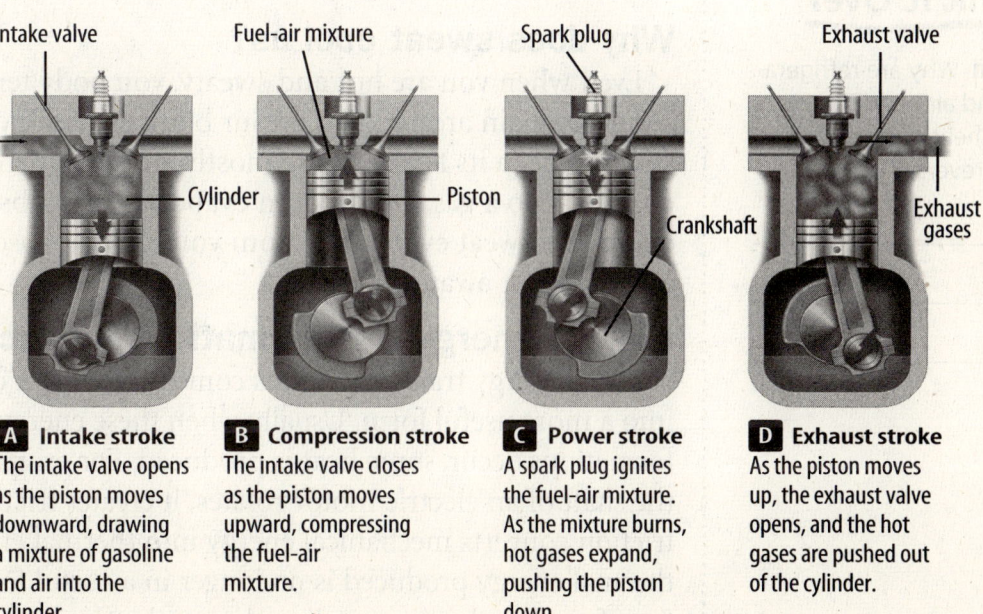

A **Intake stroke** The intake valve opens as the piston moves downward, drawing a mixture of gasoline and air into the cylinder.

B **Compression stroke** The intake valve closes as the piston moves upward, compressing the fuel-air mixture.

C **Power stroke** A spark plug ignites the fuel-air mixture. As the mixture burns, hot gases expand, pushing the piston down.

D **Exhaust stroke** As the piston moves up, the exhaust valve opens, and the hot gases are pushed out of the cylinder.

Heat Movers

The second law of thermodynamics says that it is impossible for heat to flow from a cool object to a warmer object unless work is done. So how does the inside of a refrigerator stay cold? A refrigerator is a heat mover. It does work to move heat from a cooler temperature to a warmer temperature. The energy to do the work is provided by an electric outlet.

How does a refrigerator transfer heat?

A refrigerator contains a substance called a coolant. Liquid coolant is pumped through an expansion valve and changed into a gas. When a coolant changes to a gas, it cools. The cold gas is pumped through pipes around the inside of the refrigerator. It absorbs thermal energy, and the inside of the refrigerator cools. The warm gas is then pumped to a compressor. The compressor does work by compressing the gas, making the gas even warmer. Thermal energy flows from the warm gas to the room. As the gas gives off heat, it cools and changes back into a liquid, and the cycle is repeated.

How does an air conditioner work?

An air conditioner is another heat mover. It is similar to a refrigerator, except that the warm air from the room is forced to pass over tubes containing the coolant. The warm air is cooled. The cooled air is forced back into the room. The thermal energy that is absorbed by the coolant is transferred to the air outdoors. Refrigerators and air conditioners are heat engines working in reverse. They use mechanical energy supplied by the compressor motor to move thermal energy from cooler areas to warmer areas.

Why does sweat cool us?

Even when you are hot and sweaty, you body temperature should remain around 37°C. Your body uses evaporation of sweat to keep its temperature mostly steady. When a liquid changes into a gas, as it does in evaporation, it absorbs energy. As sweat evaporates from your skin, it absorbs heat and carries it away.

How do energy transformations produce heat?

Many energy transformations convert one form of energy into a more useful form. Usually when these energy transformations occur, some heat is produced. For example, when the shaft of an electric motor rotates, it creates friction. The friction converts mechanical energy into thermal energy. The thermal energy produced is no longer in a useful form. It is transferred to the surroundings by conduction and convection.

Reading Check

7. Identify When the liquid coolant in a refrigerator passes through an expansion valve, it changes into _____.

Think it Over

8. Explain Why are refrigerators and air conditioners called heat engines working in reverse?

After You Read

Mini Glossary

first law of thermodynamics: the increase in thermal energy of a system equals the work done on the system plus the heat transferred to the system

heat engine: a device that changes heat into work

internal combustion engine: an engine in which the fuel burns inside the engine

second law of thermodynamics: heat cannot flow from a cool object to a warmer object unless work is done

solar collector: a device that absorbs radiant energy from the Sun

thermodynamics: the study of the relationship among thermal energy, heat, and work

1. Review the terms and definitions in the Mini Glossary. On the lines below, write a sentence that describes the relationship between a heat engine and an internal combustion engine.

   _____

   _____

2. Complete the chart below that describes the parts of the four-stroke cycle.

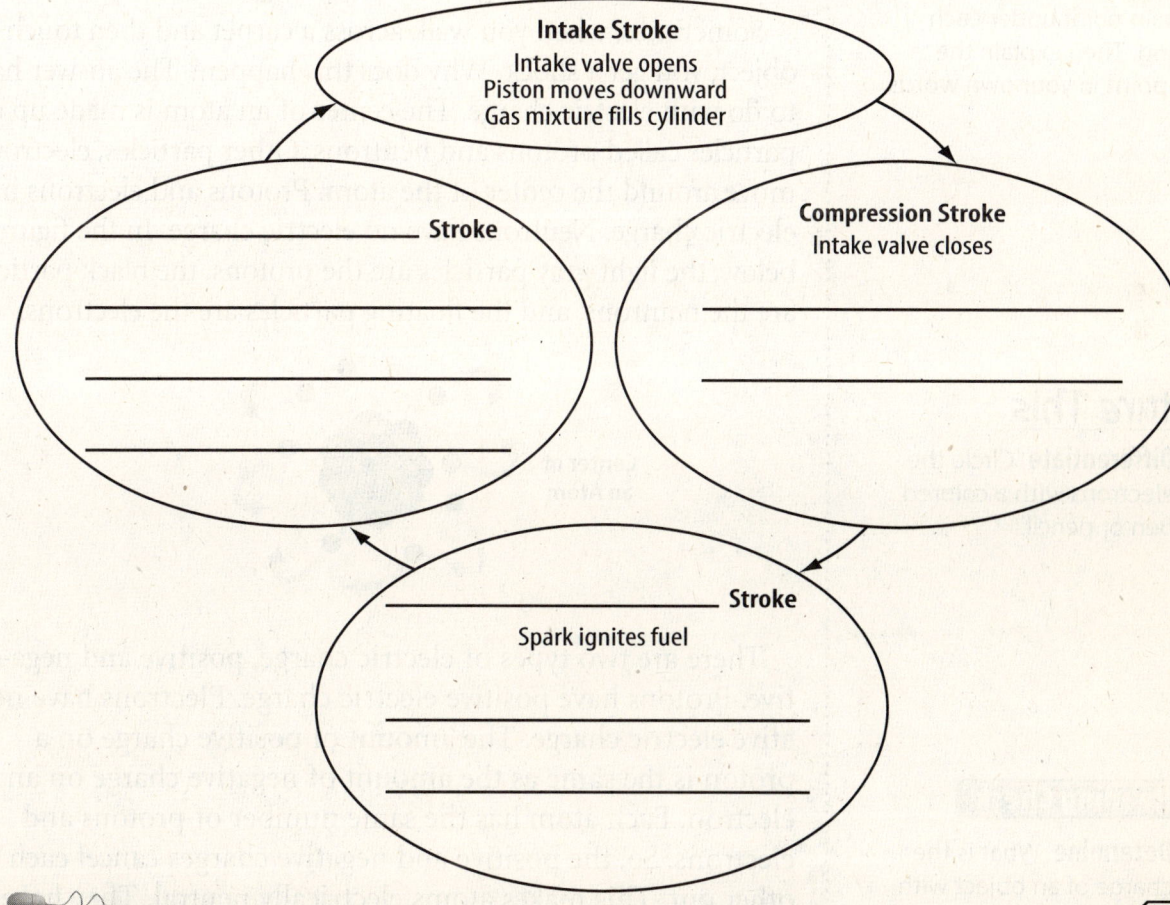

Science online Visit gpscience.com to access your textbook, interactive games, and projects to help you learn more about using heat.

End of Section

Reading Essentials **101**

chapter 7 Electricity

section ❶ Electric Charge

What You'll Learn
- how electric charges exert forces
- about conductors and insulators
- how things become electrically charged

Mark the Text

Identify Main Ideas Highlight the main point under each heading. Then explain the main point in your own words.

Picture This
1. **Differentiate** Circle the electrons with a colored pen or pencil.

✔ Reading Check

2. **Determine** What is the charge of an object with more electrons than protons?

_____

● Before You Read

Think about some electric objects that are plugged into an outlet. But cars, cell phones, and even wristwatches also use electricity. List three things you use every day that use electricity but do not plug in.

_____

● Read to Learn

Positive and Negative Charge

Sometimes, when you walk across a carpet and then touch an object, you get a shock. Why does this happen? The answer has to do with electric charge. The center of an atom is made up of particles called protons and neutrons. Other particles, electrons, move around the center of the atom. Protons and electrons have electric charge. Neutrons have no electric charge. In the figure below, the light gray particles are the protons, the black particles are the neutrons, and the floating particles are the electrons.

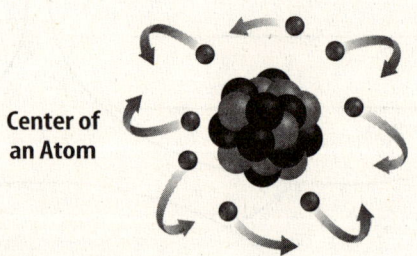

Center of an Atom

There are two types of electric charge, positive and negative. Protons have positive electric charge. Electrons have negative electric charge. The amount of positive charge on a proton is the same as the amount of negative charge on an electron. Each atom has the same number of protons and electrons. So, the positive and negative charges cancel each other out. This makes atoms electrically neutral. They have no overall electric charge. An atom becomes negatively charged if it gains extra electrons. An atom that loses electrons becomes positively charged. ✔

102 **CHAPTER 7** Electricity

How is electric charge changed?

Electrons can move from atom to atom or from object to object. For example, the electrons in the soles of your shoes are closer together than the electrons in a carpet. When you walk on a carpet, some of the looser electrons move from the carpet to the soles of your shoes.

Now, the soles of your shoes have more electrons than protons. They are negatively charged. The carpet has fewer electrons than protons. It is positively charged. The transfer of electrons changed the electric charge of each object. **Static electricity** is the buildup of electric charges on an object. When there is static electricity, the electric charges of the object are not balanced.

Is new electrical charge created?

The electrons that moved to your shoe are not new electrons. The **law of conservation of charge** states that charge can be transferred from one object to another, but it cannot be created or destroyed. An object becomes charged when electric charges have moved from one place to another.

What happens when electrical charges move?

Have you ever taken clothes out of a dryer and had them cling together? Look at the figure. Opposite electric charges attract each other. They tend to move toward each other. Electric charges that are the same repel each other. They tend to move away from each other.

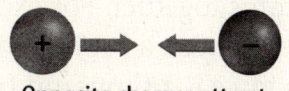
Opposite charges attract

When clothes move around in a dryer, the atoms in some of the clothes lose electrons. Those clothes become positively charged. The atoms in other clothes gain electrons and become negatively charged. The clothes have opposite charges. Objects that have opposite charges attract each other, so the clothes cling together.

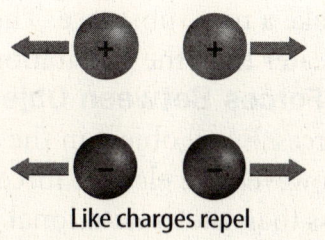
Like charges repel

How do charges exert forces?

The electric force between two charged objects depends on how far apart the objects are. The electric force between two charges decreases as the charges move farther apart.

The electric force also depends on the amount of charge on each object. When the amount of charge on one of the objects increases, the electric force increases.

FOLDABLES

A Build Vocabulary As you read this section, make the following vocabulary Foldable. Write the definition for each vocabulary word under its tab.

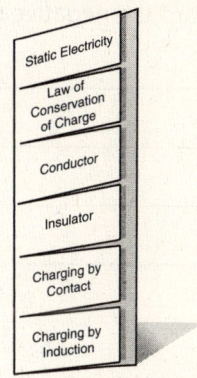

Picture This

3. **Illustrate** Look at the figure of like and unlike charges. Use two different colors of highlighters. Highlight the negative charges in one color and the positive charges in the other color. Notice the charges only attract when the colors are different.

Picture This

4. Explain Look at the figure of electric fields. Why do the arrows point outward from the positive field? Why do they point inward toward the negative field?

_____
_____
_____

What are electric fields?

When you rub a balloon on your hair, your hair will move and touch the charged balloon. What makes your hair move? The rubbing causes electrons from your hair to move to the balloon. The balloon is now negatively charged. Your hair and the balloon are attracted to each other because they have opposite electric charges.

There is an electric field, or area, around every electric charge. The electric field exerts a force. If a charge is in an electric field, it will be pushed or pulled by the field. The figure shows two electric fields. The arrows show the direction a positive charge moves in each electric field. Your hair should move to the balloon because it has the positive charge.

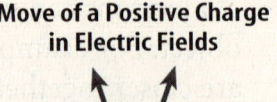
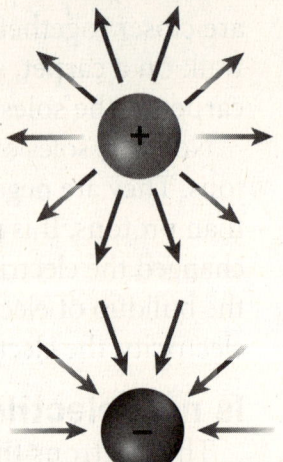

Move of a Positive Charge in Electric Fields

How do electric and gravitational forces compare?

The force of gravity between you and Earth seems very strong. However, electric forces are much stronger. Electric forces between the protons and the electrons hold the particles in atoms together. For example, there is an attractive electric force between a proton and an electron in a hydrogen atom. This force is a thousand trillion times greater than the attractive gravitational forces between the same proton and electron.

Forces Between Atoms Atoms are also are held together by electric forces. Electric forces between atoms cause chemical bonds. Chemical bonds form when two atoms join to make a new substance. These electric forces are also much greater than the gravitational forces between the atoms.

Forces Between Objects Every object exerts electric forces. Every object in the universe also produces gravity. However, the electric forces between most objects are much less than the gravitational forces between them. Most objects are close to being electrically neutral, neither positive nor negative. Because of this, there is usually very little electric force between two objects. But, if even a small amount of charge moves from one object to another, the electric force between the two objects may be noticed.

Remember the hair moving to the charged balloon. If a few electrons in just one hair are transferred to the balloon, an electric force is created. This electric force is strong enough to overcome the force of gravity on the one hair.

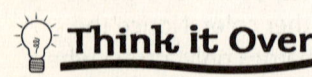 **Think it Over**

5. Explain in your own words why usually there is very little electric force between two objects.

_____
_____
_____

Conductors and Insulators

Remember the example of electrons moving from the carpet to your shoe? If you reach for a metal doorknob after walking on carpet, you might see a spark. Electrons moving from your hand to the doorknob cause the spark. How did those electrons move from your shoe to your hand?

What is a conductor?

Electrons can move more easily in some materials than in others. A **conductor** is material in which electrons can move easily. Your skin is a better conductor than your shoes. Electrons move from your shoes to your skin, spreading to your hands. The best electric conductors are metals. Atoms in metals have electrons that are able to move easily through the metal. Copper is one of the best conductors.

What is an insulator?

An **insulator** is material in which electrons cannot move easily. In insulators, electrons are held tightly to atoms. The plastic coating around an electric wire keeps you from getting a dangerous electric shock when you touch the wire. Wood, rubber, and glass are other good insulators.

Charging Objects

Just like the clothes in the dryer, when two materials are rubbed together, electrons can be transferred between them. One object will have a negative charge. The other will have the same amount of positive charge. Transferring charges by touching or rubbing is called **charging by contact**.

How can something be charged at a distance?

Remember, electric forces change when objects move closer together. If a charged object is moved near a neutral object, electrons on the neutral object will move around.

Think about the balloon that was charged by rubbing it on your hair. The charged balloon doesn't need to touch the hair to make the hair move toward it. The same is true if you hold the charged balloon close to a wall. The extra electrons on the balloon repel the electrons in the wall. The electrons in the wall move away from the balloon. Now there is a positively charged area on the wall. The negatively charged balloon is attracted to the positively charged area of the wall. A charged object rearranging the electrons on a nearby neutral object is **charging by induction**. The wall was charged by induction. The balloon will stick to the wall. An electric force holds it there. ✓

Think it Over

6. Apply Which is a better conductor, a flagpole or a flag? Why?

_____
_____
_____
_____

Think it Over

7. Explain why it is hard for electrons to move through an insulator.

_____
_____
_____
_____
_____

Reading Check

8. Compare What is the difference between charging by contact and charging by induction?

_____
_____
_____
_____
_____

Reading Essentials **105**

What is lightning?

Have you ever seen lightning hit the ground? Lightning is a large static discharge—a transfer of charge between two objects. It happens if there is a buildup of static electricity.

A large amount of static electricity is formed when air moves around in thunderclouds. Areas of positive and negative charge build up. When enough charge builds up, there is a static discharge between the cloud and the ground. As the electric charges move through the air, they run into atoms and molecules, making the atoms and molecules light up.

What is thunder?

Lightning makes a bright light. It also creates powerful sound waves. Thunder is the sound that lightning makes. The electric energy in a lightning bolt rips electrons off atoms in the air. This causes great amounts of heat. The air temperature around the lightning bolt can be as high as 25,000°C. The heat makes the air around the lightning bolt move faster. This rapid movement of air produces the sound waves that you hear as thunder.

Why is grounding important?

Lightning can cause damage and injury because it releases a great amount of energy. One way to avoid the damage is to make the charges flow to Earth's surface. Earth is a large neutral object that is also a conductor. Because Earth is so large, it can absorb a great amount of excess charge. Grounding provides a path for electric charges to move to Earth. For example, a metal lightning rod on top of a building provides a path to move excess charges to Earth's surface.

💡 Think it Over

9. Analyze Why do you think a lightning rod is made of metal?

_____
_____
_____

Detecting Electric Charge

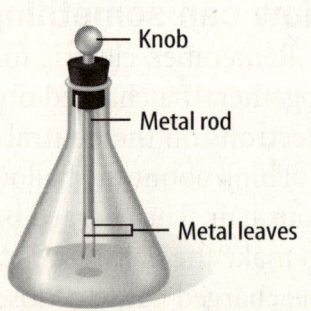

An electroscope can detect when an object has an electric charge. One type of electroscope is a glass beaker with a metal rod inside it, as shown. The metal rod connects to a knob at the top of the beaker. There are two metal branches, or leaves, at the bottom of the metal rod. The metal leaves hang down when there is no charge to the rod. When an object with a negative charge touches the knob, electrons travel down the rod to the leaves. Both leaves gain negative charges. When an object with a positive charge touches the knob, it attracts electrons that move up the rod. The leaves have a positive charge. When the leaves have a charge, they repel each other and spread apart.

Picture This

10. Draw On the figure, draw what the leaves look like if they have a charge.

After You Read

Mini Glossary

charging by contact: transferring charges by touching or rubbing

charging by induction: when electrons on a neutral object are moved by a charged object

conductor: a material in which electrons can move easily

insulator: a material in which electrons cannot move easily

law of conservation of charge: charge can be transferred from one thing to another, but it cannot be created or destroyed

static electricity: the buildup of electric charges on an object

1. Read the definitions of an insulator and a conductor in the Mini Glossary above. Use the words in a sentence that shows that you understand them.

   _____

   _____

2. Column 1 lists some of the concepts you learned about in this section. Column 2 gives a fact about each concept. Write the letter of the fact on the line next to the concept that matches it.

Column 1	Column 2
_____ 1. transferring charge	a. static electricity is discharged between a cloud and the ground
_____ 2. conservation of charge	b. electrons cannot move easily in some materials
_____ 3. insulator	c. electrons can move from one thing to another
_____ 4. lightning	d. charge cannot be created or destroyed

3. You highlighted the main points to help you understand electric charge. How did you decide what the main points were?

   _____

   _____

 Visit **gpscience.com** to access your textbook, interactive games, and projects to help you learn more about electric charge.

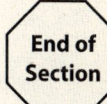

Electricity

section ❷ Electric Current

What You'll Learn
- what makes current flow
- how batteries work
- what causes electric resistance
- what Ohm's law says

Mark the Text

Identify Details Use one color to highlight each question heading. Then use another color to highlight the answer to the question.

Applying Math

1. **Convert** If a million has 6 zeroes, and a billion has 9 zeroes, how do you write 6,250 million billion using numbers only?

● **Before You Read**

Have you noticed that one end of a battery is marked with a plus sign and the other end with a minus sign? What happens if you put the batteries in a flashlight in the wrong direction?

_____

● **Read to Learn**

Current and Voltage Difference

You have read about the ways electric charges move. One example is the spark that can jump between your hand and a metal doorknob. **Electric current** is the net movement of electric charges in one direction.

To understand net movement, compare the movement of electrons in all materials to the movement of electrons in an electric current. In all materials, electrons are always moving in every direction. Since the electrons are not moving in the same direction, there is not an electric current. When an electric current flows in a wire, the electrons still move in all directions, but they also drift in the direction that the current flows. The drifting of the electrons is the net movement in one direction.

Think of a ball bouncing down a staircase. The ball bounces up and down as it hits the stairs. But the ball is always moving toward the bottom of the stairs. The force of gravity causes the ball's motion to be downward. The flow of electric current is similar.

Electric current is measured in units called amperes. Amperes are also called amps. The symbol for amperes is the letter A. Amperes measure the electrons that flow past one point. One ampere is equal to 6,250 million billion electrons moving past a point every second.

What is voltage difference?

Even though the electrons are moving in all directions, an electric force acts on the charges to make them flow in one direction. Voltage is the electric force that makes charges move.

Voltage is also like the force that acts on water in a pipe. Water flows from higher pressure to lower pressure. In the same way, electric charge flows from higher voltage to lower voltage. A **voltage difference** is related to the force that makes electric charges flow. Voltage difference is measured in units called volts. The symbol for volts is V.

What is an electric circuit?

Look at the figure. It shows an electric current doing work by lighting a lightbulb. Electric current must have a closed "loop-like" path to follow. If there is no closed path to follow, the current stops. A **circuit** is a closed path that electric current follows. If the circuit in the figure is broken by taking away one part, such as the battery or the lightbulb, current will not flow. It will also not flow if a wire is broken or cut. The lightbulb will not light.

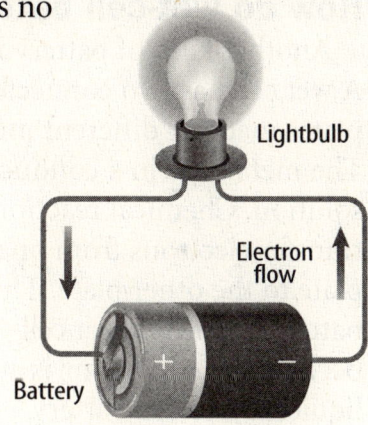

Batteries

A circuit needs a voltage difference to keep electric current flowing in it. A battery can provide the voltage difference that keeps electric current flowing. Look at the figure of the circuit again. The positive end and the negative end of a battery are called the terminals. When a closed path connects the terminals, the current will flow.

How do dry-cell batteries work?

The batteries used in a flashlight are called dry-cell batteries. Look at the figure of a dry-cell battery. The battery has two electrodes. One electrode is a carbon rod. The other electrode is a zinc container.

Around the electrodes is a moist paste. The paste is called an electrolyte. The electrolyte contains chemicals that are conductors. The electrolyte lets charges move from one electrode to the other electrode. This kind of battery is called a dry cell because the electrolyte is a paste, not a liquid.

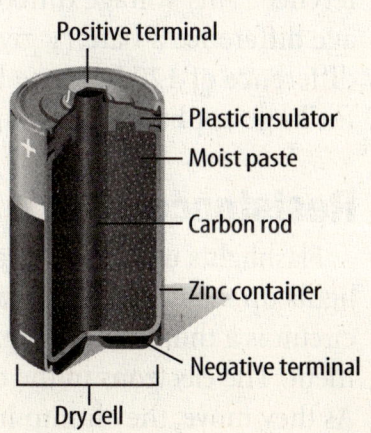

Picture This

2. Describe Look at the figure of a circuit. Which direction do electrons flow in the circuit, away from the negative terminal or away from the positive terminal?

_____

_____

FOLDABLES

B Build Vocabulary Make a Foldable as shown. Write the definitions under the tabs and add information as you read this section.

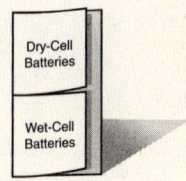

Picture This

3. Locate What is used as an electrolyte in the dry-cell battery?

_____

_____

Reading Essentials **109**

Think it Over

4. Describe when the chemical reactions in a dry-cell battery happen.

_____

_____

Picture This

5. Compare What two things are shown in both the figure of the dry-cell battery and the figure of the wet-cell battery?

_____

_____

When the two terminals are connected in a circuit, there is a reaction between the zinc and the chemicals in the electrolyte. Electrons move between some of the compounds in this chemical reaction. This makes the carbon rod become positive. The positive terminal is marked with a plus sign (+). Electrons build up on the zinc, making it the negative terminal. The negative terminal is marked with a minus sign (−). The voltage difference between the terminals makes current flow through a closed circuit.

How do wet-cell batteries work?

Another kind of battery is the wet-cell battery shown below. A wet cell has two connected plates made of different metals. The metals are in a conducting solution. Chemical reactions transfer electrons from one plate to the other plate. This battery is called a wet cell because the conductor is a liquid. A wet-cell battery contains several wet cells that are connected.

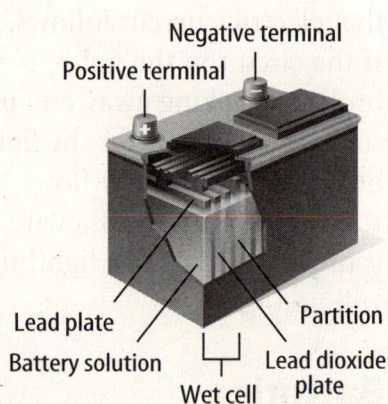

What is a lead-acid battery?

Lead-acid batteries are wet-cell batteries. They are usually used in cars. A lead-acid battery has six separate wet cells that are connected together. The cells are made of lead and lead dioxide plates. The plates are in a sulfuric acid solution. A chemical reaction gives a voltage difference of about 2 V in each cell. There are six cells, so the total voltage difference is 12 V.

How are electric outlets different from batteries?

Electric outlets, such as wall sockets, also give a voltage difference. This voltage difference usually is higher than the voltage difference a battery gives. Most wall sockets give a voltage difference of 120 V. Some have a voltage of 240 V that is used for large appliances, such as electric ovens and clothes dryers.

Resistance

Flashlights use dry-cell batteries to make the current that lights up the lightbulb. What makes a lightbulb glow? Part of the circuit is a thin wire through the bulb. The wire is called a filament. The electrons in the current flow through the filament. As they move, they bump into the metal atoms in the filament.

The electrons bump into the metal atoms, turning some of their *electric* energy into *thermal* energy. The metal filament gets hot enough to glow. The radiant energy lights up the room. ✓

How do materials resist current?

Electric current loses energy when it moves through material because of resistance. **Resistance** is the tendency for a material to oppose, or go against, the flow of electrons. Resistance turns electric energy into heat and light.

Almost all materials have electric resistance. Materials that are electric conductors have less resistance than materials that are electric insulators. Resistance is measured in units called ohms. The symbol for ohms is the Greek letter omega, or Ω.

Copper is a good conductor. It has low resistance to the flow of electrons. Copper is used to make electric wires because only a small amount of electric energy is changed to thermal energy when current flows in copper wires. ✓

What can affect resistance?

The temperature, length, and thickness of a material can affect the electric resistance of the material. Usually, the hotter something is, the more resistance it has. The resistance of an object also depends on its length and thickness. The longer the circuit is, the more resistance it has. Resistance also increases as the wire gets thinner. There is not as much space for the electrons to move around, so they bump into atoms more often.

A lightbulb filament is a thin piece of wire made into a short coil. The wire is a metal called tungsten. The uncoiled wire is about 2 m long and very thin. Tungsten is a good conductor, but since the wire is so long and thin, it has resistance. The resistance makes the filament get hot and glow. The more resistance a filament has, the brighter it glows.

The Current in a Simple Circuit

A simple electric circuit has three main parts. First, it has a source of voltage difference, such as a battery. Second, it has a device that has resistance, like a lightbulb. Third, it has conductors, such as wires. The conductors connect the resistance device to the battery terminals. When the wires are connected to the battery terminals, the path is closed and current flows.

Look at the figures of two electric circuits on the next page. Each circuit is a battery connected to a lightbulb by wires and a rod. The circuit on the right is shorter because the wires are closer together on the rod. That circuit has less resistance. The lightblub glows brighter in the circuit with less resistance.

Reading Check

6. List the three energies used to make a lightbulb glow.

_____
_____
_____

Reading Check

7. Explain Why is copper a good conductor?

_____
_____
_____

Think it Over

8. Recognize Cause and Effect How does the size of a filament affect its glow?

_____
_____
_____
_____
_____

Picture This

9. Compare In which circuit will the light be less bright, the one on the left or the one on the right?

_____

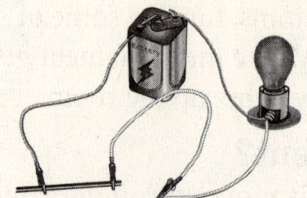 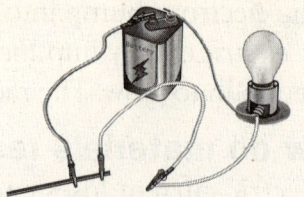

More resistance   **Less resistance**

The voltage difference, current, and resistance in a circuit are related. If the voltage difference stays the same as the resistance decreases, the current in the circuit increases. If the wire is short, the lightbulb will be brighter. If the resistance doesn't change, increasing the voltage difference increases the current. If you use a larger battery, the lightbulb will be brighter.

What is Ohm's law?

There is a relationship between voltage difference, current, and resistance in a circuit. **Ohm's law** states that the current in a circuit equals the voltage difference divided by the resistance. If I stands for electric current, Ohm's law can be written as:

Reading Check

10. Identify What do R, V, and I stand for in the equation for Ohm's law?

_____
_____
_____

$$\text{current (in amperes)} = \frac{\text{voltage difference (in volts)}}{\text{resistance (in ohms)}}$$

$$I = \frac{V}{R}$$

Ohm's law can also be used to measure resistance. Change the equation so that resistance, R, is alone on one side. Do this by multiplying both sides of the equation by R. The new equation is:

$$R = \frac{V}{I}$$

Applying Math

11. Solve the equation to show how $I = \frac{V}{R}$ becomes $R = \frac{V}{I}$.

_____
_____
_____
_____

Suppose a current of 0.5 A flows in a 75-W lightbulb. The voltage difference between the ends of the filament is 120 V. Find the resistance of the filament.

$$R = \frac{V}{I}$$
$$= \frac{120}{0.5}$$
$$= 240$$

The resistance is 240 Ω.

● After You Read
Mini Glossary

circuit: a closed path that electric current follows
electric current: the rate at which electric charges move in one direction past one point
Ohm's law: the current in a circuit equals the voltage difference divided by the resistance

resistance: the tendency for a material to oppose the flow of electrons, changing electric energy into heat and light
voltage difference: something related to the force that makes electric charges flow

1. Read the definition of resistance in the Mini Glossary. Rewrite the definition in your own words on the lines below.

2. Complete the table below to describe a simple circuit. The first column lists the parts of a circuit. In the second column, give an example of each part of a circuit. Under the heading Function, write a short description of what job each part does in the circuit.

Parts of a Simple Circuit		
Part	**Example**	**Function**
Voltage difference		
Source of resistance		
Conductors		

3. As you read this section, you highlighted the question headings and their answers. Why was using two colors helpful?

 Visit **gpscience.com** to access your textbook, interactive games, and projects to help you learn more about electric current.

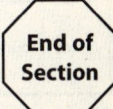

Reading Essentials 113

Chapter 7 Electricity

section ③ Electrical Energy

What You'll Learn
- the difference between series and parallel circuits
- why circuit breakers and fuses are important
- how electric power is calculated

Study Coach

Create a Quiz As you read this section, think of five quiz questions. Write them down. After you read the section, answer the quiz questions you wrote.

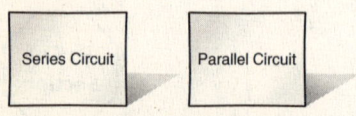

Gather Information Use two quarter-sheets of notebook paper to organize information about series circuits and parallel circuits. Include terms and calculations.

● Before You Read

Do adults ever remind you to turn off the lights when you leave a room? Like many things we use, electricity is not free. How does the electric company know how much electric energy you use? Write your answer on the lines below.

_____

_____

● Read to Learn

Series and Parallel Circuits

Think of your home. How many things are plugged into electric outlets? You might think of lamps, stereos, televisions, and clocks.

As you read in the last section, a circuit includes three parts. The first part is something that provides a voltage difference. It can be a battery or an electric outlet. The second part is something that uses electric energy and provides resistance. Lightbulbs and hair dryers are two examples. The third part is a conductor that connects the other parts. An example of a conductor is a wire. These three parts form a closed path for the electric current to travel on.

Think about using a hair dryer. The dryer needs to be plugged into an electric outlet. A generator at a power plant produces the voltage difference that ends up at the wall outlet. The voltage difference makes electric charges move when the circuit is complete. The dryer and the circuit in your house have conducting wires. The wires carry the current.

When you turn the hair dryer on, you close the circuit. The hair dryer turns electrical energy into thermal energy and mechanical energy. Mechanical energy is the energy that moves the fan in the hair dryer.

When you turn the hair dryer off, you open the circuit. This breaks the path of the current. To use electrical energy, you need a complete circuit. There are two kinds of circuits, series circuits and parallel circuits.

What is a series circuit?

One kind of circuit is called a series circuit. In a **series circuit**, the current has only one loop to flow through. Series circuits are used in flashlights. ☑

How does an open circuit affect a series circuit?

Some older strings of holiday lights will not work if just one lightbulb is burned out. The lights are connected in a series circuit. In a series circuit, the parts are wired one after another. The amount of current is the same through every part. When any part of a series circuit is disconnected, no current flows through the circuit. This is called an open circuit. One burned-out bulb makes the string of lights an open circuit.

What about parallel circuits?

What would happen if your home were wired in a series circuit? If you turned off one light, the circuit would be open. All the other lights and appliances in your house would go off, too. This is why houses are wired with parallel circuits. **Parallel circuits** have at least two paths for current to move through.

Look at the parallel circuit in the figure. The parallel circuit breaks up current into two separate paths. This lowers the resistance. Remember Ohm's law from the last section. More current flows through the paths that have lower resistance.

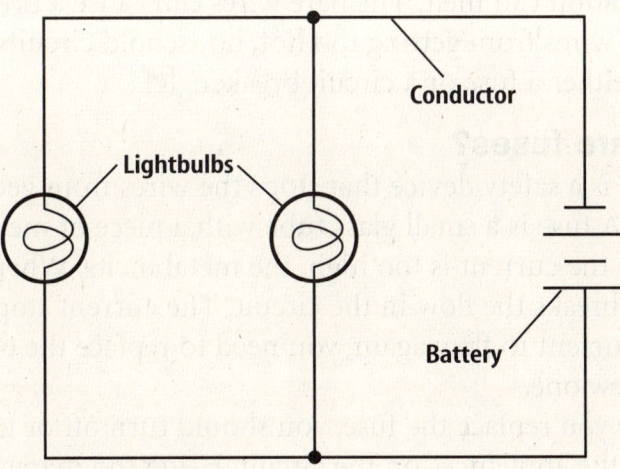

Houses, cars, and most electric systems use parallel circuits. When one path is opened, the current still flows through the other paths. One part can be turned off without turning off the whole circuit.

✓ Reading Check

1. Identify How many loops does a series circuit have?

_____

Picture This

2. Illustrate In the space below, draw a parallel circuit that has three branches. Label the parts of the circuit.

Picture This

3. Highlight In the figure, use a highlighter to trace a path of a circuit from the meter to the wall socket on the far side of the room.

Reading Check

4. Describe What happens to the current flow in a circuit when more appliances are added to the circuit?

_____

_____

Household Circuits

Many things in your house use electric energy. You don't see all the wires, because they are hidden behind the walls, ceiling, and floors. The wiring is mostly a combination of parallel circuits. The circuits are connected in an organized way.

Look at the figure below showing the wiring in a house. There is a main switch and a circuit breaker or fuse box. These are like the electric headquarters for the house. Parallel circuits branch out from the circuit breaker or fuse box. The circuits run to wall outlets, appliances, and lights.

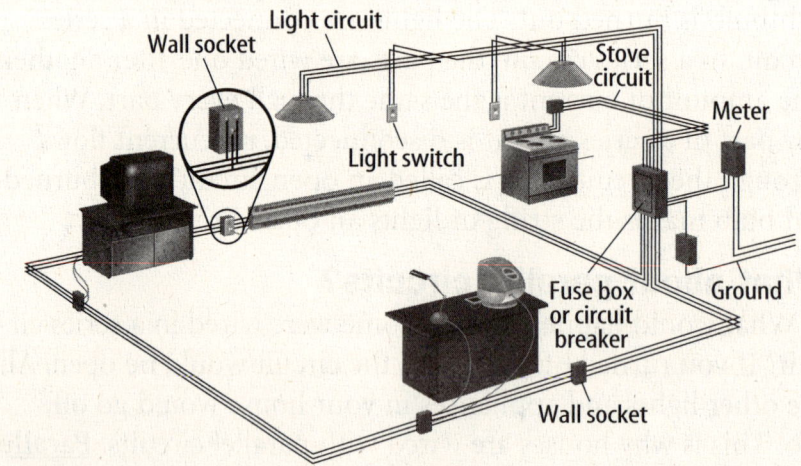

Many appliances use current from the same circuit. If more appliances are plugged in on a circuit, more current will flow through the wires. As more current flows through the wires, more heat is produced in the wires. If the wires get too hot, the insulation can melt. The bare wires can cause a fire. To keep the wires from getting too hot, household circuits include either a fuse or a circuit breaker.

What are fuses?

A fuse is a safety device that stops the wires from getting too hot. A fuse is a small glass tube with a piece of metal inside. If the current is too high, the metal melts. When it melts, it breaks the flow in the circuit. The current stops. To get the current to flow again, you need to replace the old fuse with a new one.

Before you replace the fuse, you should turn off or unplug some of the appliances on the circuit. Using too many appliances at once is the main cause for a blown fuse.

How does a circuit breaker work?

A circuit breaker is another device that keeps a circuit from overheating. A house usually has a metal box, called a breaker box, which contains many circuit breakers. A circuit breaker is a switch, like a light switch, that has a piece of metal inside. If the current in the circuit is too high, the metal warms up and bends. When the metal bends, it flips the switch and opens the circuit. The flow of current stops.

You can usually reset the circuit breaker by flipping the switch inside the breaker box back to its original position. But, before you flip the switch, you should turn off or unplug some of the appliances on the circuit. Otherwise, the circuit breaker will flip the switch off again.

Electric Power

Electrical energy is useful because it is easy to change into other kinds of energy. For example, it can be changed to thermal energy in a hair dryer. It can also be turned into light, or radiant energy, in a lamp or mechanical energy in a fan. **Electric power** is the rate at which electrical energy is changed into another form of energy.

Different appliances use different amounts of electric power. Appliances are usually marked with a power rating. The power rating tells how much power the appliance uses. Appliances that have electric heating elements, such as ovens and hair dryers, usually use the most power.

How is electric power calculated?

The amount of electric power something uses depends on the voltage difference and the current. You can use the following equation to calculate electric power.

electric power = current × voltage difference
(in watts)   (in amperes)   (in volts)

$$P = IV$$

The unit for power is the watt. The abbreviation for watt is W. The watt is a small unit of power. Because of this, electric power usually is measured in kilowatts. *Kilo-* means "thousand." One kilowatt equals 1,000 watts. The abbreviation for kilowatt is kW.

Think it Over

5. **Apply** Think about two clocks. One is a digital clock with a lighted display. The other is an analog clock with a lighted face and hands that move. Which clock would use more electric power? Why?

_____
_____
_____

Applying Math

6. **Calculate** The current in an electric clothes dryer is 15 A when it is plugged into a 240-V outlet. How much power does the clothes dryer use? Show your work. Show your answer in kW.

_____
_____
_____

Think it Over

7. Infer Why do you think electric companies charge by the amount of electric energy used as opposed to the amount of electric power used?

_____
_____
_____

How is electrical energy calculated?

Using electric power costs money. However, electric companies charge by the amount of electrical energy used, not the amount of electrical power. Electrical energy usually is measured in units of kilowatt hours. The abbreviation for kilowatt hours is kWh. Kilowatt hours can be calculated using this equation:

electrical energy = electric power × time
(in kWh)   (in kW)   (in hours)

$$E = Pt$$

How much does it cost to use electric energy?

You can figure out how much it costs to use an appliance. You do this by multiplying the electric energy used by the cost of each kilowatt hour. Suppose you leave a 100-W lightbulb on for 5 h. The amount of electric energy it uses is

$$E = Pt = (0.1 \text{ kW})(5 \text{ h}) = 0.5 \text{ kWh}$$

If the power company charges $0.10 per kWh, the cost of using the light for 5 h is

cost = (kWh used)(cost per kWh)
= (0.5 kWh)($0.10/kWh) = $0.05

So, in this example, it costs five cents to use a 100-W lightbulb for 5 h.

The cost of using some household appliances is given in the table. The cost of $0.09 per kWh was used in the calculations.

Picture This

8. Observe Which appliance has the greatest monthly cost?

_____

Cost of Using Home Appliances

Appliance	Hair Dryer	Stereo	Color Television
Power rating	1,000	100	200
Hours used daily	0.25	2.0	4.0
kWh used monthly	7.5	6.0	24.0
Cost per kWh	$0.09	$0.09	$0.09
Monthly cost	$0.68	$0.54	$2.16

After You Read

Mini Glossary

electric power: the rate at which electric energy is changed into another form of energy

parallel circuit: a circuit with at least two paths for current to move through

series circuit: a circuit with just one loop for current to move through

1. Read the terms and definitions in the Mini Glossary above. On the lines below, write a sentence that shows your understanding of the difference between a series circuit and a parallel circuit.

   _____

   _____

2. Complete the graphic organizer.

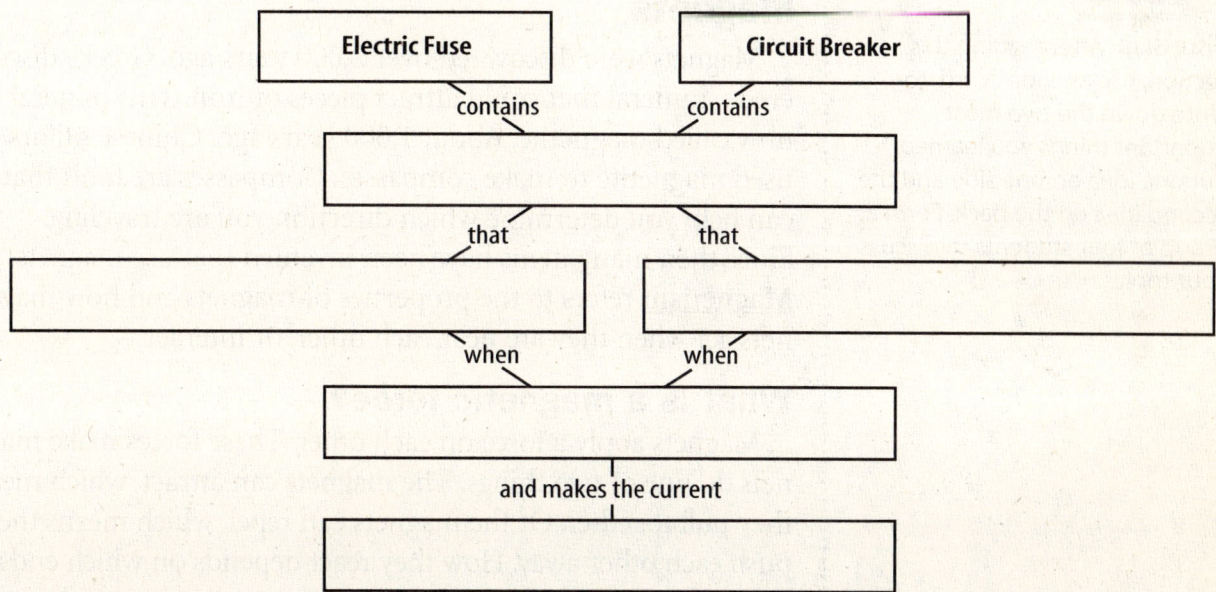

3. You used the create-a-quiz strategy as you read this section. Look at the quiz questions you wrote. How many of them can you answer correctly? Did this strategy help you understand and remember what you read?

   _____

   _____

 Visit gpscience.com to access your textbook, interactive games, and projects to help you learn more about electrical energy.

Magnetism and Its Uses

section ❶ Magnetism

What You'll Learn
- how a magnet applies force
- how temporary and permanent magnets act
- magnetic materials and magnetic domains

Study Coach

Discussion After reading this section, use an index card to write down the two most important things you learned. Put one idea on one side and the second idea on the back. Form a group of four students to discuss your topics.

● Before You Read

Think about a magnet that you have used. Tell what it looked like and the kinds of materials it attracted.

_____

_____

● Read to Learn

Magnets

Magnets were discovered over 2,000 years ago. Greeks discovered a mineral that could attract pieces of iron. This mineral is now called magnetite. About 1,000 years ago, Chinese sailors used magnetite to make compasses. Compasses are tools that can help you determine which direction you are traveling. Since then many items have been invented that use magnets. <u>Magnetism</u> refers to the properties of magnets and how magnets act when they are near each other, or interact.

What is a magnetic force?

Magnets apply a force on each other. These forces make magnets do one of two things. The magnets can attract, which means they pull together. Or the magnets can repel, which means they push each other away. How they react depends on which ends of the magnets are close together. Two magnets interact with each other even before they touch. As the magnets move closer together, the force between them increases. As the magnets move farther apart, the force decreases.

What is a magnetic field?

The way magnetic forces interact with each other is caused by magnetic fields. The <u>magnetic field</u> exerts a force on other magnets and objects that are made of magnetic material. The magnetic force is strongest close to the magnet.

✓ Reading Check

1. **Infer** What causes the ways magnetic forces interact with each other?

_____

Lines can represent a magnetic field. The figures below show what magnetic fields might look like. The arrows show that magnetic fields have direction.

What are magnetic poles?

<u>Magnetic poles</u> are the places on a magnet that exert, or put forth, the strongest magnetic force. All magnets have a north and a south pole. As shown, the north and south poles are at the opposite ends of a bar magnet. The lines that represent the magnetic field are closest together at the poles.

The next figures show the magnetic poles of two magnets with different shapes. A horseshoe-shaped magnet has its north and south poles at its two ends. The magnetic field lines start at the north pole and end at the south pole. Look at the disk magnet and the bar magnet. Like all magnets, the magnetic field lines of the disk magnet and the bar magnet go from north pole to south pole.

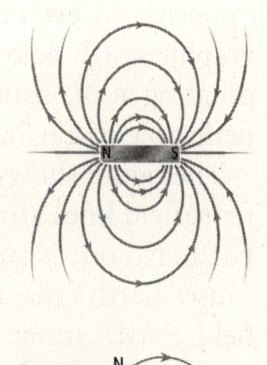

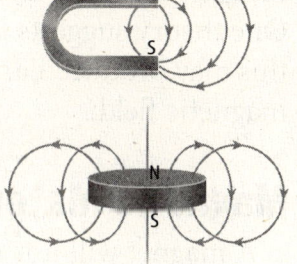

Magnetic Fields

Picture This
2. **Highlighting** Using a highlighter, trace the lines of the magnetic fields for all three types of magnets shown on this page. At which magnetic pole did you always start?

_____

How do magnetic poles interact?

Remember, two magnets can either attract or repel each other. This depends on which poles of the magnets are placed close together. Two north poles will repel each other. The same is true for two south poles. However, a north pole and a south pole always attract each other. Like magnetic poles repel each other and unlike poles attract each other.

How do magnets affect compasses?

A compass needle is a small bar magnet. The force exerted by another bar magnet will make the compass needle turn. The needle turns until it lines up with magnetic field lines. The south pole in a compass needle is attracted to the north pole of a magnet.

How does Earth act like a magnet?

Earth is like a giant bar magnet. It is surrounded by a magnetic field and has a north and a south magnetic pole. Earth also has a north and a south geographic pole. The geographic poles are at opposite ends of Earth—one in the north and one in the south. These are different from the magnetic poles. The south magnetic pole is near the geographic north pole.

FOLDABLES

A **Organizing Information** Make the following Foldable to help you organize information about magnetic fields, magnetic poles, and how magnets repel and attract.

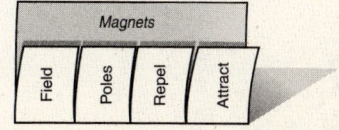

Why does a compass point north?

A compass needle lines up with Earth's magnetic field lines. The needle always points toward Earth's geographic north pole. Remember that magnetic poles are attracted only to their opposite. So, even though the compass is pointing at the geographic north pole of Earth, the north pole of the compass is pointing at the south magnetic pole of Earth.

The figure shows the magnetic field lines around Earth. No one is sure what causes Earth's magnetic field. Earth's inner core is made of iron and nickel. One theory suggests that this may produce Earth's magnetic field.

Picture This

3. **Determine** Write an N on the figure to show where Earth's north magnetic pole is. Write an S on the figure to show where Earth's south magnetic pole is.

Magnetic Materials

A magnet will not attract all metal objects. For example, a magnet will not attract aluminum foil. Only a few metals, such as iron, cobalt, and nickel are attracted to magnets. These metals can be made into permanent magnets. Permanent magnets keep their magnetism even after they have been removed from a magnetic field. Think back to what you have learned about electrons. Recall that electrons have magnetic properties. In most elements, the magnetic properties of the electrons cancel out. But in iron, cobalt, and nickel, they don't cancel out. Each atom in these metals acts like a small magnet with its own magnetic field. ☑

Even though the atoms in iron, cobalt, or nickel have magnetic fields, objects made from them do not always act like magnets. For example, a nail is made from iron. If you hold an iron nail close to a refrigerator door and then let it go, it will fall to the floor. However, you can make the nail act like a magnet temporarily.

Reading Check

4. **Explain** Why do the atoms in nickel act like a magnet?

_____

_____

What are magnetic domains?

In magnetic materials, the magnetic field made by each atom exerts a force on other nearby atoms. This causes the atoms to rotate and form a magnetic domain. A **magnetic domain** is a large group of atoms with their magnetic poles lined up in the same direction. Because the atoms are lined up, a domain acts like a magnet. A domain has a north pole and a south pole.

How do domains line up?

An iron nail has a large number of magnetic domains that act like magnets. So why doesn't a nail act like a magnet? The poles of the domains point in different directions. Since the domains do not line up, the magnetic fields cancel each other out. Therefore, the nail does not act like a magnet. The figure below shows the magnetic domains of a nail.

One way to make the domains line up is to touch a bar magnet to the nail. The domains will rotate and point in the same direction because of the magnetic field of the magnet. Now the nail acts like a magnet. The second figure shows domains that have lined up.

Reading Check

5. **Explain** Why doesn't a nail act like a magnet?

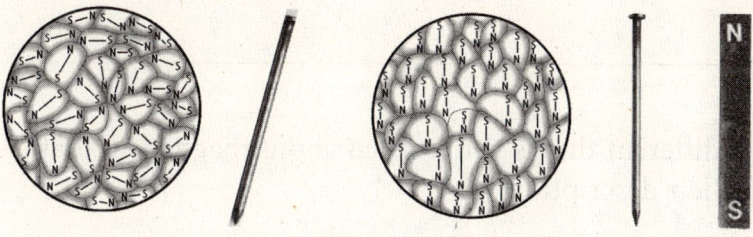

Picture This

6. **Observe** Look at the second figure. How are the north and south poles lined up?

If the bar magnet is taken away, the atoms in the nail move around and bump each other. This causes the domains to move out of line. This is why a nail is not always a magnet.

How can you make a permanent magnet?

Permanent magnets can be made by placing a magnetic material, such as iron, in a strong magnetic field. The strong magnetic field causes the magnetic domains to line up and combine to make a strong magnetic field inside the material. This field keeps the atoms from bumping the domains out of line. The material becomes a permanent magnet.

Heating a permanent magnet causes it to lose its magnetism. Heat causes the atoms in the magnet to move faster. This moves the domains out of line. The permanent magnet will then lose its magnetic field.

Can a pole be isolated?

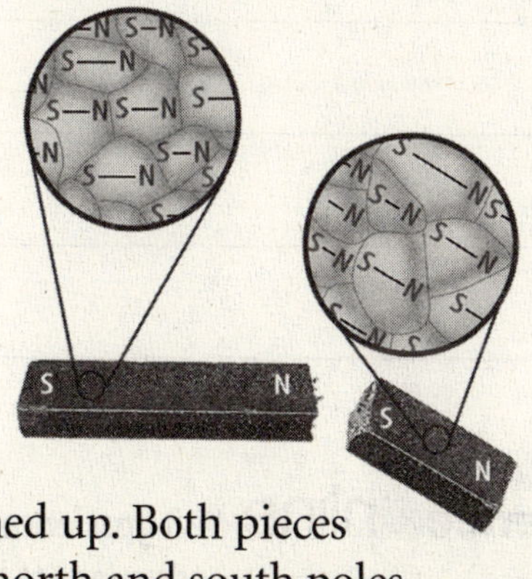

Suppose a magnet is broken into two pieces as shown in the figure. Is one piece a north pole and the other piece a south pole? Remember, each atom in a magnetic material acts like a tiny magnet. So, every magnet is made up of many smaller magnets that are lined up. Both pieces of a broken magnet have their own north and south poles.

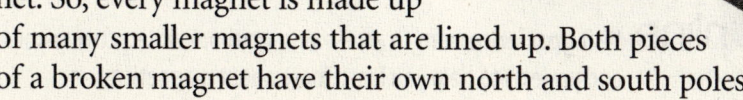

Picture This

7. **Infer** If the two pieces of magnet were put back together, what would happen to the north and south poles at the broken edge?

Reading Essentials 123

After You Read
Mini Glossary

magnetic domains: groups of atoms with poles that line up in the same direction

magnetic field: the lines of force around a magnet

magnetic poles: the north and south pole where the forces of the magnet are strongest

magnetism: the properties of magnets and how magnets interact with each other

1. Review the definitions of the vocabulary words. Write a sentence that explains how Earth is a magnetic field.

   _____

   _____

2. Complete the chart below. List the different things you learned about magnetism that have the word *magnetic* as part of their description.

Magnetism
1. magnetic force
2.
3.
4.
5.

3. Review the ideas your group wrote on the index cards. Write one idea that you all agreed was important. How did this idea help you to understand magnetism?

   _____

   _____

   _____

 Visit gpscience.com to access your textbook, interactive games, and projects to help you learn more about magnetism.

124 CHAPTER 8 Magnetism and Its Uses

Chapter 8: Magnetism and Its Uses

section 2 Electricity and Magnetism

● Before You Read

List some machines that use motors to make them work.

_____

_____

What You'll Learn
- how an electric current and electromagnet produce magnetic fields
- about electromagnets and electric motors

● Read to Learn

Electric Current and Magnetism

In 1820, a Danish physics teacher discovered that there is a connection between electricity and magnetism. He was doing a demonstration using an electric current. There was a compass near the electric circuit. He noticed that the compass needle changed direction depending on the flow of the electric current. This made the teacher hypothesize that an electric current produces a magnetic field around the wire. He also hypothesized that the direction of a magnetic field changes with the direction of the electric current.

Mark the Text

Highlighting As you read this section, use a highlighter to mark the most important ideas in each paragraph.

How do magnetic field lines change?

The teacher's hypotheses have been proven to be true. We now know that moving charges produce a magnetic field. For example, when an electric current flows through a wire, a magnetic field forms around the wire. The direction of the magnetic field depends on the direction of the current in the wire. The magnetic field lines form circles around a wire carrying electric current. The figure shows that when the current in the wire changes direction, then the direction of the magnetic field lines also change. If the current in the wire gets stronger, the magnetic field gets stronger. As you move farther away from the wire, the magnetic field gets weaker.

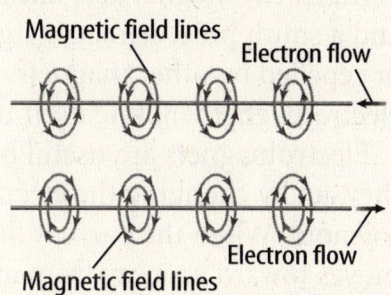

FOLDABLES

B Note Cards As you read this section, make note cards out of half sheets of paper to write notes about the three main topics.

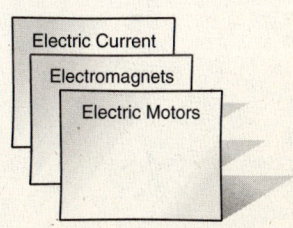

Reading Essentials **125**

✓ Reading Check

1. **Describe** a solenoid.

_____
_____
_____
_____

💡 Think it Over

2. **Explain** How is a solenoid used to make an electromagnet?

_____
_____

Picture This

3. **Draw** In the second figure, change the drawing of the electromagnet so that it will be stronger than the one shown.

Electromagnets

You can make a magnetic field stronger. This is done with a solenoid. A **solenoid** (SOH luh noyd) is a single wire carrying electric current that is wrapped into a cylinder-shaped coil. It looks like a spring. The figure to the right is a solenoid. ✓

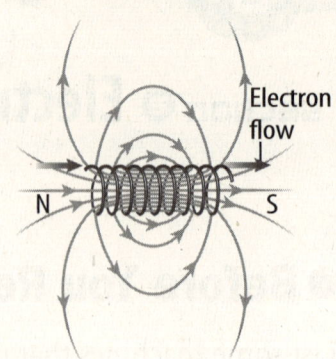

What does a solenoid do?

A solenoid makes a magnetic field stronger. Suppose there is an electric current flowing through a wire. If a loop is formed out of the wire, the magnetic fields inside the loop combine. This makes the magnetic field inside a loop stronger than the magnetic field around a straight wire. Since a solenoid has many loops, the magnetic field inside it is even stronger.

The solenoid is used to make an electromagnet. An **electromagnet** is a solenoid wrapped around an iron core. The figure to the right is an electromagnet. The solenoid's magnetic field magnetizes the iron core. The magnetic field inside the electromagnet can be 1,000 times stronger than the field inside a solenoid without an iron core.

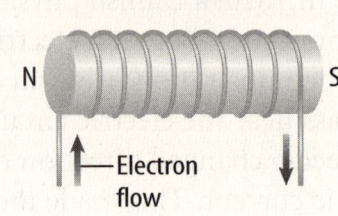

How do electromagnets behave?

Electromagnets are temporary magnets. The magnetic field is present only when current is flowing in the solenoid. The magnetic field of an electromagnet can be made stronger by adding more coils of wire or by adding more current.

An electromagnet acts like any other magnet. It has a north and a south pole, attracts magnetic materials, and is attracted or repelled by other magnets. If put in a magnetic field, an electromagnet will line itself up along the magnetic field lines.

Electromagnets are useful because you can control how they act by changing the electric current flowing through the solenoid. When the current flows in the electromagnet and it moves toward or away from another magnet, electric energy is changed into mechanical energy. The mechanical energy will do work. Electromagnets make mechanical energy to do the work in many devices, such as stereo speakers and electric motors.

How do electromagnets make sound?

There is an electromagnet in the speaker you use when you listen to a CD. The electromagnet is connected to a speaker cone made from paper, plastic, or metal. A permanent magnet surrounds the electromagnet.

Changing Electric Current Recall that increasing the current passing through a wire increases the strength of a magnetic field. A CD player produces a voltage, a measure of electrical potential energy that can be changed into other forms of energy. As voltage increases, more electrical potential energy is ready to be changed into other forms of energy. The CD's voltage produces an electric current in the electromagnet next to the speaker cone. The CD contains information that changes the amount of electric current and its direction. The changing electric current causes the direction and the strength of the magnetic field around the electromagnet to change. A change in direction causes the electromagnet to attract or repel the permanent magnet. This makes the electromagnet move back and forth causing sound.

The electromagnet changes electrical energy to mechanical energy. The mechanical energy vibrates the speaker cone so it reproduces the sound that was recorded on the CD.

Reading Check

4. **Determine** What causes the speaker cone to vibrate?

_____
_____
_____
_____

What makes an electromagnet rotate?

A permanent magnet can apply forces on an electromagnet to make it rotate. The figure below shows an electromagnet between the north and south poles of a permanent magnet. The north and south poles of the electromagnet are attracted to the opposite poles of the permanent magnet. This causes a downward force on the left side of the electromagnet in the figure and an upward force on the right side. These forces make the electromagnet rotate until opposite poles are lined up.

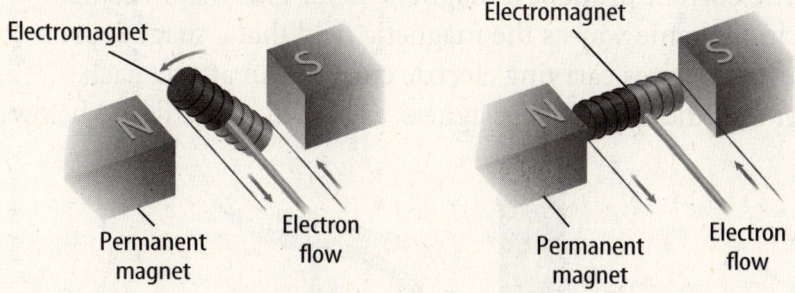

Picture This

5. **Identify** Which pole of the electromagnet is attracted to the north pole of the permanent magnet?

_____

The electromagnet continues to rotate until its poles are next to the opposite poles of the permanent magnet. Once the north and south poles of the electromagnet are in opposite directions, the electromagnet stops rotating.

How does a fuel gauge work?

A <u>galvanometer</u> is a device that uses an electromagnet to measure electric current. Look at the first figure. It is a galvanometer. An electromagnet is located between the poles of a permanent magnet and is connected to a small spring. The electromagnet rotates until the force applied by the spring is balanced with the magnetic forces on the electromagnet. A needle is attached to the electromagnet, so it turns also.

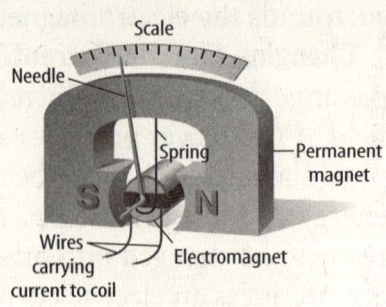

The second figure is a fuel gauge from a car's dashboard. The fuel gauge is a galvanometer. When the amount of gasoline in the car's fuel tank changes, the needle in the gauge moves. This movement is started by a sensor in the car's fuel tank that tells

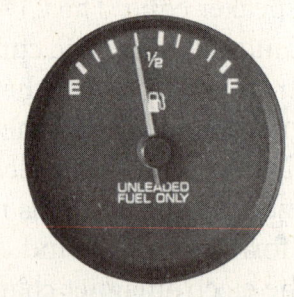

when the fuel level changes. This sensor sends an electric current to the galvanometer. The current changes as the amount of fuel changes. The current change causes the electromagnet to turn. This makes the needle move to different positions on the gauge. The gas gauge is set so that when the fuel tank is full, the needle moves to the full mark on the gauge.

Picture This

6. Analyze What would make the needle in the fuel gauge turn to the right?

_____

_____

Electric Motors

On hot days, do you use a fan to keep cool? A fan uses an electric motor. An <u>electric motor</u> is a machine that changes electrical energy into mechanical energy. The wires carrying electric current produce a magnetic field. This magnetic field acts in the same way as the magnetic field that a magnet produces. Two wires carrying electric current can attract each other as if they were two magnets, as shown in the figure below.

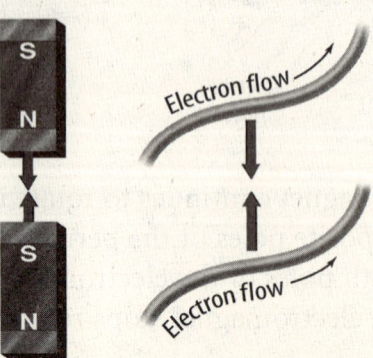

Picture This

7. Trace the arrows between both the magnets and the wires that show the attractions.

128 CHAPTER 8 Magnetism and Its Uses

Where are electric motors used?

Electric motors are used in many types of machines. You can find these machines in industry, agriculture, and transportation. Objects as large as airplanes and cars use electric motors. Objects as small as personal-sized fans also use motors. You can find machines with electric motors in most rooms of your house. Almost every appliance with a moving part contains an electric motor. Electric motors are used in CD and DVD players, computers, hair dryers, electric fans, and many other appliances.

How does a simple electric motor work?

A magnet and a wire carrying electric current exert forces on each other, just like two magnets do. The magnetic field around the wire causes it to be moved by a magnet. The wire is either pushed or pulled depending on the flow of the electric current. As a result, some of the electric energy from the current is changed into kinetic energy of the moving wire, as shown in the figure on the left. Any machine that changes electric energy into kinetic energy is a motor. To keep a motor running, the wire carrying electric current is formed into a loop, or closed circuit. This allows the magnetic field to make the wire spin without stopping as shown in the figure below. ✓

Reading Check

8. Identify What is a machine that changes electric energy into kinetic energy called?

_____

Simple Electric Motors

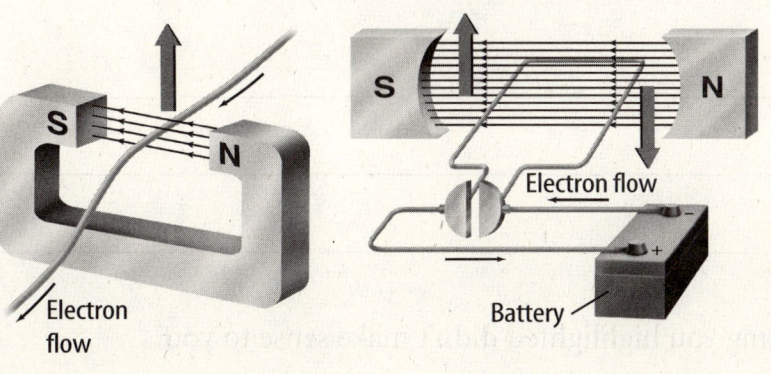

Picture This

9. Determine In the figure on the right, which direction will the loop spin, clockwise or counterclockwise?

_____

After You Read

Mini Glossary

electric motor: a device that changes electrical energy into mechanical energy

electromagnet: a single wire carrying an electric current that is wrapped around an iron core

galvanometer: a device that uses an electromagnet to measure electric current

solenoid: a single wire carrying electric current that is wrapped into a coil shaped like a cylinder

1. Review the vocabulary words and their definitions in the Mini Glossary. Write a sentence that shows how a solenoid and an electromagnet are related.

2. Complete the diagram to list three items that use electromagnets to make them work.

Use electromagnets to make them work

3. Look at the parts of the text that you highlighted. How did this help you to learn about electricity and magnetism?

4. What could you do if something you highlighted didn't make sense to you?

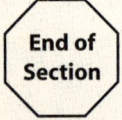

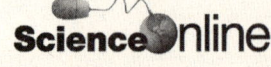

End of Section

ScienceOnline Visit **gpscience.com** to access your textbook, interactive games, and projects to help you learn more about electricity and magnetism.

Magnetism and Its Uses

section ❸ Producing Electric Current

Before You Read

Name three things you used today that use electrical energy to make them work. Where do you think the electrical energy came from?

_____

_____

What You'll Learn
- what electromagnetic induction is
- how a generator produces an electric current
- the difference between direct and alternating current
- how to change the voltage of an alternating current

Read to Learn

From Mechanical to Electrical Energy

In 1831, scientists discovered that if they moved a loop of wire through a magnetic field, it caused an electric current to flow in the wire. They also found that moving a magnet through a loop of wire produced a current in the wire. In both cases, the movement caused an electric flow in the wire. In other words, mechanical energy was changed into electrical energy.

The loop of wire or the magnet has to move to make an electric current. This makes the magnetic field inside the loop change. When the magnetic field changes, it causes an electric current to flow in the wire. The current change in the wire can start a current in a nearby coil. **Electromagnetic induction** is the making of a current by a changing magnetic field.

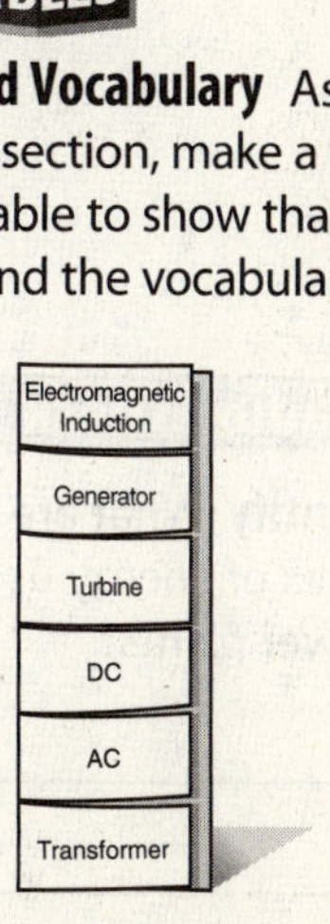

How do electric generators work?

Generators use electromagnetic induction to change mechanical energy into electrical energy. The figure shows a hand turning the handle of a simple generator. The turning handle provides mechanical energy to rotate the coil between the poles of a permanent magnet producing a current in the coil.

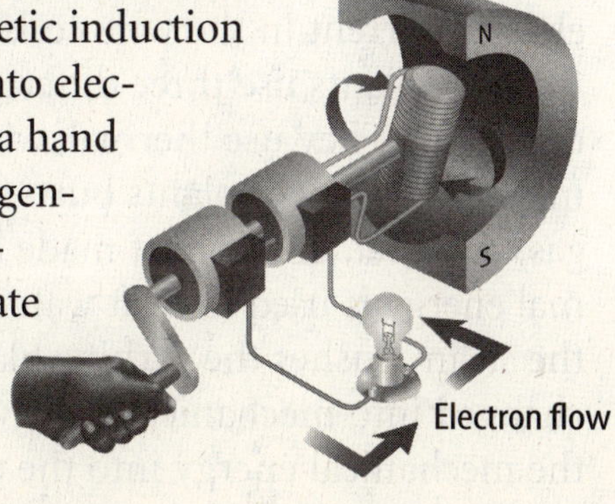

Electron flow

Study Coach

Sticky-Note Discussions Place sticky notes at parts of the section you find interesting or that you have a question about. Write the question on the sticky note.

FOLDABLES

C Build Vocabulary As you read this section, make a vocabulary Foldable to show that you understand the vocabulary terms.

- Electromagnetic Induction
- Generator
- Turbine
- DC
- AC
- Transformer

Reading Essentials **131**

Picture This

1. Explain What provides the original mechanical energy for the simple generator in the figure? *(Hint: It is not the handle.)*

_____

_____

Current Flow A simple generator with a turning wire coil is shown in the figure on the previous page and this page. The coil rotates through the magnetic field of the permanent magnet. This makes a current flow through the coil. Each time the coil makes half of a turn, the ends of the coil move past opposite poles of the permanent magnet. This causes the current to change direction. The current in the coil changes direction twice each time it makes one full turn. You can control how often the current changes direction by controlling how fast the generator rotates. In the United States, generators rotate 60 times per second to produce electric current. This is equal to 3,600 turns per minute.

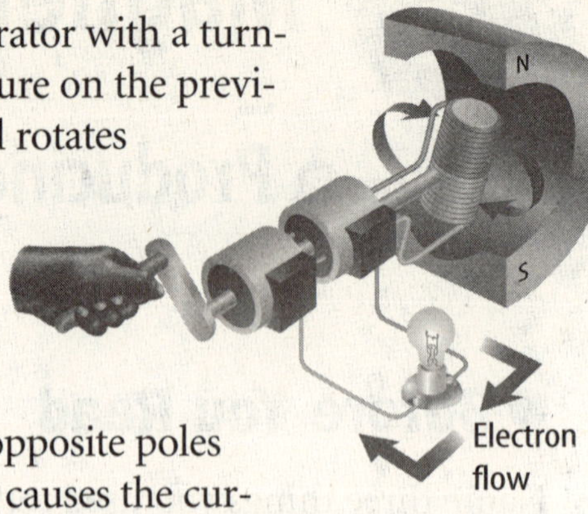

How are electric generators used?

Generators are used in cars. They are called alternators and provide electrical energy for the car's lights. They also provide electrical energy to the spark plugs in the car's engine. The spark plugs ignite (cause a spark that burns) the fuel in the cylinders of the engine. Once the engine is running, the fuel provides the mechanical energy to turn the coil in the alternator.

Electric power plants produce most of the electrical energy. The huge generators in electric power plants operate in a different way from the alternator. The coil does not rotate. Instead, the permanent magnet rotates. Mechanical energy rotates the magnet, and electrical current is produced in the coil.

How is electricity produced for your home?

The electrical energy you use in your home comes from a power plant with huge generators. These generators have many coils of wire wrapped around huge iron cores. The magnets are connected to a <u>turbine</u> (TUR bine), a large windmill-like wheel. The turbine rotates when it is pushed by steam, water, or wind, and the rotating magnets produce the electric current in the wire coils.

Power plants use three different kinds of energy to make electricity. They use thermal, wind, or water energy. For thermal energy, power plants burn fossil fuels such as oil, natural gas, and coal, or use heat made by nuclear reactors. The thermal energy is used to heat water and produce steam. When the steam pushes the turbine blades, the thermal energy is changed into mechanical energy. The generator then changes the mechanical energy into the electrical energy you use. ✓

Reading Check

2. Identify What are three types of energy used by power plants?

_____

_____

Wind Energy Some power plants use the mechanical energy in falling water to turn the turbines. Other power plants use wind energy. Fields of windmills, like those in the figure below, use mechanical energy in wind to turn the generators. The propeller on each windmill is connected to an electric generator. The turning propeller rotates a coil or a permanent magnet.

Picture This
3. **Make Connections** What do the propellers rotate?
   _____
   _____

Direct and Alternating Currents

Have you ever had a power outage in your house? The electrical devices do not work because electrical energy is not coming into your house. Some electrical devices use batteries as a primary source of energy. A battery-operated radio is an example of this type of device. The current produced by a battery is different than the current from an electric generator.

A battery produces a direct current. A **direct current** (DC) is a current that flows in only one direction through a wire. A CD player or any other appliance that plugs into a wall outlet uses alternating current. An **alternating current** (AC) reverses or changes the direction of the current twice during each rotation of the coil. Electronic devices that use batteries for backup energy, such as a radio, usually need direct current to operate. When the radio is plugged into a wall outlet, electronic parts in the radio change the alternating current to direct current. ☑

Think it Over
4. **Apply** List two devices that use DC only and two devices that use AC only.
   _____
   _____
   _____
   _____

Transmitting Electrical Energy

Have you ever seen power lines along a highway? They carry electrical energy from a power plant to buildings. When electrical energy travels through wires, some of the electrical energy changes into heat because of electrical resistance in the wires. As wires get longer, there is more electrical resistance and the heat increases. One way to reduce the heat produced in a power line is to send the electrical energy at high voltage. This voltage in the lines is too high for appliances to use. A transformer is used to decrease the voltage before it enters your home.

✓ Reading Check
5. **Explain** Why can some radios work on both batteries and when plugged into a wall outlet?
   _____
   _____
   _____

Reading Essentials **133**

Reading Check

6. Identify What causes the magnetic fields in the primary coil to change direction?

_____

_____

Picture This

7. Identify In the two figures, circle the labels that are the same.

Transformers

A <u>transformer</u> is used to increase or decrease the voltage of an alternating current. A transformer has a primary coil and a secondary coil. Both coils are wrapped around the same iron core. An input voltage of alternating current passes through the primary coil. This causes the coil's magnetic field to magnetize the iron core. When the current in the primary coil changes direction, this causes the magnetic fields in the primary coil and the iron core to also change directions. This causes an output voltage in the secondary coil. ☑

The figures show two kinds of transformers, a step-up transformer and a step-down transformer.

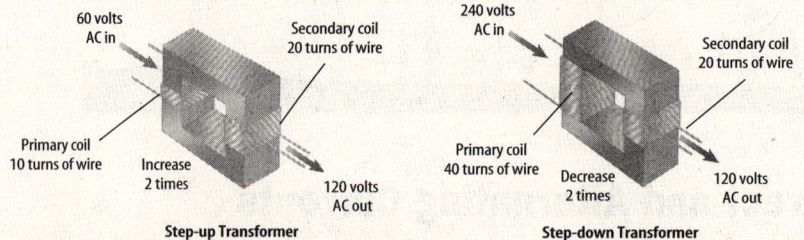

Step-Up Transformer A step-up transformer increases voltage so that the output voltage is greater than the input voltage. The secondary coil then has more turns than the primary coil. In the first figure, an input voltage of 60 volts in the primary coil provides an output voltage of 120 volts in the secondary coil. The secondary coil has twice as many turns as the primary coil has. Therefore, the output voltage is twice as large as the input voltage.

Step-Down Transformer A step-down transformer decreases voltage so that the output voltage is less than the input voltage. The secondary coil then has fewer turns than the primary coil. In the second figure, the input voltage of 240 volts in the primary coil is changed to an output voltage of 120 volts in the secondary coil. The secondary coil has half as many turns as the primary coil has. Therefore, the output voltage is one-half of the input voltage.

What path does an alternating current follow?

Power plants usually make alternating current because the voltage can be increased or decreased with transformers. As the electrical energy leaves the power plant, a step-up transformer increases the voltage. This electrical energy is carried along power lines. When the electrical energy leaves the power lines to enter a building, a step-down transformer decreases the voltage. Even though the voltage is changed, the amount of electrical energy is not changed.

● **After You Read**

Mini Glossary

alternating current (AC): electrical current that changes its direction twice during each rotation of a coil

direct current (DC): electrical current that flows in only one direction through a wire

electromagnetic induction: a changing magnetic field producing an electric current in a wire

generator: a device that uses electromagnetic induction to change mechanical energy into electrical energy

transformer: a device that increases or decreases the voltage of an alternating current

turbine: a large wheel that rotates when it is pushed by water, wind, or steam

1. Review the definitions of the vocabulary words in the Mini Glossary. Choose one of the words and write what it means in your own words.

   _____
   _____

2. Complete the chart below to organize information you have learned about electric current.

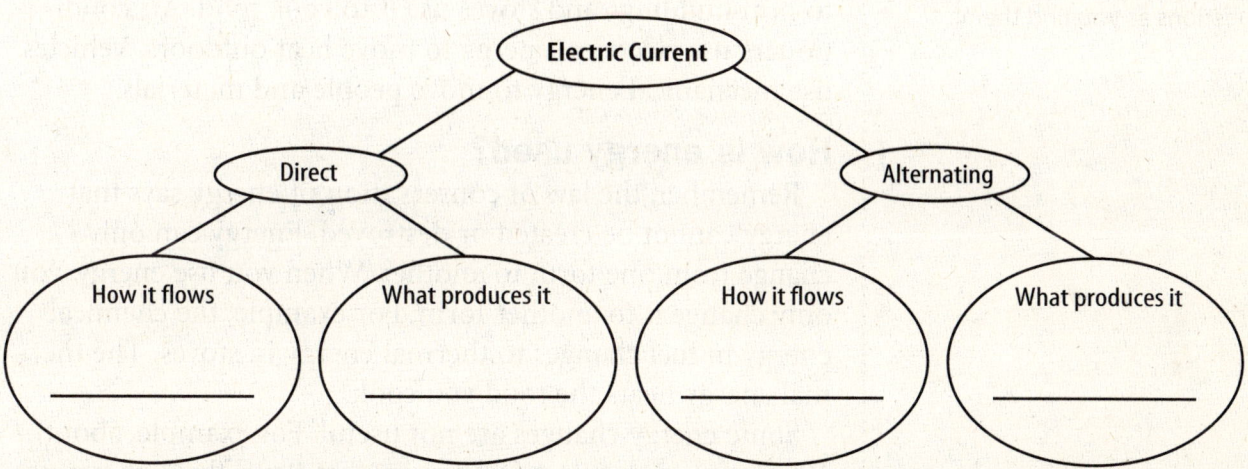

3. Think about what you have learned in this section. Look at the parts you marked with sticky notes. How did these notes help you learn?

   _____
   _____

Science Online Visit **gpscience.com** to access your textbook, interactive games, and projects to help you learn more about producing electric current.

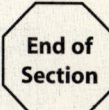

End of Section

chapter 9 Energy Sources

section ❶ Fossil Fuels

What You'll Learn
- what fossil fuels are
- how fossil fuels are formed

Mark the Text

Locate Information Underline every heading in the section that asks a question. Then, highlight the answers to those questions as you find them.

FOLDABLES

A Describe Make the following Foldable to help describe fossil fuels and how they are converted to energy.

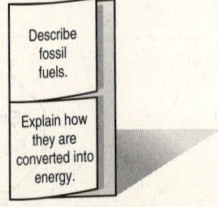

● Before You Read

Write what you think the energy source is for the heating system, water heater, and stove in your home.

● Read to Learn

Using Energy

Energy is used in many ways. Furnaces use thermal energy to heat buildings and stoves use it to cook food. Air conditioners use electrical energy to move heat outdoors. Vehicles use mechanical energy to move people and materials.

How is energy used?

Remember, the law of conservation of energy says that energy cannot be created or destroyed. Energy can only change from one form to another. When you use energy, you only change it to another form. For example, the chemical energy in fuel changes to thermal energy in stoves. The thermal energy heats the food you cook.

Some energy changes are not useful. For example, about 10 percent of the electrical energy that flows through power lines changes to thermal energy in the lines. This thermal energy is the result of friction between the flowing electrons and the atoms in the wire.

How is energy used in the United States?

The United States uses more energy than any other country. Look at the figure at the top of the next page. The first circle graph shows energy use in the United States. Energy is used in homes for heating and cooling, running appliances, lighting buildings, and heating water. Energy used for transportation powers cars, trucks, and airplanes. Businesses use energy to heat, cool, and light stores and offices. Factories and farms use energy to make products and grow crops.

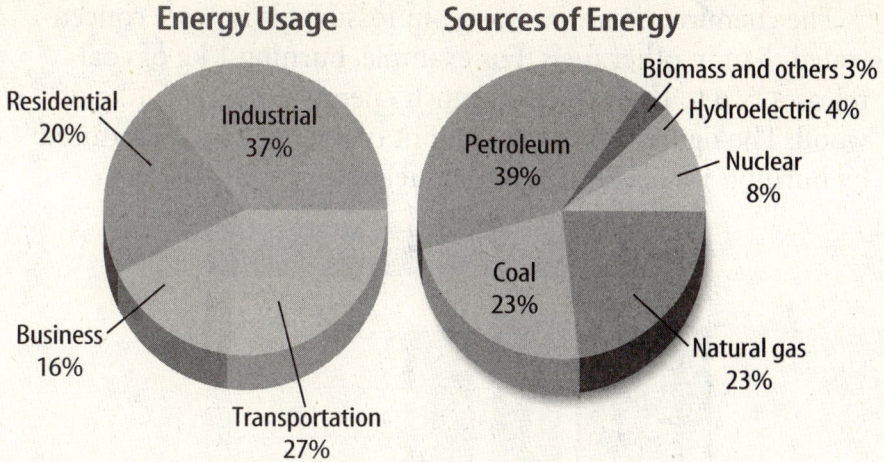

The second circle graph shows the main sources of the energy used in the United States. Almost 85 percent of the energy used comes from burning petroleum, natural gas, and coal. Nuclear power plants provide about eight percent of the energy used in the United States.

Making Fossil Fuels

A car might use several gallons of gasoline in one hour of freeway driving. Did you know that it takes millions of years to make fuels like oil, natural gas, and coal? These fuels are called fossil fuels. **Fossil fuels** are formed from the decayed remains of ancient plants and animals.

Petroleum, or oil, and natural gas form when matter made up of dead plants and animals piles up on the ocean floor. The matter is buried slowly under layers of sediment. Chemical reactions caused by heat and great pressure change the matter into oil and natural gas. The oil and gas may bubble up to the surface or be trapped beneath a layer of rock. Coal is formed from dead plants in ancient swamps in much the same way.

Why are fossil fuels concentrated energy sources?

Fossil fuels are energy sources. They store chemical energy in the bonds between the atoms. When a fossil fuel burns, a chemical reaction takes place. The carbon and hydrogen atoms in the fossil fuel combine with oxygen in the air to form carbon dioxide and water. Heat and light are produced when chemical potential energy is converted in the chemical reaction. Chemical bonds in fossil fuels are the source of the potential energy.

Applying Math

1. **Interpret a Graph** What percentage of energy used in the United States is used by businesses and industries together?

   _____

Reading Check

2. **Infer** What is necessary to change matter into fossil fuels?

   _____
   _____
   _____

The chemical potential energy in fossil fuels is more concentrated than in other fuels. For example, burning 1 kg of coal releases two to three times as much energy as burning 1 kg of wood. The figure shows the amount of energy that is produced by burning 1 g of different fossil fuels.

Applying Math

3. Compare About how much more energy is made by one gram of gasoline than one gram of wood?

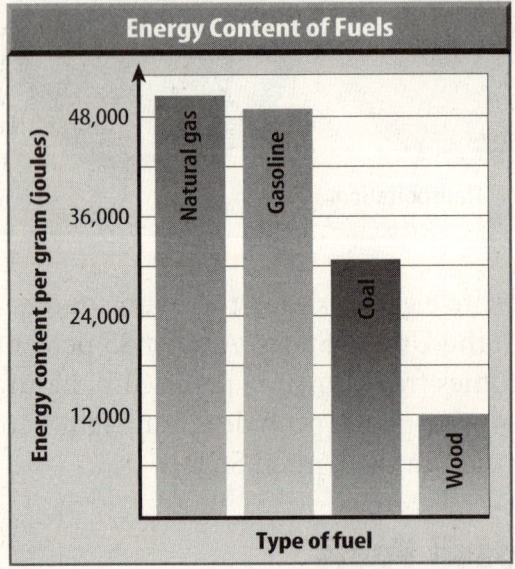

Reading Check

4. Identify What two types of atoms do petroleum molecules contain?

Petroleum

<u>Petroleum</u> is a liquid fuel formed by decayed organisms. It is a mixture of thousands of chemical compounds. Most of these compounds are hydrocarbons, which means their molecules contain only carbon atoms and hydrogen atoms. Petroleum, or crude oil, is pumped from wells deep in Earth's crust. Millions of gallons of petroleum are pumped every day.

How are hydrocarbons separated?

Petroleum is a mixture of many different hydrocarbon compounds. Some are thick and heavy, like asphalt. Others are very light, like gasoline. Heavy hydrocarbon compounds have large molecules with many carbon and hydrogen atoms. Light hydrocarbon compounds have small molecules with few carbon and hydrogen atoms.

Petroleum compounds must be separated or refined in a process called fractional distillation. At a refinery, crude oil is heated in the bottom of a tall, slender tank. The compounds boil and turn to vapor. Hydrocarbons with the lowest boiling points rise to the top of the tank. Those with high boiling points remain liquid and are drained from the bottom of the tower.

Most of the hydrocarbons that are separated at a refinery end up as fuels such as gasoline, jet fuel, diesel fuel, and heating oil. The rest are used as ingredients in other chemical products.

What are other uses for petroleum?

Not all of the products we get from petroleum are turned into fuels. In the United States, about 15 percent of petroleum-based substances are used for other products. Plastics and synthetic fabrics are made from the hydrocarbons found in petroleum. Lubricants, such as grease and motor oil, and the asphalt used in roads also are made from petroleum.

Natural Gas

Natural gas and petroleum are formed by the same process. However, natural gas is a gas and petroleum is a liquid. Underground, natural gas is usually found above petroleum deposits. Natural gas is made up mostly of the hydrocarbon, methane, CH_4. It also contains other hydrocarbon gases such as propane, C_3H_8 and butane, C_4H_{10}. Natural gas provides energy for cooking, heating, and manufacturing. About one-fourth of the energy used in the United States comes from natural gas. Natural gas contains more energy per kilogram than petroleum or coal. It also burns more cleanly than other fossil fuels and produces fewer pollutants.

Coal

Coal is a solid fossil fuel. It is found in underground mines. From 1900 to 1950, coal provided more than half of the energy that was used in the United States. Now, almost two-thirds of the energy used comes from petroleum and natural gas. Only about one-fourth comes from coal. About 90 percent of all the coal that is used in the United States is burned by power plants to generate electricity. ✓

Where does coal come from?

Coal deposits, the places where coal can be found, were once ancient swamps. Over thousands of years, plants that lived in the swamps died. Coal formed from the dead plants. The amount of coal worldwide is estimated to be 20 to 40 times greater than the supply of petroleum.

Coal is a solid mixture of hydrocarbons and other chemicals, such as sulfur compounds. Because of the chemical compounds it contains, burning coal often produces more pollutants than burning petroleum or natural gas. The sulfur compounds in coal, when burned, produce sulfur dioxide. Sulfur dioxide is a major cause of acid rain.

💡 Think it Over

5. Identify Look around you. List three products that are made from petroleum.

_____
_____
_____

✓ Reading Check

6. Use Percentages About what percent of all the coal that is used in the United States is used to produce electricity?

_____

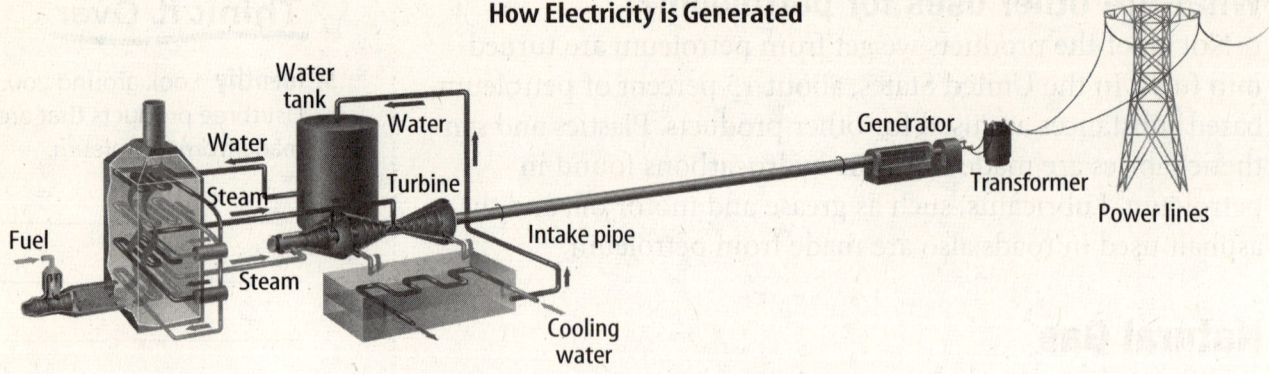

How Electricity is Generated

Picture This

7. Observe Circle the parts of the figure that show where the electric current is produced.

Generating Electricity

Burning fossil fuels produces almost 70 percent of the electrical energy used in the United States. The chemical energy contained in fossil fuels is converted or changed to electrical energy in electric power stations. Follow the process in the figure above. First, the fuel is burned and converted to thermal energy. The thermal energy heats water and produces steam under high pressure. Then the steam spins the blades of a turbine. The turbine is connected to an electric generator. The spinning turbine turns magnets inside the generator. This produces electric current. Finally, the electric current is sent to homes, schools, and businesses through power lines.

Efficiency of Power Plants

When fossil fuels are burned, not all the chemical energy in the fuel changes into electrical energy. Some of the energy is converted into thermal energy. Efficiency tells how energy is delivered. You can find the overall efficiency of a power plant by multiplying the efficiencies of each stage of the process. The overall efficiency is only about 35 percent. This means that only about 35 percent of the energy contained in the fossil fuels is delivered as electrical energy. The rest of the energy is converted to thermal energy as the chemical energy is changed into the electrical energy that is delivered to energy users.

Applying Math

8. Calculate What percent of energy is converted to thermal energy during fossil fuel conversion?

_____

_____

The Costs of Using Fossil Fuels

Fossil fuels are useful for making electricity. They also provide power for transportation. But they have some harmful side effects. Petroleum products and coal give off smoke when they are burned. The smoke contains small particles called particulates. The particulates can cause breathing problems for some people.

140 CHAPTER 9 Energy Sources

Burning fossil fuels also releases carbon dioxide. The amount of carbon dioxide in Earth's atmosphere has increased greatly over the past hundred years. The increase is due to burning fossil fuels. An increase of carbon dioxide in the atmosphere could cause Earth's surface temperature increase.

What are the problems with using coal?

There is more coal than any other fossil fuel. However, burning coal releases more pollutants than oil or natural gas. Many power plants that burn coal remove some of these pollutants before they are released into the air. Mining coal also can be dangerous. Miners risk being killed or injured. Some miners suffer from lung diseases that are caused by breathing coal dust for long periods of time.

Nonrenewable Resources

Fossil fuels are nonrenewable resources. **Nonrenewable resources** are resources that cannot be replaced by natural processes as quickly as they are used. Fossil fuel reserves are decreasing. Reserves are the amounts of fossil fuels remaining in the ground. At the same time, the population and industrial needs for fuel are increasing. The graph below shows that over the next 50 years, oil production might decrease from more than 25 billion barrels to about 5 billion barrels per year.

> **Reading Check**
>
> 9. **List** two harmful side effects of using fossil fuels.
>
> _____
>
> _____
>
> _____

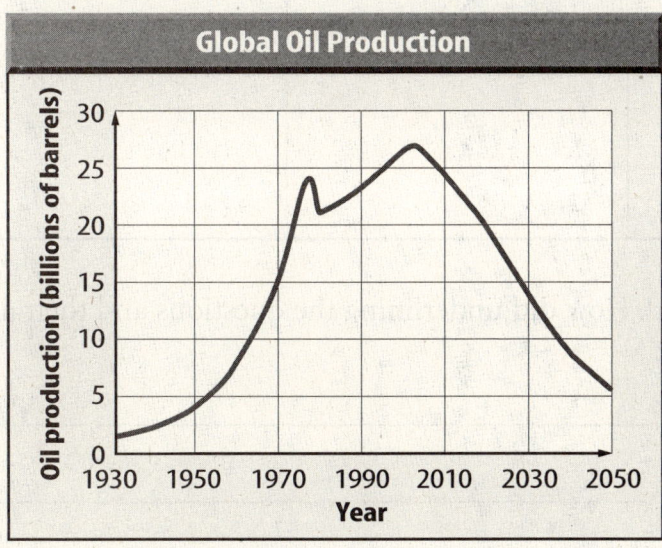

> **Applying Math**
>
> 10. **Reading Graphs** In what year was global oil production the most?
>
> _____

Conserving Fossil Fuels

Since reserves of fossil fuels are decreasing, the price of fossil fuels is increasing. However, the need for energy keeps increasing because the population of the world is increasing. One way to meet energy needs and to reduce the use of fossil fuels is to get energy from other sources.

After You Read
Mini Glossary

fossil fuel: fuels that are formed from the decayed remains of ancient plants and animals

nonrenewable resource: resources that cannot be replaced by natural processes as quickly as they are used

petroleum: a liquid fuel formed by decayed organisms

1. Review the terms and their definitions in the Mini Glossary. Explain why fossil fuels are called nonrenewable resources.

   _____

   _____

2. Complete the graphic organizer below to organize the information you learned in this section about fossil fuels.

	Comes From	Mainly Used For . . .	Effects of Burning
Petroleum			
Natural Gas			
Coal			

3. Think about what you have learned. How did underlining the questions and their answers help you as you read the section?

   _____

   _____

   _____

 Visit **gpscience.com** to access your textbook, interactive games, and projects to help you learn more about fossil fuels.

142 **CHAPTER 9** Energy Sources

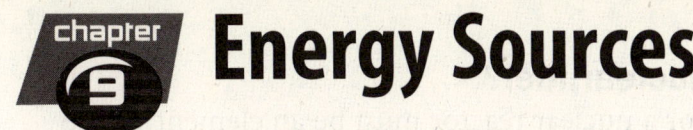

Energy Sources

section ❷ Nuclear Energy

● Before You Read

What do you know about nuclear energy? Write all the information you know about nuclear energy on the lines below.

_____

_____

● Read to Learn

Using Nuclear Energy

Electric power plants that do not burn fossil fuels have been developed. Some of these are nuclear power plants. Since fossil fuels are not burned, nuclear power plants do not cause pollution or release carbon dioxide. Nuclear power plants convert nuclear energy to electrical energy through a process called nuclear fission. Nuclear fission is the process of breaking apart the nucleus of an atom. This releases energy. In the process, an extremely small amount of matter is converted into a huge amount of energy. Today almost 20 percent of the electricity produced in the United States comes from nuclear power plants. Nuclear power plants produce about eight percent of all the energy used in the United States. In 2003, there were 65 nuclear power plants in the United States. These plants contained 104 nuclear reactors.

Nuclear Reactors

A **nuclear reactor** uses the energy from controlled nuclear reactions to generate electricity. Nuclear reactors can have different designs, but they all have the same major parts. Nuclear reactors contain a fuel that can undergo nuclear fission. They also have control rods that can speed up or slow down the nuclear reactions. Nuclear reactors have a cooling system that keeps the reactor from being damaged by the heat that is produced. The actual fission of the atoms of the fuel happens in a small part of the reactor known as the core.

What You'll Learn
- how a nuclear reactor works
- advantages and disadvantages of nuclear energy

Study Coach

Make Flash Cards For each heading in this section, think of a question your teacher might ask on a test. Write the question on one side of a flash card. Then write the answer on the other side. Quiz yourself until you know the answers.

Reading Essentials 143

What is nuclear fuel?

The fuel for a nuclear reactor must be an element whose nucleus can undergo fission, or split. One element that is often used is uranium. Uranium dioxide is the fuel that is used most often in nuclear reactors. One isotope of uranium can split. It is called the U-235 isotope. However, only about 0.7 percent of the naturally occurring uranium is the U-235 isotope. Natural uranium must be enriched to increase the amount of U-235. Uranium nuclear fuel for reactors is enriched so that it contains three to five percent U-235.

What is in the reactor core?

The reactor core contains small, enriched uranium dioxide pellets that are sealed inside metal fuel rods. The pellets look like pencil erasers. They are placed end to end in a tube. The tubes are bundled up into what are called fuel rods. Then they are covered with a metal alloy. A typical reactor core contains about 100,000 kg of uranium in hundreds of fuel rods. For every kilogram of uranium that undergoes fission in the core, 1 g of matter is converted into energy. The energy released by this one gram of matter is equal to the energy released by burning more than 3 million kg of coal.

How does nuclear fission happen?

The figure shows a uranium nuclear fission reaction. When a neutron hits the nucleus of a U-235 atom, the nucleus splits into two smaller nuclei. Two or three neutrons are released at the same time. The smaller nuclei are called fission products. The neutrons from the first reaction hit other U-235 nuclei. Every uranium nucleus that splits apart releases neutrons that cause other uranium atoms to split apart. This process is called a nuclear chain reaction.

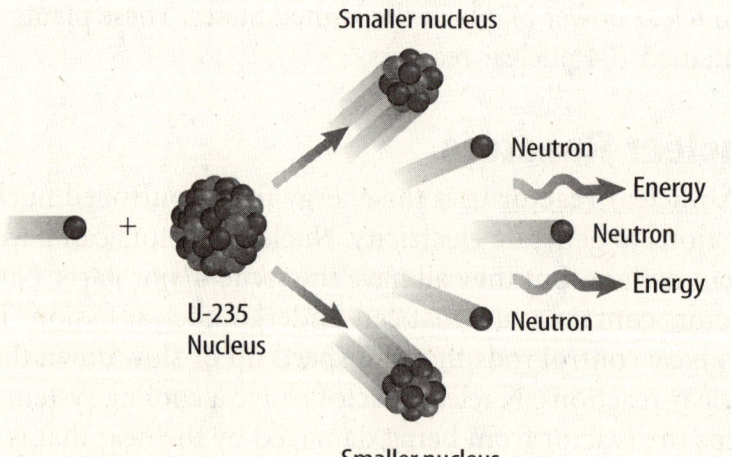

Applying Math

1. **Calculate** Suppose that over a period of time, 100 kg of nuclear fuel is converted to energy in a nuclear power plant. How many kilograms of coal would have to be burned to release the same amount of energy? Write your answer in both numerals and words.

Picture This

2. **Locate** Circle the fission products.

The number of nuclei that split in a nuclear chain reaction can more than double at each step of the reaction. So, a huge number of nuclei can be split after only a small number of reactions. Nuclear chain reactions take place in less than a second. If nuclear chain reactions aren't controlled, the chain reaction could create an explosion instead of a controlled release of energy.

How is the chain reaction controlled?

To control a nuclear chain reaction, some of the neutrons that are released when U-235 splits apart must be stopped from hitting other U-235 nuclei. Control rods containing boron or cadmium metals are used to absorb some of the neutrons from the reactions. The rods are inserted into the reactor core. Moving the control rods deeper into the reactor makes them absorb more neutrons. This slows down the chain reaction. Eventually, only one of the neutrons released in the fission of each of the U-235 nuclei strikes another U-235 nucleus. When this happens, energy is released at a constant rate.

Nuclear Power Plants

Nuclear fission reactors produce electricity in almost the same way as other power plants. Thermal energy from the reactor is used to boil water and produce steam. The figure below shows a nuclear reactor. You can see that cool water, or coolant, is pumped through the reactor. The coolant absorbs heat from the core and then is pumped to a heat exchanger in the boiler. The thermal energy is transferred from the coolant and boils water to make steam. This steam then drives a turbine. The turbine rotates an electric generator.

The overall efficiency of nuclear power plants is about 35 percent. This is similar to the efficiency of fossil fuel power plants.

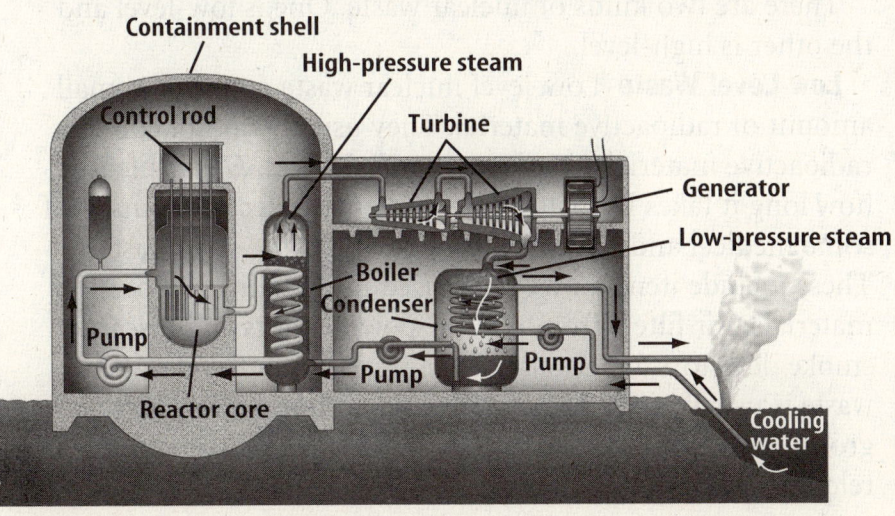

FOLDABLES

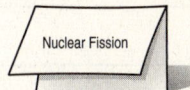

B Summarize Use a sheet of notebook paper to make the following Foldable and summarize what you have learned about the process of nuclear fission.

Picture This

3. **Point Out** In the figure, circle where the coolant absorbs heat, put a box around where the heat is exchanged, circle the turbine, and put a box around the generator.

Think it Over

4. Draw Conclusions What would you say is the main risk from nuclear power plants and nuclear fuel?

_____
_____
_____
_____

Reading Check

5. Explain What can happen if a person is exposed to too much radiation?

_____
_____
_____

Think it Over

6. Draw Conclusions Why is it important to find a more permanent way to store or dispose of radioactive waste?

_____
_____
_____

The Risks of Nuclear Power

Nuclear power plants don't use fossil fuels. They don't release pollutants into the atmosphere. They don't produce carbon dioxide. These are all advantages over fossil fuel power plants. However, there are disadvantages to producing energy from nuclear fission. The mining of uranium can damage the environment. Also, water that is used as a coolant in the reactor core must be cooled before being released into streams and rivers. If the water is not cooled properly, its excess heat could harm fish and other animals and plants in the water.

What happens when radiation is released?

A serious risk of nuclear power is the escape of harmful radiation from power plants. If released, radiation could damage living organisms. To prevent accidents, nuclear reactors and plants have many safeguard systems, strict safety precautions, and highly trained workers. Yet, accidents have happened.

A reactor overheated in 1986 in Chernobyl, Ukraine. The building was partially destroyed, and radioactive materials escaped. As a result, 28 people died of radiation sickness. Up to 260,000 people may have been exposed to radiation that was carried by winds. Power plants in the United States are built to prevent accidents such as Chernobyl. However, many people worry that an accident like this could happen again.

The Disposal of Nuclear Waste

The U-235 in nuclear fuel is used up after about three years. The used rods contain radioactive fission products and the remaining uranium. **Nuclear waste** is any radioactive by-product that is created when radioactive materials are used.

What happens to nuclear waste?

There are two kinds of nuclear waste. One is low-level and the other is high-level.

Low-Level Waste Low-level nuclear waste contains a small amount of radioactive material. They usually do not contain radioactive materials that have long half-lives. A half-life is how long it takes for half of the material to decay. Products of some medical and industrial processes are low-level wastes. These include items of clothing used in handling radioactive materials. Air filters from nuclear power plants and discarded smoke detectors are also low-level wastes. Low-level nuclear waste is usually sealed in containers and buried deep in the ground. When dilute enough, low-level waste is sometimes released into the air or water.

146 CHAPTER 9 Energy Sources

High-Level Waste High-level nuclear waste contains larger amounts of radioactive material. High-level waste is made by nuclear power plants and nuclear weapons programs. Some of this material has long half-lives and will remain radioactive for tens of thousands of years. High-level waste is usually stored in deep pools of water or disposed of in extremely strong and long-lasting containers.

Nuclear Fusion

A different type of nuclear reaction, thermonuclear fusion, is the source of the Sun's energy. In fusion, two very small nuclei collide and stick, or fuse, together to form a larger nuclei. In the fusion reaction shown in the figure below, two isotopes of hydrogen come together to form a helium nucleus. This reaction makes even more energy than fission. A small amount of mass is converted into energy. Fusion is the most concentrated energy source known.

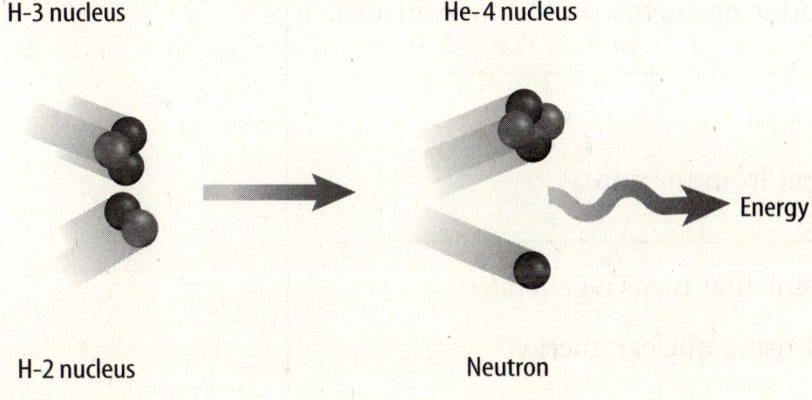

Picture This

7. **Determine** In the first figure, circle the four particles that come together to form the He-4 nucleus.

What are the advantages and disadvantages of fusion?

Fusion has advantages and disadvantages. One advantage is it uses hydrogen as fuel. Hydrogen is abundant on Earth. Another advantage is that fusion reaction produces helium, which is not radioactive. One disadvantage is that fusion can take place only at extremely high temperatures. A great deal of energy needs to be used to reach these temperatures. The amount of energy used in this process can be more than the energy produced in the reaction. Another problem is how to contain a reaction that happens at such high temperatures.

Reading Check

8. **List** the advantages of fusion.

After You Read
Mini Glossary

nuclear reactor: a system that uses the energy from controlled nuclear reactions to generate electricity

nuclear waste: any radioactive by-product that is created when radioactive materials are used

1. Review the terms and their definitions in the Mini Glossary. Tell how nuclear reactors and nuclear waste are connected.

   _____

2. Complete the outline to help you organize what you learned about nuclear energy.

   Nuclear Energy
   I. How does nuclear fission happen?
      A. _____
      B. The nuclei split producing energy, and more neutrons.
      C. _____
   II. How does a reactor work?
      A. Coolant absorbs heat from the core.
      B. _____
      C. Steam drives a turbine that turns a generator.
   III. What are advantages of using nuclear energy?
      A. _____
      B. Doesn't cause air pollution or release carbon dioxide.
   IV. What are disadvantages of using nuclear energy?
      A. _____
      B. Harmful radiation could escape.

3. How could you use the flash cards you made to study for a test?

   _____

   _____

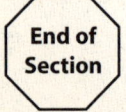
End of Section

 Visit **gpscience.com** to access your textbook, interactive games, and projects to help you learn more about nuclear energy.

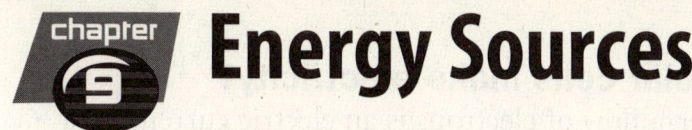

Energy Sources

section ❸ Renewable Energy Sources

● Before You Read

Have you ever heard the term alternative? An alternative is a choice or option that is different from the usual one. Write the name of an alternative energy source on the line below.

_____

What You'll Learn
- why alternate energy sources are needed
- alternate methods for generating electricity

● Read to Learn

Energy Options

The demand for energy keeps increasing, but supplies of fossil fuels are decreasing. Nuclear reactors produce nuclear waste that has to be disposed of safely. To solve these problems, other sources of energy are being developed. A **renewable resource** is an energy source that is replaced almost as quickly as it is used. Some alternative energy sources are renewable resources.

Energy from the Sun

You can feel energy from the Sun if you step outside on a clear day. The amount of solar energy that falls on the United States in one day is more than the total amount of energy used in the United States in one year. The Sun is expected to produce energy for several billion more years. Because of this, solar energy will not run out. It is considered a renewable resource.

What are solar cells?

Many objects, such as solar-powered calculators, use solar energy for power. They use a **photovoltaic cell** that converts radiant energy from the Sun directly into electrical energy. Photovoltaic cells are also called solar cells.

What are solar cells made of?

Solar cells are made of two layers of semiconductor materials. The top layer has many electrons, but the bottom layer does not. The semiconductors are sandwiched between two layers of conducting metal.

Mark the Text

Identify Main Points Highlight each kind of alternative energy source as it is discussed in this section. Circle the main points of each energy source.

FOLDABLES

ⓒ Compare and Contrast Make the following one sheet Foldable out of notebook paper. Compare and contrast information about *alternative* energy sources. Label the columns on the right "Advantages" and "Disadvantages."

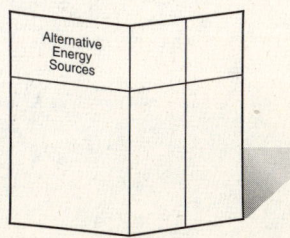

Reading Essentials **149**

How do solar cells make electricity?

Remember, a flow of electrons is an electric current. In a solar cell, energy from sunlight knocks electrons away from their atoms in the top semiconductor. These electrons flow through the top metal layer, then through a circuit to the bottom metal layer, and then to the bottom semiconductor. An electric current is produced. Only about 7 to 11 percent of the radiant energy from the Sun that strikes a solar cell is converted to electrial energy.

How is solar energy used?

It is cheaper to make electricity from fossil fuels than with solar cells. But in some areas, where it is too expensive to run power lines, solar cells are a good source of electrical power.

Another way to generate electricity from solar energy is through a solar concentrating power plant. In a solar concentrating power plant, mirrors focus sunlight onto a receiver. Heat from the sunlight boils water to produce steam. The steam turns a turbine, making electricity. The world's largest concentrating solar power plant is in the Mojave Desert in California. The plant generates over 350 megawatts of power. This is enough electrical power to meet the needs of about 500,000 people.

Energy from Water

Water behind a dam has gravitational potential energy. When water is released from a dam, its potential energy is changed to kinetic energy because the water is in motion. The kinetic energy of water flowing from a dam can be used to spin a turbine and generate electricity. The figure shows how a dam makes electricity.

Applying Math

1. **Use Statistics** How many megawatts of electricity would be needed for a city of three million people?

Picture This

2. **Draw Conclusions** What moves the turbine in hydroelectric power plants?

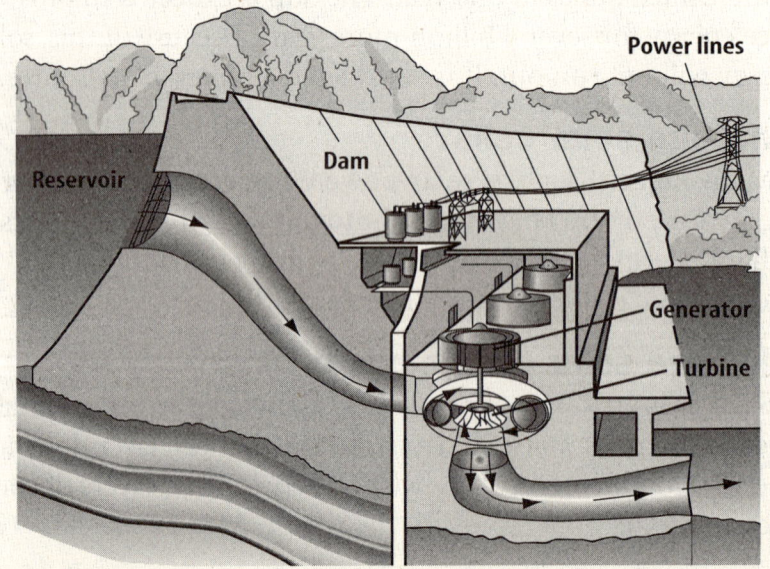

What is hydroelectricity?

Electricity produced from the energy of moving water is called **hydroelectricity**. A dam that produces hydroelectricity is a hydroelectric dam. About 8 percent of the electrical energy used in the United States is produced by hydroelectric power plants. Not all of the kinetic energy of flowing water is changed to electrical energy in a hydroelectric dam. Some of the energy is lost to friction, and some water escapes without moving the turbine. But, because no heat must be produced to turn the turbine, hydroelectric dams are almost twice as efficient as fossil fuels or nuclear power plants.

What are advantages and disadvantages of hydroelectricity?

The lakes formed by dams can provide water for drinking and irrigation. These lakes also may be used for boating and swimming. After a hydroelectric dam is paid for, the cost of the electricity it produces is relatively cheap.

However, artificial dams can disturb the balance of natural ecosystems. For example, dams can prevent fish from moving up rivers to lay their eggs. Also, the lakes formed by dams destroy land habitats when the land is submerged. Some rivers that would make great sources for hydroelectric power are often too far from cities that need the power.

Energy from the Tides

Gravity from the Sun and Moon causes bulges in Earth's oceans that move westward as Earth rotate. The movement of the bulges causes the height of the ocean to rise and fall continually. These rises and falls are ocean tides that can generate hydroelectric power. As the tide comes in, the moving water spins a turbine in a hydroelectric dam. The water then is trapped behind the dam. As the tide lowers, the water behind the dam flows back out to the ocean. This spins the turbines again and generates more electrical power.

Hydroelectricity from tides causes almost no pollution. It is about as efficient as hydroelectricity from dams. However, there aren't many places on Earth where tides are great enough for generating hydroelectricity. There is only one power plant in North America that makes hydroelectricity from tides. It is in Annapolis Royal, Nova Scotia in Canada.

Reading Check

3. **Explain** Why are hydroelectric dams almost twice as efficient as fossil fuels or nuclear power plants?

_____
_____
_____

Reading Check

4. **Identify** What causes tides?

_____
_____

💡 Think it Over

5. Think Critically Which do you think would work best near your community, a hydroelectric dam or a windmill farm? Explain.

_____
_____
_____
_____

💡 Think it Over

6. Draw a Conclusion Why might burning biomass as a source of energy be harmful to the environment?

_____
_____
_____

Harnessing the Wind

Windmills have been used for centuries to pump water. Now, windmills are made to generate electricity from wind's energy. A windmill has a propeller that is turned by the wind. The propeller then spins a generator. Windmills often are placed together in large groups called windmill farms in areas that are regularly windy. Some windmill farms contain several hundred windmills.

Because the wind is never constant, electricity from windmills is not always reliable. Windmills are also only 20 percent efficient. Researchers are trying to increase this efficiency. Wind energy has some disadvantages. Windmills change the landscape and they make noise. They can also confuse migrating birds. However, windmills do not use nonrenewable natural resources and they do not pollute the atmosphere or water.

Alternative Fuels

The use of fossil fuels could be reduced if cars used other energy sources. Cars have been developed that use electrical energy supplied by batteries. Hybrid cars use both electric motors and gasoline engines. Hydrogen gas is another possible alternative fuel. It produces only water vapor when it burns and creates no pollution.

What are biomass fuels?

Could any other materials be used to heat water and produce electricity like fossil fuels and nuclear fission? Biomass can be burned to convert stored chemical energy to thermal energy. **Biomass** is renewable organic matter, such as wood, sugarcane fibers, rice hulls, and animal manure. Burning biomass for energy is probably the oldest use of natural resources to meet human energy needs.

Energy from Inside Earth

The interior of Earth is hot from the decay of radioactive elements. This heat is called geothermal heat. Geothermal heat causes the rock beneath Earth's crust to soften and melt. This hot molten rock is called magma. The thermal energy that is contained in hot magma is called **geothermal energy**.

In some places, Earth's crust has cracks or thin spots that allow magma to rise near the surface. Active volcanoes are places where magma escapes from Earth's crust. In other areas, magma heats water underground forming hot springs and geysers, such as the Old Faithful in Yellowstone National Park. In some areas, water heated by geothermal energy is pumped to houses to provide heat.

How does a geothermal power plant generate electricity?

Geothermal energy also can be used to make electricity. A geothermal power plant is shown in the figure below. Where magma is close to the surface, the surrounding rocks are also hot. Water is pumped into the hot rocks through a well. The hot rocks boil the water to produce steam. The steam then returns to the surface. It rotates the turbines that spin generators.

Geothermal power plants generate electricity without using fossil fuels. The also create very little pollution. But, their efficiency is only about 16 percent. Also, there aren't many places where magma is close enough to Earth's surface to use the geothermal energy.

Think it Over

7. Infer Where must geothermal power plants be located?

_____
_____

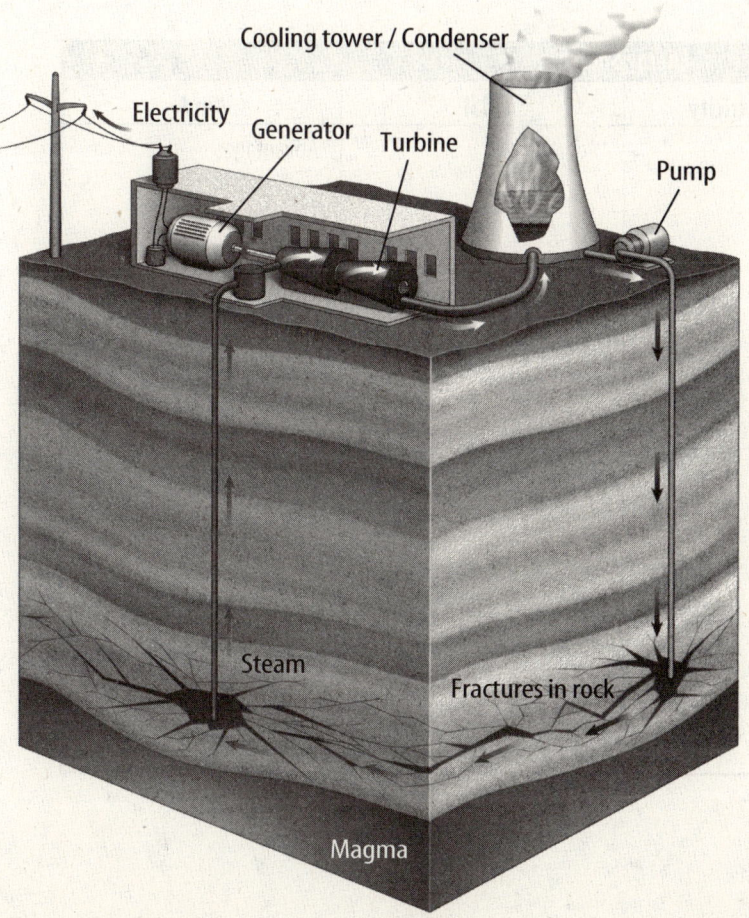

Picture This

8. Highlight Use a highlighter to follow the path of the water and steam in the geothermal power plant. Be sure to start at the pump.

Reading Essentials **153**

After You Read

Mini Glossary

biomass: renewable organic matter, such as wood and plant fibers

geothermal energy: thermal energy contained in hot magma

hydroelectricity: electricity produced from the energy of moving water

photovoltaic cell: a device that converts radiant energy from the Sun into electrical energy

renewable resource: an energy source that is replaced almost as quickly as it is used

1. Review the terms and their definitions in the Mini Glossary. Is hydroelectricity a renewable resource? Why or why not?

___

___

___

2. Complete the chart below to summarize what you know about these alternative energy sources.

Energy Sources			
Solar	**Hydroelectricity**	**Tidal**	**Wind**
Advantage:	Advantage:	Advantage:	Advantage:
Disadvantage:	Disadvantage:	Disadvantage:	Disadvantage:

154 CHAPTER 9 ENERGY SOURCES

3. Fill in each empty box with either the action or energy in each step of producing hydroelectricity.

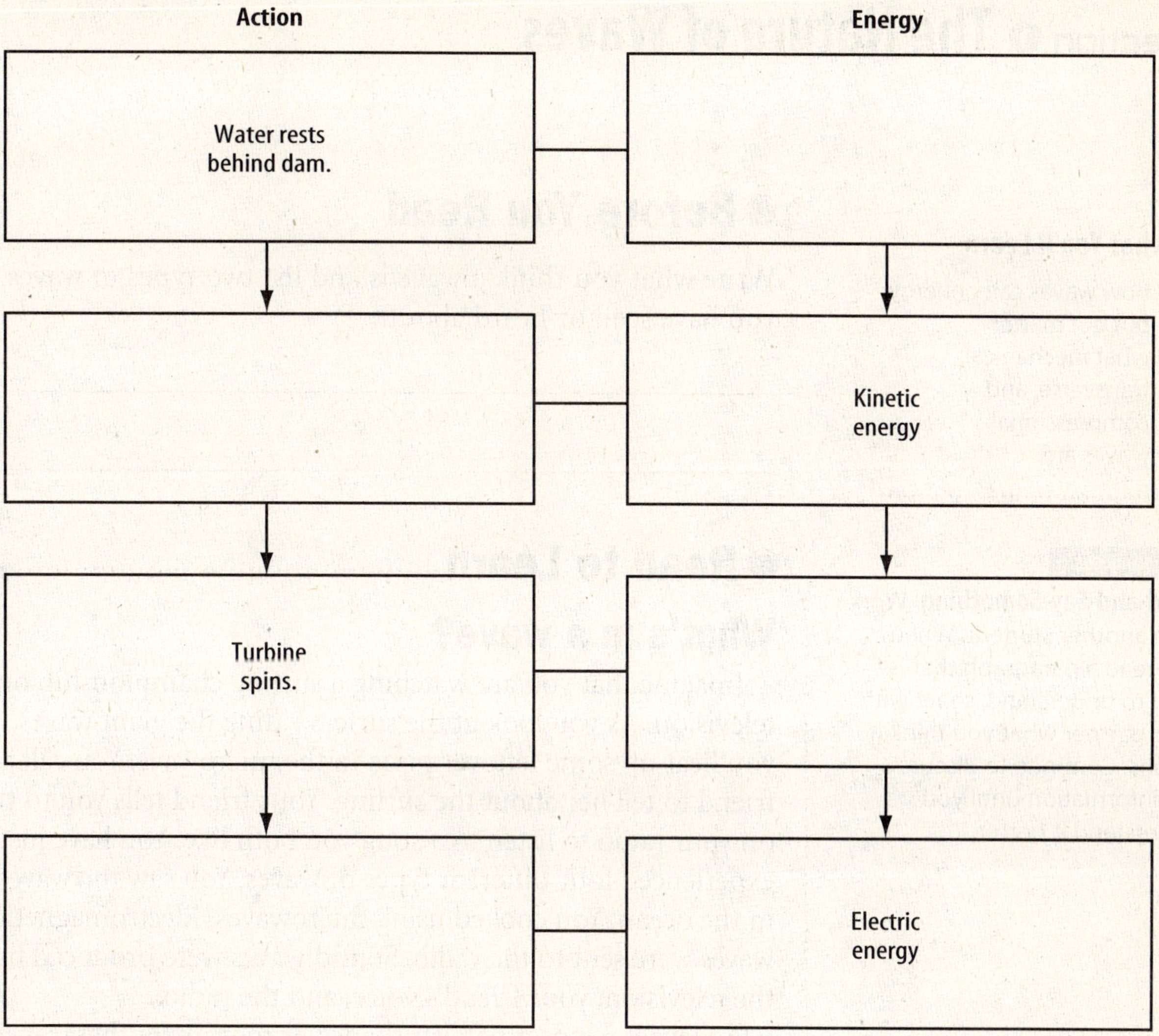

Action	Energy
Water rests behind dam.	
	Kinetic energy
Turbine spins.	
	Electric energy

4. You identified main points as you read this section. Would you recommend this strategy to a friend who is reading this chapter? Why or why not?

_____

_____

_____

 Visit gpscience.com to access your textbook, interactive games, and projects to help you learn more about renewable energy sources.

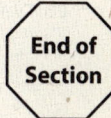
End of Section

Reading Essentials **155**

chapter 10 Waves

section ❶ The Nature of Waves

What You'll Learn
- how waves carry energy but not matter
- what mechanical, transverse, and compressional waves are

Study Coach

Read-and-Say-Something Work with another student. When you read a paragraph that is hard to understand, share with your partner what you think it means. Continue to discuss the information until you understand it better.

✔ Reading Check

1. Identify What do waves carry with them?

_____

● Before You Read

Write what you think a wave is and list two types of waves you have seen or heard about.

_____

_____

● Read to Learn

What's in a wave?

Imagine that you are watching a surfing championship on television. As you look at the surfers riding the giant waves, you heat up some leftover pizza in the microwave. You call a friend to tell her about the surfing. Your friend tells you to turn on your radio to listen to a song you both like. You have just experienced four different types of waves. You saw the waves in the ocean. You cooked using microwaves. Electromagnetic waves were sent to the radio. Sound waves were produced by the television, your friend's voice, and the radio.

A **wave** is a repeating disturbance or movement that transfers energy through matter or space. For example, ocean waves disturb the water and transfer energy through it. In an earthquake, powerful waves transfer energy through Earth. Microwaves, radio waves, and sound waves all transfer energy.

Waves and Energy

Have you ever watched a pebble fall into a pool of water and seen ripples form? The pebble causes a disturbance in the water. Some of the energy from the falling pebble transfers to the water molecules that are close by. The water molecules pass the energy along to other water molecules that are next to them. In this way, the energy passes from molecule to molecule until it is farther and farther away from where the pebble first fell. A wave is formed. It carries the energy along the surface of the water. ✔

Do waves carry matter?

Suppose you are in a boat on a lake. The waves in the water bump against your boat. You notice that the boat moves up and down and maybe even back and forth a little. But, after the waves have moved past, your boat has not moved to a different place. The waves don't carry the water along with them. They only carry energy as they move. This is true for all waves. They carry energy without moving matter from place to place.

How can you make a wave?

A wave will travel as long as there is both energy and a medium to carry it. Think of the pebble and the pool of water. The ripples eventually stop and the water is smooth again. There is no more energy to carry.

The figure shows a hand making a wave with a rope. As the hand moves up and down, a wave of energy begins in the rope. The wave of energy keeps moving down the rope until it gets to the end. Once it reaches the end, there is no more rope to carry the energy, so the rope is still again.

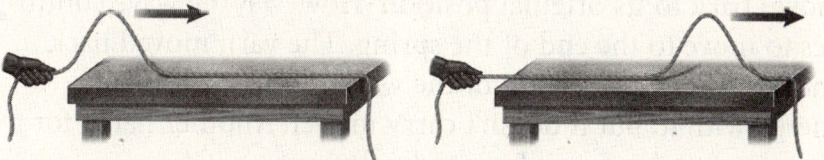

Producing a Wave

The hand in the figure is moving up and down. Anything that moves up and down or back and forth repeatedly is vibrating. The hand moving up and down is a vibrating movement. Vibrations cause all waves.

Mechanical Waves

Sound waves travel through the air to reach your ears. Ocean waves travel through water to reach the shore. Both kinds of waves move their energy through a medium. A **medium** is the matter through which a wave travels. The medium can be a solid, a liquid, or a gas. It can also be a combination of these forms of matter. The medium for sound waves is air. The medium for ocean waves is water. Not all waves need a medium in order to travel. Light waves and radio waves are examples of waves that do not need a medium. They can travel through space. Waves that travel only through a medium are called mechanical waves. There are two types of mechanical waves—transverse waves and compressional waves.

Picture This

2. **Mark the Figure** Use a pen or a pencil to circle the parts of the rope that show where the wave is.

Reading Check

3. **Identify** What must a mechanical wave travel through?

_____

FOLDABLES

A Compare and Contrast
Use two quarter-sheets of notebook paper to make a Foldable that compares and contrasts transverse waves and compressional waves.

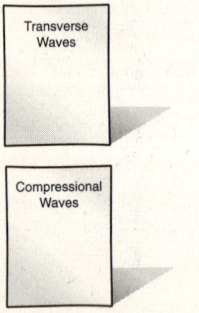

Picture This

4. **Explain** Why are the coils closer together at the middle of the spring in the figure?

_____
_____
_____

Think it Over

5. **Analyze** When a wave travels through a spring, why does it look like the whole spring is moving?

_____
_____
_____

How does matter move in a transverse wave?

In a <u>transverse wave</u>, the matter in the medium moves at a right angle to the direction the wave travels. An ocean wave moves across, horizontally, while the water it passes through moves up and down, vertically. The wave and the matter in the medium move at right angles to each other.

How does matter move in a compressional wave?

In a <u>compressional wave</u>, matter in the medium moves back and forth in the same direction that the wave travels. The figure below shows how a compressional wave moves along a coiled spring. Suppose you hold onto both ends of a spring. You squeeze some coils together at one end of the spring, then let go of them. A wave travels along the spring. If you tie a piece of yarn to one of the coils of the spring, you can see the back and forth movement in the spring. As the wave passes through the coil with the yarn, the yarn moves in the direction of the wave. After the wave passes the yarn, it moves back to its original position. However, the wave continues to move to the end of the spring. The yarn moved back and forth in the direction of the wave. The wave carries energy with it, but it doesn't carry matter. Another name for compressional waves is longitudinal waves.

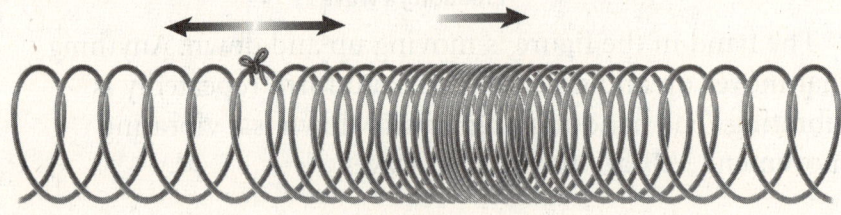

Compressional Wave

How do sound waves move?

Sound waves are compressional waves. Some sound waves travel through air. If you pick a string on a guitar, the string vibrates. The vibration pushes nearby air molecules close together. The air molecules squeeze together like the coils in the spring. Then the compressions travel through the air to make a wave. Sound waves also can travel through mediums such as water and wood. The particles in those mediums also squeeze together and move apart when sound waves travel through them.

When a sound wave reaches your ear, it causes your eardrum to vibrate. Your inner ear sends signals to your brain. Your brain understands these signals as sound.

How do water waves move?

Water waves look like transverse waves. They are actually a combination of transverse and compressional waves. The figures below show the movement of water in a wave. The small arrows show the direction of the wave and the large arrows show the movement of the water. As a wave goes by, the water moves up and down. The water also moves back and forth for a short distance in the same direction that the wave moves. Waves have both high and low points. Water pushes forward or backward toward the high part of the wave. This causes the low part of the wave to form as the first figure shows. Then as the wave passes, the water that was pushed forward or backward moves back to where it was.

Water is pushed aside.

Water returns to where it was.

Picture This

5. **Highlighting** Use your highlighter to trace the arrows that show the movement of water when it is pushed forward or backward toward the high part of the wave.

The up-and-down and back-and-forth motion causes the water to move in circles. An object that floats on the surface of the water takes in some of the energy from the waves. This causes it to bob in a circular motion.

How are ocean waves formed?

Wind blowing across the ocean surface causes most ocean waves to form. The changing speed of the wind acts like a vibration on the water. The size of the waves depends on the speed of the wind, the length of time the wind blows, and how far it travels over the water.

What are seismic waves?

Forces in Earth's crust can cause parts of the crust to shift, bend, or even break. When this happens, Earth's crust vibrates and releases energy. This creates seismic (SIZE mihk) waves that carry energy outward and cause an earthquake.

Seismic waves are a combination of transverse and compressional waves. They travel through Earth and along Earth's surface. Objects on Earth's surface move and shake when they take in some of the energy from the seismic waves. The more Earth's crust moves during an earthquake, the more energy is released.

Reading Check

6. **Identify** What usually starts the movement of a wave in an ocean?

_____

_____

After You Read
Mini Glossary

compressional wave: a wave in which the matter in the medium moves back and forth in the same direction that the wave travels

medium: the matter through which a wave travels

transverse wave: a wave in which matter in the medium moves back and forth at a right angle to the direction the wave travels

wave: a repeating disturbance or movement that transfers energy through matter or space

1. Review the definitions of the vocabulary words. What is the difference between a transverse wave and a compressional wave?

   _____

   _____

2. Complete the chart below to list what you learned about mechanical waves.

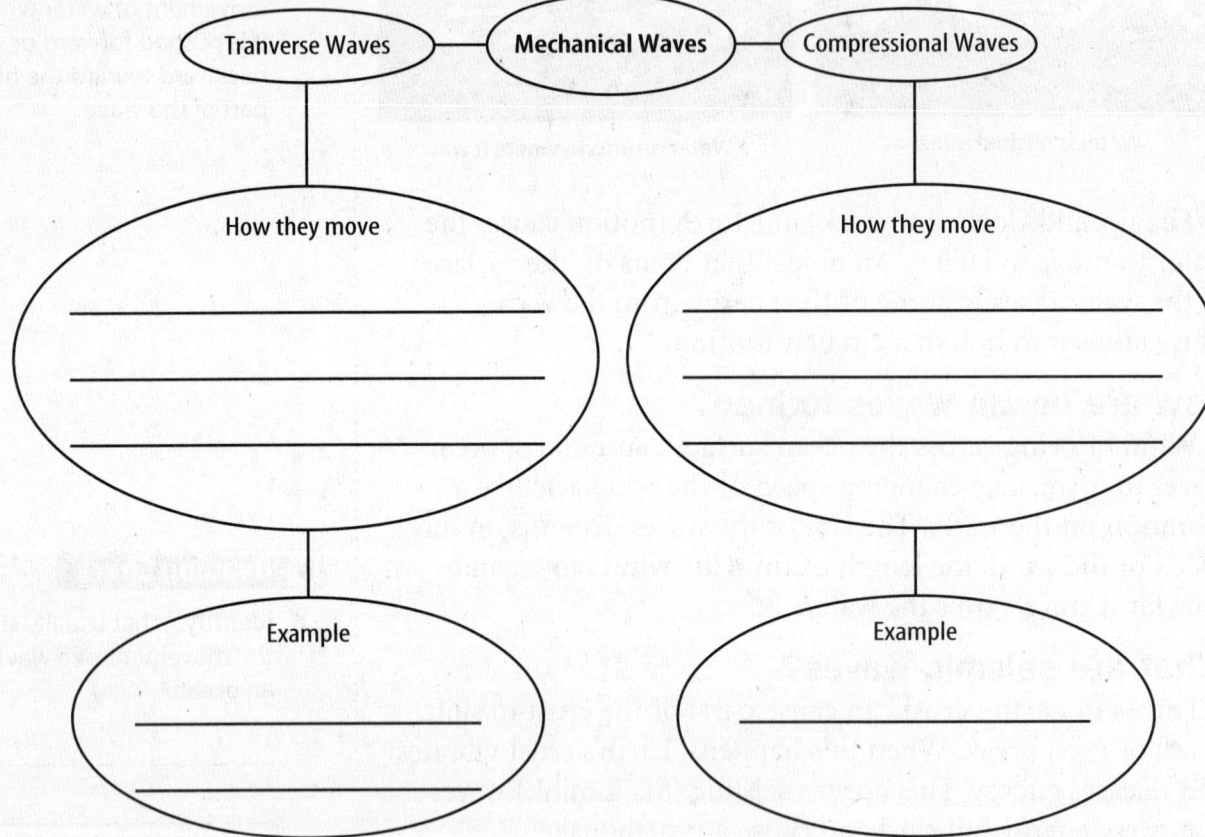

Science Online Visit gpscience.com to access your textbook, interactive games, and projects to help you learn more about producing electric current.

End of Section

Chapter 10 Waves

section 2 Wave Properties

Before You Read

Have you ever sat up high in the stadium at a baseball game and heard the sound of a ball being hit by a bat? Did you noticed that you heard the sound after the ball was hit? Write why you think this happened.

_____

_____

What You'll Learn
- what is wavelength, frequency, period, and amplitude
- how frequency and wavelength are related
- how a wave's energy and amplitude are related
- how to calculate wave speed

Read to Learn

Mark the Text

Selective Underlining As you read this section, look for important information about the properties of waves. Underline all of the ideas that you think are important to understand and remember.

The Parts of a Wave

How are sound waves, water waves, and seismic waves different? Some waves have more energy than others. Some travel faster than others. There are other ways that waves are different.

Remember that transverse and compressional waves act differently as they travel through a medium. The first figure below shows a transverse wave. Notice that the medium (the rope) has alternating high points and low points. A **crest** is the high point of a transverse wave. A **trough** is the low point of a transverse wave. The second figure shows a compressional wave in a coiled spring. It does not have crests or troughs. Instead, the wave creates areas where the coils are close together. This area is called a compression. In other areas, the coils are spread apart. A **rarefaction** (RAYR uh fak shun) is the area where the medium is more spread out.

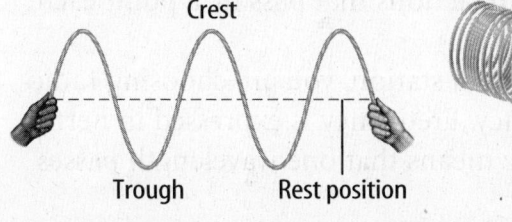

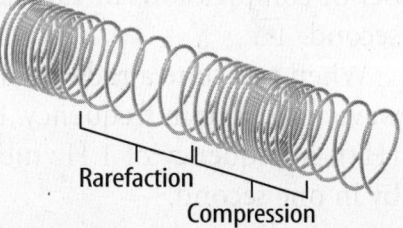

Picture This

1. **Describe** the pattern shown in the medium by the transverse wave.

_____

_____

Reading Essentials **161**

FOLDABLES

B Differentiate Make a 4-tab Foldable to help you organize notes that will help you learn the difference between wavelength, frequency, speed, and amplitude and energy.

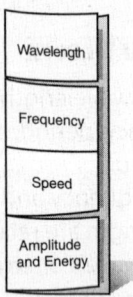

Picture This

2. Analyze What do you notice about the sizes of the wavelengths in the figure showing the transverse wave?

_____
_____
_____

3. Explain How is the frequency of a wave measured?

_____
_____
_____

Wavelength

Another way to describe a wave is by its wavelength. A <u>wavelength</u> is the distance between one point on a wave and the nearest point just like it. The first figure below shows a transverse wave. The wavelength is the distance from the top of one crest to the top of the next crest. Or you could measure the wavelength from the bottom of one trough to the bottom of the next trough.

The second figure below shows a compressional wave. The wavelength is the distance from the center of one compression to the center of the next compression. Or you could measure the wavelength from the center of one rarefaction to the center of the next rarefaction.

Humans cannot hear all sounds. We can only hear sounds that have wavelengths with measures between a few centimeters and about 15 m. We cannot hear sound with wavelengths that are smaller than a few centimeters. Nor can we hear sounds with wavelengths that are much greater than 15 m.

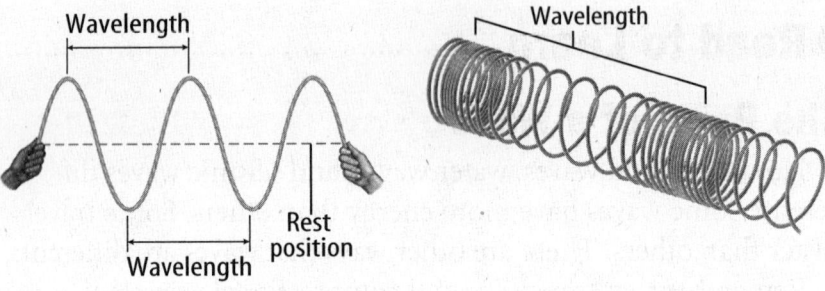

Frequency and Period

Have you ever heard someone say the word frequency? It usually means how often something happens in a given period of time. Frequency has a similar meaning when referring to waves. The <u>frequency</u> of a wave is the number of wavelengths that pass by a point each second. To find the frequency of a transverse wave, count the number of crests or troughs that pass by a point each second. In the same way, to find the frequency of a compressional wave, count the number of compressions or rarefactions that pass by a point each second.

When you tune a radio to a station, you are choosing radio waves of a certain frequency. Frequency is expressed in hertz (Hz). A frequency of 1 Hz means that one wavelength passes by in one second.

The **period** of a wave is the amount of time it takes one wavelength to pass a point. A period is measured in units of seconds. If it takes two seconds for an ocean wave, from one crest to the next, to pass a point, the wave has a period of 2.

How are frequency and wavelength related?

There is a relationship between frequency and wavelength. If you make transverse waves with a rope, you can increase the frequency by moving the rope up and down faster. The frequency increases because more crests or troughs pass by a point in one second. Moving the rope faster also makes the wavelength shorter. The distance from crest to crest or trough to trough is shorter. This relationship is always true—as the frequency of waves increases, their wavelength decreases.

Look at the two figures below. The waves in the ropes show that as the frequency of waves increases, their wavelength decreases. The first figure shows one wavelength passing by in one second. The frequency of the first wave is 1 Hz. The second figure shows two wavelengths passing by in one second. The frequency of the second wave is 2 Hz. As more wavelengths pass by in one second, the wavelengths get shorter. In the two figures, the frequency increases from 1 Hz to 2 Hz, so the wavelength decreases. If you move a rope up and down 5 times in 1 s, the frequency of the wave is 5 Hz. However, its wavelength would be even shorter than the rope with a frequency of 2 Hz.

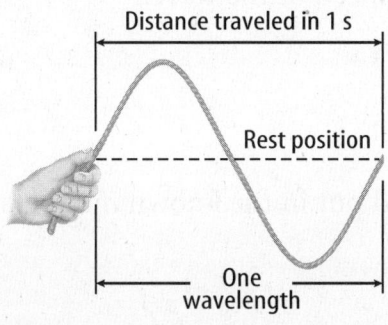
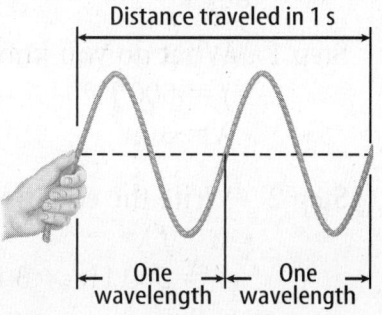

Wave Speed

Look back at the question at the beginning of this section. It asks if you have ever sat up high in the stadium at a baseball game, heard the sound of a ball being hit by a bat, and realize that you heard the sound after you saw the ball being hit. You saw the baseball being hit before you heard it because light waves travel through gases much faster than sound waves. Air is a gas. The light waves reflected from the ball reached your eyes before the sound waves created by the bat hitting the ball reached your ears.

Picture This

4. **Identify** What kind of waves are shown in the figure?

_____

Think it Over

5. **Explain** If you are watching fireworks from a distance, why do you see fireworks explode before you hear the noise?

_____
_____
_____

Reading Essentials **163**

What determines wave speed?

The speed of a wave depends on the medium it is traveling through. Even though light waves travel faster than sound waves through gases, sound waves usually travel faster than light waves through liquids and solids. Sound waves are compression waves. Compression waves travel faster in liquids and solids than they do in gases. Also, sound waves usually travel faster in a material if the temperature of the material is increased. For example, sound waves travel faster in air at 20°C than in air at 0°C.

How do you calculate wave speed?

You can calculate the speed of any wave by multiplying its frequency times its wavelength. The Greek letter lambda (λ) represents the wavelength, f represents the frequency, and v represents the speed of the wave. The wave speed equation is:

speed (in m/s) = **frequency** (in Hz) × **wavelength** (in m)
$$v = f\lambda$$

Why does multiplying the frequency unit Hz by the distance unit m result in the unit for speed, m/s? Recall the SI unit Hz is the same as 1/S. So multiplying m × Hz equals m × 1/S. This equals m/s.

Using the equation, you can calculate the speed of a wave traveling in water. If the wave has a frequency of 500 Hz and a wavelength of 3 m, what is the speed of the wave?

Step 1   What do you know?
f = 500 Hz
λ = 3 m

Step 2   Write the equation and put in the known numbers.
$v = f\lambda$
v = 500 Hz × 3 m

Step 3   Solve the equation.
v = 1,500 m/s
The speed of the wave is 1,500 m/s.

Amplitude and Energy

Why do some earthquakes cause terrible damage, while others are hardly felt? This is because waves can carry different amounts of energy. **Amplitude** is a measure of the energy that a wave carries. The greater a wave's amplitude, the more energy the wave carries. Amplitude is measured differently for compressional and transverse waves.

Reading Check

6. **Determine** Do light waves or sound waves travel faster through gases?

Applying Math

7. **Calculate** What is the speed of a sound wave that has a frequency of 150 Hz and a wavelength of 0.00002 mm? Show your work.

How is amplitude measured for compressional waves?

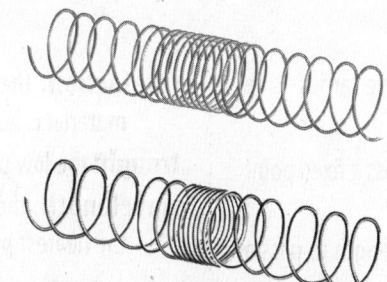

The amplitude of a compressional wave depends on how tightly the medium is pushed together at the compressions. The tighter the medium is pushed together at the compressions, the greater its amplitude. The greater the amplitude is, the more energy the wave carries. Think of a coiled spring toy. It takes more energy to push the coils tightly together than it does to barely move them.

Compare the compressions in the first figure with the compressions in the second figure. The coils in the compressions of the second figure are closer together. The second wave has greater amplitude and more energy compared to the first wave.

Also look at the rarefactions in the two figures. The closer the coils are in the compression, the farther apart they are in the rarefaction. So the less dense a medium is at the rarefactions, the more energy the wave carries.

How is amplitude measured for transverse waves?

Have you ever been knocked over by an ocean wave? If so, you know that the higher waves carry more energy. Remember that the greater a wave's amplitude, the more energy the wave carries. So a tall ocean wave has a greater amplitude than a short ocean wave. The amplitude of a transverse wave is measured differently than the amplitude of a compressional wave. The figure below shows that the amplitude of a transverse wave is the distance from the crest or trough of the wave to the rest position. The greater this distance is, the greater the amplitude is.

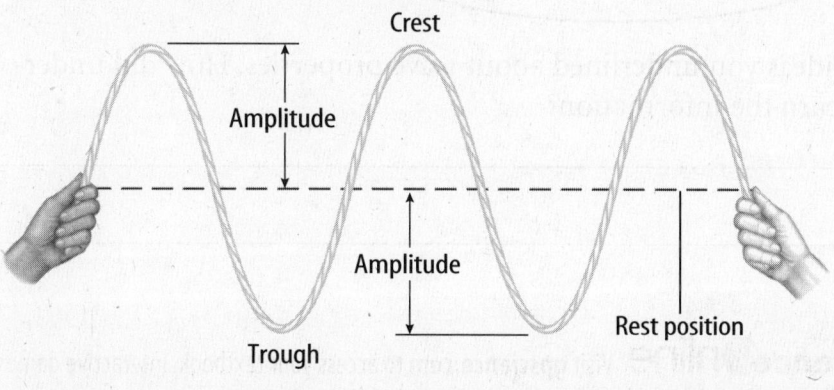

Picture This

8. Identify In the second coil, are the coils closer together or farther apart in the rarefactions?

_____

_____

Picture This

9. Draw Using the same rest position, draw a wave with a greater amplitude than the one shown. Then draw a wave with a smaller amplitude. Use your own paper.

After You Read
Mini Glossary

amplitude: a measure of the energy that a wave carries
crest: the high point of the transverse wave
frequency: the number of wavelengths that pass a fixed point in one second
period: the amount of time it takes one wavelength to pass a fixed point

rarefaction: the section in a compression wave where the material is less crowded and more spread out
trough: the low point of the transverse wave
wavelength: the distance between one point on a wave and the nearest point just like it

1. Review the vocabulary terms and their definitions. Explain how a transverse wave's wavelength is different from its amplitude.

   _____

   _____

   _____

2. Complete the concept map below to list the properties of waves that you learned about in this section.

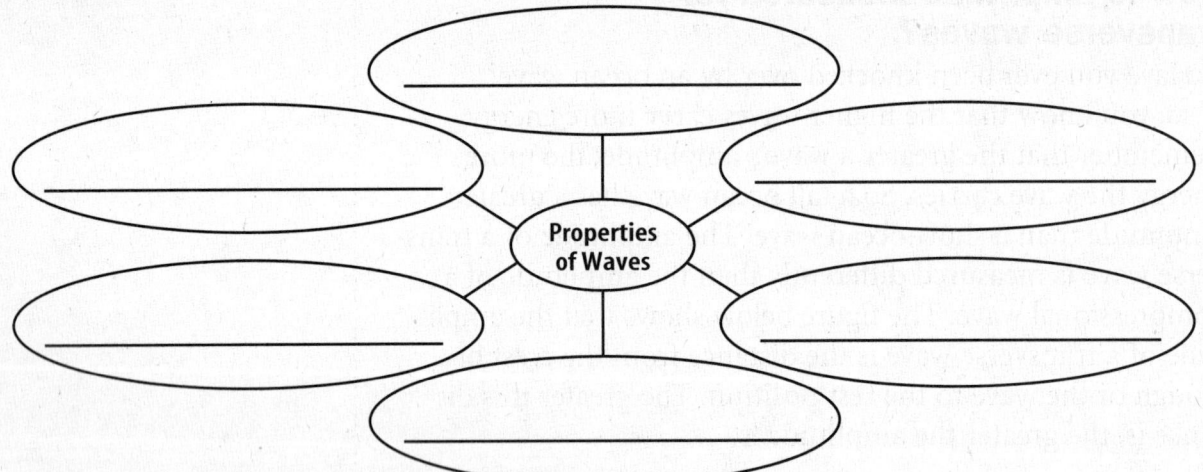

3. Look back at the important ideas you underlined about wave properties. How did underlining these ideas help you learn the information?

   _____

   _____

End of Section

 Visit gpscience.com to access your textbook, interactive games, and projects to help you learn more about wave properties.

Waves

section ❸ The Behavior of Waves

● Before You Read

Think about a time that you walked down an empty hallway and heard the echo of your footsteps. Write what you think caused the echo.

___

What You'll Learn
- the law of reflection
- how waves change direction
- what are refraction and diffraction
- how waves interfere with each other

● Read to Learn

Reflection

Suppose you and a friend are the last students to leave your school building. You shout to your friend at the other end of the hallway. Your voice echoes throughout the hallway. You also notice your reflection in one of the glass windows. These are both examples of wave reflection. Wave reflection causes the echo you hear and the image you see of yourself. Wave reflection happens when a wave strikes an object or surface and bounces off of it. All types of waves—including sound, water, and light waves—can be reflected.

How do light waves reflect?

What happens when you see your face in a mirror? First, light waves strike your face and bounce off. Then, the reflected light strikes the mirror and reflects back to your eyes.

What are echoes?

Echoes are reflected sound waves. When you called to your friend in the school building, your voice echoed around the hall. Sound waves formed when you shouted. The waves traveled through the air to your ears and to other objects. The waves reflected off the walls, floor, and ceiling and then came back to your ears. You could hear your voice again, a few seconds after you first heard your voice. This caused the echo. Dolphins and bats use echoes to determine where objects are. Dolphins make clicking noises and listen to the echoes.

Study Coach

Create a Quiz Write questions on index cards as you read this section. After you read, form a group of three students. Take turns asking each other your questions and answering them.

FOLDABLES

C Build Vocabulary Make the following Foldable. It will help you understand the content of this section by defining the vocabulary terms.

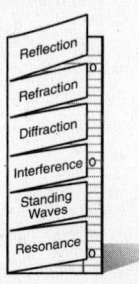

Reading Essentials 167

Picture This

1. **Labeling** On the figure, label the incident beam as *i* and the reflected beam as *r*.

Picture This

2. **Recall** Describe a time when you saw an object in water that looks like the pencil in the figure.

_____
_____
_____

What is the law of reflection?

The figure below shows a flashlight beam striking a mirror. The light beam that strikes the mirror is called the *incident beam*. The light beam that bounces off the mirror is called the *reflected beam*. The line that is at a right angle to the mirror is called the *normal*. The angle made by the incident beam and the normal is the *angle of incidence*. The angle made by the reflected beam and the normal is the *angle of reflection*. The law of reflection says that the angle of incidence is equal to the angle of reflection.

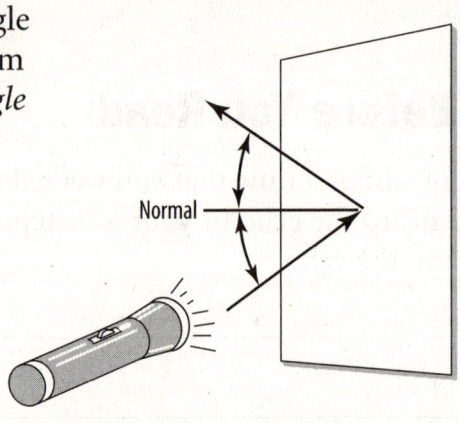

Refraction

Suppose you put a pencil in a glass of water and then look at it from the side of the glass. The pencil looks like it is broken into two pieces at the water line. But if you pull the pencil out of the water, you see that it is not broken! What causes the pencil to appear broken?

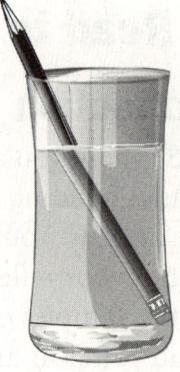

What is refraction?

Remember that a wave's speed depends on the medium it is moving through. When a wave moves from one medium to another, such as from water to air, it changes speed. If this wave is traveling at an angle when it passes from one medium to another, it changes direction, or bends, as it changes speed. **Refraction** is the bending of a wave caused by a change in its speed as it moves from one medium to another. Two things happen to the light waves as they move from the air to the water in the glass. The light waves change speed and direction. They are refracted so the pencil appears to be broken.

What happens to light waves as they change speed?

When a light wave passes into a material in which it slows down, it bends toward the normal. When a light wave passes into a material in which it speeds up, the wave bends away from the normal.

The figures show what these refractions look like. The first figure shows light waves traveling from air to water. The light waves slow down when they enter the water. This causes them to change direction and bend toward the normal. The second figure shows light waves traveling from water to air. The waves speed up and bend away from the normal.

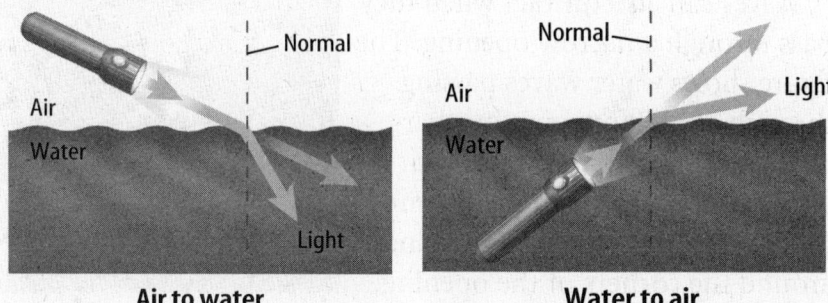

Air to water   **Water to air**

Picture This

3. **Highlighting** Look at the figure on the right. Use your highlighter to trace the angle that shows the light bending away from the normal.

How does light bend in water?

Have you ever noticed that the feet of someone standing in a swimming pool look closer to the surface than they really are? Your brain wants to think that the light waves travel in a straight line. But refraction causes the light waves from the feet to bend away from the normal as they pass from the water to the air.

This is similar to the pencil in the glass of water. The pencil looks broken at the surface of the water. The light waves coming from the part of the pencil above the water are not bent. The light waves that move from the air to the water in the glass change speed and bend. This makes the part of the pencil that is underwater look like it has shifted.

Diffraction

Suppose you are in a classroom and you hear music coming from another room. The sound waves bend around corners and travel from the room down the hall to where you are. Refraction does not cause sound waves to bend. Instead, they bend because of diffraction. **Diffraction** takes place when an object causes a wave to change direction and bend around it.

Light waves can diffract, too. Light waves do not diffract as much as sound waves do. Suppose you walk toward the room where you hear the music. As you walk toward the open door, you can see light coming out of the room. Light waves bend around the edges of the open door. But the amount of light that bends is not enough for you to be able to see around the corner and into the room. Yet, you can hear the all the music that is being played in the room.

Reading Check

4. **Identify** Circle the term that describes the bending of a wave around an object.

   a. reflection

   b. refraction

   c. diffraction

   d. rarefaction

Picture This

5. Trace Using a pencil, trace along the waves in the figure that have passed through the opening. What happens to the lengths of the waves as they get farther from the opening?

_____

_____

_____

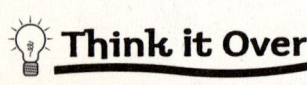

Think it Over

6. Explain Why is it hard to see around a corner?

_____

_____

_____

When do water waves diffract?

Ocean waves refract when they strike an island. The waves change direction and bend around the island. Diffraction and refraction both cause waves to bend, but there is a difference. Waves refract when they pass through an object. They diffract when they pass around an object.

Waves can also diffract when they pass through a narrow opening. The figure shows water waves passing through a small opening in a barrier. They diffract and spread out after they pass through the opening. In this case, the waves are bending around the corners of the opening.

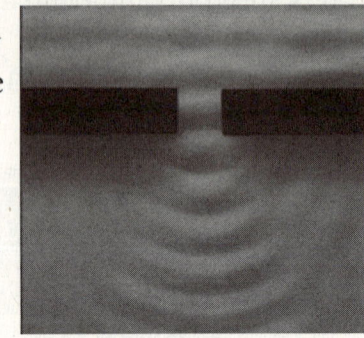

How much will a wave bend?

To find out how much a wave will bend when it strikes an object, compare the size of the object to the wavelength. When an object is smaller than the wavelength, the waves bend around it. Suppose you shine a large spotlight on a very thin tree. The object is smaller than the wavelength, so the light waves bend around the tree. The shadow is narrower than the tree. A large amount of diffraction takes place. Suppose you shine a small flashlight on a very large tree. The object is larger than the wavelength, so the light waves will not bend around the tree easily. The shadow is wider than the tree. Almost no diffraction takes place.

How do sound waves bend around corners?

Think back to the example of the music coming from another room in your school. You can hear the sounds before you reach the door. The wavelengths of sound waves are about the same size as the door opening. The sound waves diffract around the door and spread out into the hallway.

How is this different from the light waves in the room? You aren't able to see into the room because light waves have a much shorter wavelength than sound waves. So, the light waves are not diffracted by the door as much as the sound waves are.

How do radio waves diffract?

AM radio waves have longer wavelengths than FM radio waves. Because the AM radio wavelengths are longer, AM radio waves can diffract around objects like buildings and mountains much more than FM radio waves. You may have noticed that you can get more AM radio stations than FM stations when you are around tall buildings and hills.

Interference

Suppose you throw two pebbles into a still pond. Each pebble causes ripples to form around it. The waves of ripples travel toward each other. When the two waves meet, they pass right through each other and continue moving. **Interference** is the point where two waves meet each other and overlap to form a new wave. The new wave lasts only as long as the two waves continue to overlap. There are two kinds of interference, constructive and destructive.

What is constructive interference?

The figure shows constructive interference. In constructive interference, the waves meet at the same point and add together. This happens when the crests of transverse waves overlap each other. The troughs of the waves also overlap. The amplitudes of the two waves add together to make a larger wave. So, the new wave has a higher crest and a lower trough. The amplitude of the new wave is the sum of the amplitudes of the other two waves. Constructive interference also happens when two compressional waves overlap. If they are sound waves, the new wave is louder.

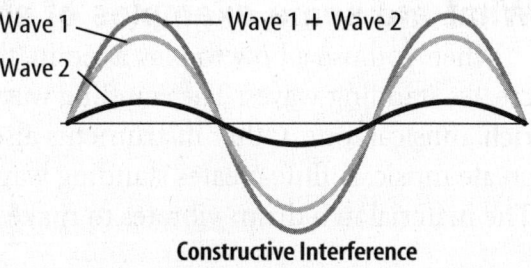
Constructive Interference

What is destructive interference?

In destructive interference, the overlapping waves subtract from each other. This happens when the crest of one transverse wave overlaps with the trough of another transverse wave. The amplitudes of the two waves combine to make a new wave with a smaller amplitude. The figure shows the new wave that forms during destructive interference.

The same is true for compressional waves. When the compression of one wave overlaps the rarefaction of the other wave, this causes destructive interference. The compressions and rarefactions combine to form a smaller wave. In other words, the sound gets softer.

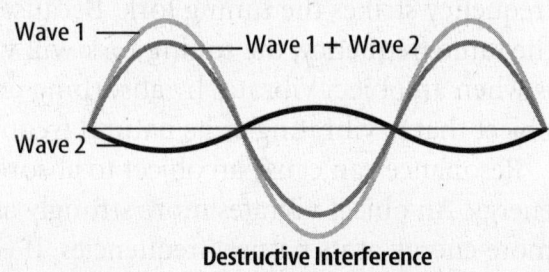
Destructive Interference

Think it Over

7. Describe a time when you heard or used the word "interference" or "interfere."

Picture This

8. Highlighting In both figures, use a highlighter to trace over the new wave that forms during either constructive or destructive interference.

Standing Waves

Suppose that you and a friend are holding the ends of the same rope. You both shake the rope and make waves the same size that travel toward each other. Interference happens when the waves from one end overlap waves from the other end. A new wave forms when a crest of one of your waves meets a crest of one of your friend's waves. The new wave has a larger amplitude.

When a crest of one wave meets a trough of another wave, the waves cancel each other out. Then there is no movement. A **standing wave** is a wave pattern that forms when two equal-sized waves travel in opposite directions and continuously interfere with each other. The interference of these two waves makes the rope vibrate and creates a pattern of crests and troughs. This makes it look like the rope is standing still. Nodes are the places where the two waves cancel each other. The nodes always stay in the same place on the rope. The wave pattern vibrates between the nodes.

What are some examples of standing waves?

When you use a bow to play a violin, the string vibrates and creates standing waves. The standing waves in the string make a rich, musical tone. Other instruments also use standing waves to create music. A flute creates standing waves in a column of air. The material in a drum vibrates to make standing waves.

Resonance

Bells of different sizes and shapes make different sounds. When you strike a bell, it vibrates at its own natural frequencies. All objects have their own natural frequencies of vibration. The frequencies depend on the size and shape of the object. Frequency also depends on the kind of material that the object is made of.

There is another way to make an object vibrate at its natural frequencies. Suppose you have a tuning fork with a single natural frequency. Imagine that a sound wave of the same frequency strikes the tuning fork. Because the sound wave has the same frequency, the tuning fork will vibrate. **Resonance** is when an object vibrates by absorbing energy from another object that is vibrating at its natural frequencies.

Resonance can cause an object to absorb a large amount of energy. An object vibrates more strongly as it keeps absorbing more energy at its natural frequencies. If enough energy is absorbed, the object can vibrate too much and break apart.

Reading Check

9. **Describe** What happens when a crest of one wave meets a trough of another wave?

Think it Over

10. **Summarize** How can resonance cause an object to break?

After You Read
Mini Glossary

diffraction: the bending of a wave around a barrier
interference: the point where two waves meet each other and overlap to form a new wave
refraction: the bending of a wave caused by a change in its speed as it moves from one medium to another

resonance: an object vibrates by absorbing energy from another object that is vibrating at its natural frequencies
standing wave: a wave pattern that forms when two equal-sized waves travel in opposite directions and interfere with each other

1. Review the definitions of the vocabulary words in the Mini Glossary. How are refraction and diffraction the same? How are they different?

   _____
   _____
   _____
   _____

2. List the main topics you learned about the behavior of waves. For help, use the main headings in the section.

   The Behavior of Waves

   1. _____
   2. _____
   3. _____
   4. _____
   5. _____

3. Look at the questions you wrote on index cards during this section. How did writing these questions help you learn?

   _____
   _____

 Visit gpscience.com to access your textbook, interactive games, and projects to help you learn more about the behavior of waves.

Reading Essentials 173

chapter 11 Sound

section ❶ The Nature of Sound

What You'll Learn
- how sound travels
- what changes the speed of sound
- how your ears allow you to hear

Mark the Text

Identify the Main Point
Highlight the main idea of each paragraph. Then review what you highlighted after you finish reading the section.

Before You Read

Place your hand on your throat and hum a tune. How is the movement that you feel related to the sound you made?

_____

_____

Read to Learn

What causes sound?

An amusement park is a noisy place. Music is playing, videogames are beeping, and people on rides are screaming. It can be hard to hear your friends talking. These sounds are all different, but they all have something in common. Each sound is made by an object that vibrates.

Remember that vibration is a quick, rhythmic back-and-forth movement. The sounds of your friends' voices and people screaming are made by the vibrations of their vocal cords. Music and noises from videogames are made by vibrating speakers. All sounds are made by something that vibrates.

Sound Waves

Think of a radio speaker. When the speaker vibrates, it bumps into nearby molecules in the air. Some of the speaker's energy is transferred to air molecules. These air molecules bump into other air molecules. They pass the energy on to the new molecules. The energy that came from the vibrating speaker keeps moving from one molecule to another. The collisions and the energy they transmit make a sound wave.

Sound waves are compressional waves. Recall that there are two areas in compressional waves. One area is the compression where the molecules of the medium are very close together. The other area is the rarefaction where the molecules of the medium are more spread out. A compressional wave has alternating areas of compressions and rarefactions.

In each figure, the radio speaker is vibrating. In the first figure, the speaker is vibrating outward. When this happens, molecules in the air are pushed together to form a compression. The second figure shows the speaker moving inward. Air molecules have room to spread out. A rarefaction forms. The back-and-forth vibrations form compressions and rarefactions.

Picture This

1. Communicate Look at the figures to the left. Describe how the particles in a compression are different from the particles in a rarefaction.

_____
_____
_____

Compression

When the speaker vibrates outward, molecules in the air next to it are pushed together to form a compression.

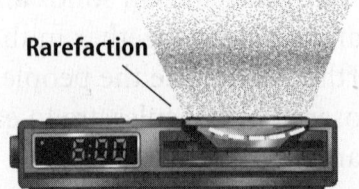

Rarefaction

When the speaker vibrates inward, the molecules spread apart to form a rarefaction.

How does a sound wave travel?

The air molecules around the speaker bump into other molecules. The speaker keeps vibrating, so a series of compressions and rarefactions forms a sound wave. This sound wave travels out from the speaker for your ear to hear.

Moving Through Materials

Did you ever swim underwater and hear sounds? Most sounds that you hear travel through air. Sound can also travel through solids, liquids, and other gases. Sound waves travel in any material, or medium. But sound waves cannot travel through empty space. There are no molecules to transmit a sound wave.

How does the speed of sound change in different materials?

How quickly a sound wave moves through a material depends on the material and whether it is solid, liquid, or gas. The table shows that sound travels slowest through gases and fastest through solids. Molecules in liquids and solids are closer than molecules in gases. When molecules are close, they transmit energy more quickly. Loud and soft sounds travel through a material at the same speed.

Speed of Sound in Different Mediums	
Materials	Speed of Sound (in m/s)
Air	347
Cork	500
Water	1,498
Brick	3,650
Aluminum	4,877

Applying Math

2. Use a Table Suppose lightning strikes two kilometers away from you. Find how long it would take the sound of the thunder to reach your ears by dividing the distance by the speed of sound through air. Round your answer to the nearest tenth of a second.

_____

Why do close molecules transmit sound more quickly?

Imagine a large group of people standing in a line. They are passing a bucket of water from person to person. If everyone stands far apart, each person has to take the time to walk the bucket to the next person. But if everyone stands close together, each person quickly can hand the bucket to the next person.

The molecules in solids and liquids are like the people standing close together in the line. The particles in gases are farther apart, like the people standing far apart in line. The closer the molecules are to each other, the faster they can transfer energy.

How does temperature affect the speed of sound?

The speed of sound waves also depends on the temperature of a medium. As temperature increases, molecules move faster. It is easier for molecules to bump into each other if they are moving quickly. If the particles in a medium are bumping into each other more often, more energy can be transferred faster. So the higher the temperature, the faster sound waves move. For example, when the air temperature is 0°C, sound travels through the air at 331 m/s. But at a temperature of 20°C, sound travels at 343 m/s.

Human Hearing

Our vocal cords and mouths produce many different kinds of compressional waves. But how does your brain make sense of sound waves? Your ears and your brain work together. They turn compressional waves into something that you can understand. Making sense of sound waves involves four steps. First, your ears gather the compressional waves. Next, your ears amplify the waves, or make them stronger. The amplified waves are changed to nerve impulses that travel to the brain. Finally, the brain makes sense of the nerve impulses.

How does the outer ear gather sound waves?

When you think of your ear, you probably picture the part that is on the outside. But the human ear has three parts: the outer ear, the middle ear, and the inner ear. Look at the figure on the next page. It shows the three parts of the human ear. The outside part of the ear together with the ear canal and the eardrum make up the outer ear. The outer ear is where sound waves are gathered. It helps catch the sound waves, then sends them into the ear canal.

✓ Reading Check

3. **Explain** Why does sound travel faster in solids and liquids than it does in gases?

_____
_____
_____

💡 Think it Over

4. **Explain** Hearing aids are placed in a person's outer ear. This makes it easy for the person to put them in and take them out. What do hearing aids do when they are placed in a person's outer ear?

_____
_____
_____

The sound waves travel along the ear canal to the eardrum. The **eardrum** is a tough membrane, or tissue, about 0.1 mm thick. The eardrum is stretched over the end of the ear canal. When sound waves reach the eardrum, they transfer their energy to the eardrum and it vibrates.

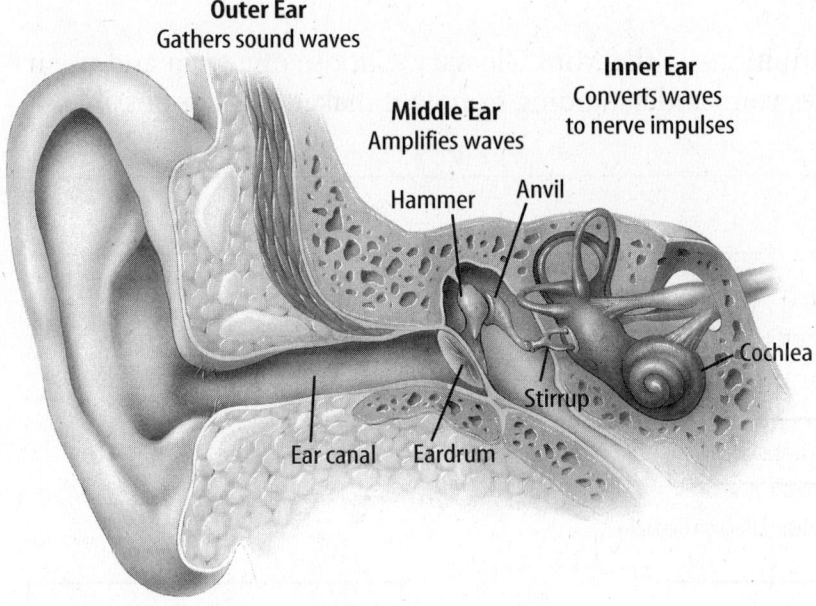

What does the middle ear do?

When the eardrum vibrates, it transfers the vibrations into the middle ear. There are three tiny bones in the middle ear that make the vibrations stronger. The bones are called the hammer, anvil, and stirrup. They are a lever system that makes the force and pressure of the sound waves stronger. The stirrup is connected to a membrane in the oval window. When the stirrup vibrates, the membrane vibrates too.

What does the inner ear do?

When the membrane in the oval window vibrates, the sound vibrations are sent into the inner ear. The inner ear contains the cochlea. The **cochlea** (KOH klee uh) is a spiral-shaped structure that is filled with liquid. It also contains tiny hair cells. When the hair cells in the cochlea begin to vibrate, nerve impulses are sent to the brain. When the nerve impulses reach the brain, they are interpreted as sounds. It is the cochlea that changes sound waves to nerve impulses.

If the hair cells in the cochlea are damaged or destroyed, a person can lose some hearing ability. This damage can happen when someone is exposed to loud sounds. Scientists are finding that the hair cells may be able to repair themselves.

Picture This

5. **Use a Scientific Illustration** Using a highlighter, draw a circle around the structures that make up the outer ear. Then use two different colors of highlighters to draw circles around the structures of the middle ear and of the inner ear.

FOLDABLES

A **Drawing and Identifying** On the inside of a 3-tab book Foldable, draw a human ear. Identify the three sections of the ear and label all the parts found in each section.

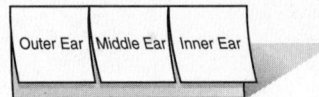

Think it Over

6. **Think Critically** Write two things you can do to help protect your hearing.

_____
_____
_____
_____

After You Read
Mini Glossary

cochlea: a spiral-shaped structure in the inner ear that turns vibrations into nerve impulses

eardrum: a tough membrane, or tissue, at the end of the ear canal that transmits vibrations to the middle ear

1. Review the terms and their definitions in the Mini Glossary. Choose one term and use it in a sentence that demonstrates your understanding of how it makes hearing possible.

_____

_____

2. Complete this graphic organizer to outline how vibrations travel through the air to be interpreted as sounds by the brain.

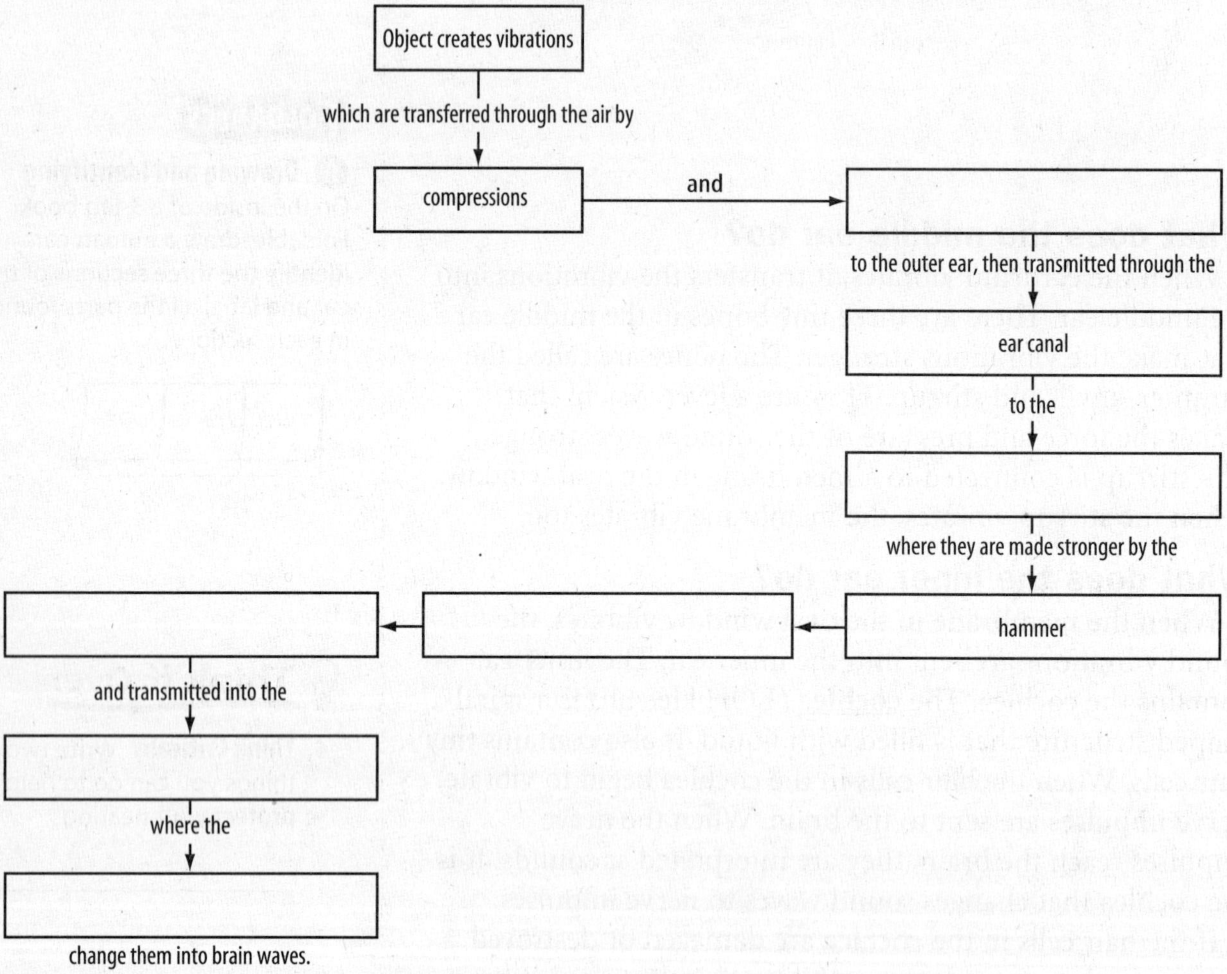

```
          Object creates vibrations
                    |
       which are transferred through the air by
                    ↓
             compressions  ──and──→  [          ]
                                          |
                          to the outer ear, then transmitted through the
                                          ↓
                                    [ ear canal ]
                                          |
                                        to the
                                          ↓
                                    [          ]
                                          |
                          where they are made stronger by the
                                          ↓
      [          ]  ←──────────────  [ hammer ]
           |
    and transmitted into the
           ↓
      [          ]
           |
        where the
           ↓
      [          ]
    change them into brain waves.
```

Science online Visit **gpscience.com** to access your textbook, interactive games, and projects to help you learn more about the nature of sound.

178 CHAPTER 11 Sound

Chapter 11 Sound

section 2 Properties of Sound

Before You Read

Do your parents ever ask you to turn down the volume on the television? Why do you think some sounds might seem too loud to one person and just right to another?

_____

_____

Read to Learn

Intensity and Loudness

Suppose you turn down the volume on your radio. The notes sound the same, but the sound is not as loud. What happens to the sound waves? The quieter sound waves do not carry as much energy as louder sound waves do.

The amount of energy a wave carries depends on its amplitude. For a compressional wave, amplitude is related to the density of the molecules. Density means how close together molecules are.

Compare the figures. Molecules vibrate with a lot of energy. Sound waves with dense compressions vibrate with a lot of energy. The second sound wave shown has a high-amplitude. The compressions are much denser in a high-amplitude sound wave. Molecules act in the opposite way in the rarefactions. The rarefactions are less dense in the high-amplitude sound wave.

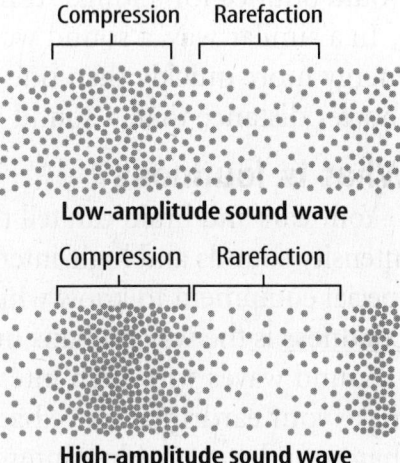

Remember that sound waves are compressional waves. The wave carries energy, but it doesn't carry matter. Matter compresses and expands as a sound wave passes through the matter.

What You'll Learn

- how amplitude, intensity, and loudness are related
- how sound is measured
- the relationship between frequency and pitch
- the Doppler effect

Mark the Text

Identify Definitions As you read the section, highlight the headings that are questions. Highlight the answers in a different color.

FOLDABLES

B Build Vocabulary Make the following vocabulary book Foldable to define the vocabulary terms of this section.

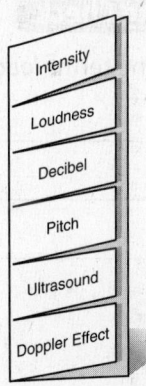

Reading Essentials 179

Think it Over

1. Apply How can you increase the intensity of the sound coming from a radio?

_____

_____

What is intensity?

Imagine sound waves moving from a radio, through a small window, and then to your ear. The amount of energy that passes through the window in 1 s is a measure of intensity. **Intensity** is the amount of energy that flows through a certain area in a specific amount of time. When you turn down the volume of your radio, you reduce the energy carried by the sound waves. So you are also reducing their intensity.

Intensity affects how far away a sound can be heard. Think about whispering with a friend. The sound waves you make have low intensity. They do not travel far. You have to sit close together to hear each other. Now think about shouting to someone. You can be much farther apart. When you shout, the sound waves have high intensity. They can travel farther.

How are intensity and distance related?

Some of a wave's energy is converted into other forms of energy when it is passed from particle to particle. So intensity affects how far a wave will travel.

What happens when you drop a basketball? The ball held above the ground has potential energy. When the ball falls, the potential energy is changed into energy of motion. When it hits the ground and bounces up, some of the ball's energy is transferred to the ground. The ball doesn't have enough energy to bounce back as high as it was. The ball transfers a small amount of energy with each bounce. Finally, the ball has no more energy and it stops bouncing. If you held the ball higher above the ground, it would have more energy and would bounce for a longer time before it stopped.

In a similar way, a sound wave with low intensity loses its energy more quickly than one with high intensity. It travels a shorter distance than a sound wave of higher intensity.

What is loudness?

Your ears and brain can tell the difference between low-intensity sounds and high-intensity sounds. You do not need special equipment to know which sounds have greater intensity. **Loudness** is the way humans understand sound intensity.

Sound waves with high intensity carry more energy. They make your eardrums move back and forth a greater distance than sound waves of low intensity do. The bones of the middle ear change the increased movement of the eardrum into increased movement of the hair cells in the inner ear. This makes you hear a loud sound instead of a quiet one. As the intensity of a sound waves increases, the loudness of the sound you hear increases.

2. Define the terms *loudness*.

_____

_____

_____

How is loudness measured?

It is hard to say how loud too loud is. Two people might not agree on whether a noise is too loud. A sound that seems fine to you may seem much too loud to your teacher. However, the intensity of sound can be measured. Each unit on the scale for sound intensity is called a **decibel** (DE suh bel). The abbreviation for decibels is dB.

The quietest sound that most people can hear is 0 dB. The intensity of the noises in an average home is 50 dB. Sounds that are louder than 120 dB can cause pain and even permanent hearing loss. During some rock concerts, sound intensity reaches this damaging level. Factories, construction sites, and other workplaces also can have noise levels that might damage hearing. Wearing ear protection, such as earplugs, around loud sounds can help protect against hearing loss. The figure below shows the intensity levels of some sounds in decibels.

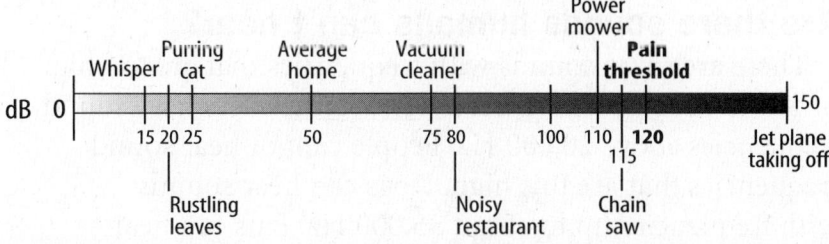

Picture This
3. **Interpret a Graph** How many decibels louder is a jet plane taking off than a vacuum cleaner?

_____

Pitch

Have you ever sung the musical scale do, re, mi, fa, so, la, ti, do? When you sing this scale, your voice starts low and becomes higher with each note. What you hear is a change in pitch. **Pitch** is how high or low a sound seems to be. The pitch of a sound is related to the frequency of the sound waves.

How are frequency and pitch related?

Frequency, being the number of wavelengths that pass a fixed point in one second, is measured in hertz (Hz). One hertz means one wavelength passes by in 1 second. The frequency of a compressional wave is the number of compressions or the number of rarefactions passing by a fixed point each second.

When a sound wave with high frequency hits your ear, many compressions hit your eardrum each second. The vibrations are transmitted to your brain. You understand them as a sound with a high pitch. As the frequency of a sound wave decreases, the pitch becomes lower. A whistle with a frequency of 1,000 Hz has a high pitch. Low-pitched thunder has a frequency of less than 50 Hz.

Think it Over
4. **Explain** Why can the frequency of a compressional wave be counted as either the number of compressions or the number of rarefactions that passes by in one second?

_____
_____
_____

The figure shows different notes and their frequencies. A healthy human ear can hear sound waves with frequencies from about 20 Hz to about 20,000 Hz. Humans can best hear sounds that are between 440 Hz and 7,000 Hz. Most people can hear much softer sounds at this range than at higher or lower frequencies.

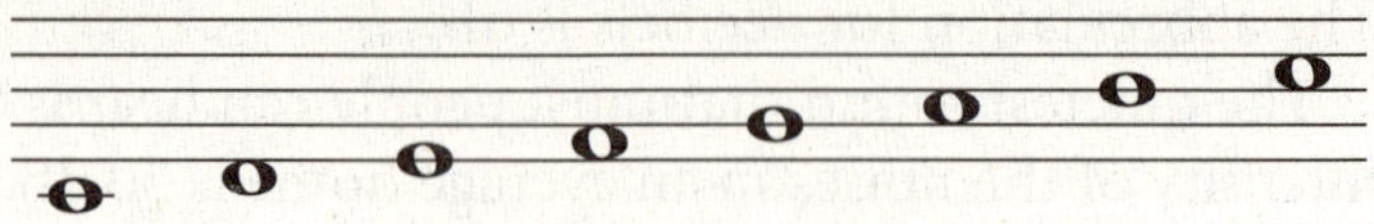

Applying Math

5. Apply Notice that the musical note C is both the first and the last note of the scale. What do you notice about the frequencies of these two notes?

Note:	C	D	E	F	G	A	B	C
Sound:	do	re	mi	fa	so	la	ti	do
Frequency:	262 Hz	294 Hz	330 Hz	349 Hz	393 Hz	440 Hz	494 Hz	524 Hz

Are there sounds humans can't hear?

There are some sounds with frequencies that are too high or too low for people to hear. **Ultrasonic** waves have sound frequencies above 20,000 Hz. People cannot hear sound frequencies that are this high. Dogs can hear sounds with frequencies up to about 35,000 Hz. Bats can hear sounds with frequencies of more than 100,000 Hz. Even though humans can't hear ultrasonic waves, they use them for many things. Doctors use ultrasonic waves to diagnose and treat illnesses. Ultrasonic waves also are used to estimate the size, shape, and depth of underwater objects.

Reading Check

6. Identify What are two ways that people use ultrasonic waves?

Sound waves with frequencies below 20 Hz are called infrasonic waves. They are also called subsonic waves. These frequencies are too low for most people to hear. Infrasonic waves are produced by sources that vibrate very slowly. Wind, heavy machinery, and earthquakes are some things that produce infrasonic waves. You probably can't hear infrasonic waves. But sometimes you can feel them as a rumble inside your body.

The Doppler Effect

Suppose that you are standing beside a racetrack. Race cars are zooming past. As a car moves toward you, the pitch of its engine becomes higher. As it moves away from you, the pitch becomes lower. The **Doppler effect** is a change in pitch or wave frequency because a wave source is moving.

What happens when a sound source moves?

As a race car moves, it sends out sound waves. The sound waves are made up of compressions and rarefactions. Look at the first figure. The race car creates a compression, labeled compression A. Compression A moves through the air toward the flagger. Now look at the second figure. By the time compression B leaves the race car, the car has moved forward. Compressions A and B are closer together than they would be if the car had stayed still. Because the compressions are closer together, more compressions pass by the flagger each second than if the car were at rest. So the flagger hears a higher pitch. Look at the second figure again. You can see that the compressions behind the moving car are farther apart. This means the sound of the car has a lower frequency and a lower pitch as the car moves away from the flagger.

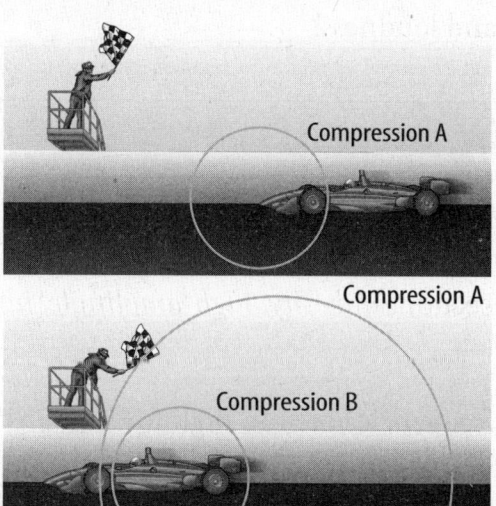

What happens when a listener moves?

You also can notice the Doppler effect when you are moving past a sound source that is standing still. Suppose you pass a building with a ringing bell. The pitch of the bell sounds higher as you get closer to the building. The bell sounds lower as you get farther from the building. The Doppler effect happens any time position changes. The change can be the position of the source of a sound or of the listener. The faster the change in position, the greater the change in frequency and pitch.

How is the Doppler effect used?

The Doppler effect also occurs with other kinds of waves, such as electromagnetic waves. Radar guns with electromagnetic waves are used to measure the speed of cars. The radar gun sends radar waves toward a moving car. The waves reflect from the car. The frequency of the waves changes, depending on the speed and direction the car is moving. The radar gun measures the change and finds the speed of the car. Weather radar also uses the Doppler effect to show the movement of winds in storms.

Picture This

7. **Draw and Label** In the space below, draw and label a diagram that explains the Doppler effect.

Reading Check

8. **Explain** When does the Doppler effect happen?

After You Read

Mini Glossary

decibel: the unit for measuring sound intensity
Doppler effect: when pitch or wave frequency changes because a wave source is moving
intensity: the amount of energy that flows through a certain area in a specific amount of time
loudness: how humans understand sound intensity
pitch: how high or low a sound is
ultrasonic: sound frequencies above 20,000 Hz

1. Review the vocabulary terms and their definitions in the Mini Glossary above. What is the difference between intensity and loudness?

   _____

   _____

2. Fill in the blanks below with the following words to make correct statements about the material you read in this section: *intensity, high-amplitude, ultrasonic, decibel, subsonic, frequency.*

   _____ sound frequencies are too high for humans to hear.

   As the _____ of a sound wave increases, the pitch increases.

   In a _____ sound wave, the compressions are dense.

   The _____ is the unit for measuring how intense a sound is.

   As the _____ of a sound wave increases, the loudness increases.

   Another name for infrasonic is _____.

3. As you read this section, you highlighted the headings that are questions. Will this strategy help you remember what you read in the section? Why or why not?

   _____

   _____

   _____

 Visit **gpscience.com** to access your textbook, interactive games, and projects to help you learn more about the properties of sound.

184 CHAPTER 11 Sound

Sound

section ❸ Music

● Before You Read

Ask a parent or teacher what kind of music he or she liked at your age. Have you ever heard that kind of music? If not, ask the person to sing part of a favorite song. How is it similar to or different from the kind of music you like?

What You'll Learn
- the difference between noise and music
- why different kinds of instruments sound differently
- how instruments make music

● Read to Learn

What is music?

Music and noise are both caused by vibrations. But there are some important differences between them. Noise has random patterns and pitches. <u>Music</u> is made up of sounds that are carefully chosen and have regular patterns.

What are natural frequencies?

Every material when hit, struck, or disturbed in some way will vibrate. The set of frequencies at which the material will vibrate is called its natural frequencies. No matter how you pluck a guitar string, you hear the same pitch because the string always vibrates at its natural frequencies. The guitar string's natural frequencies depend on its thickness and length. It also depends on how tightly the string is stretched. Each string on a guitar is tuned to different natural frequencies. This lets you play different notes on the guitar and make music. Every musical instrument contains something that vibrates at its natural frequencies to create a pitch. It could be strings, a membrane, or a column of air.

What is resonance?

In wind instruments, such as an oboe or a flute, the column of air inside the instrument vibrates. The air vibrates because of resonance. Resonance is the ability of a medium to vibrate by absorbing energy at its own natural frequencies.

Mark the Text

Vocabulary Highlight all the vocabulary terms and their definitions in this section.

FOLDABLES

❸ Identify Make the Foldable shown below to hold quarter sheets of paper that contain information on sound quality and musical instruments.

A musician plays a wind instrument by blowing air into its mouthpiece. The mouthpiece vibrates. The air in the instrument absorbs some of this energy and starts to vibrate. The resonance of the air makes the sound louder. Resonance helps amplify the sound of many musical instruments.

Sound Quality

Suppose someone played a note on a flute and then played the same note on a piano. You could tell the difference between the two instruments, even if you had your eyes closed. They would not sound the same. **Sound quality** describes the differences among sounds of the same pitch and loudness. Each instrument has its own sound quality.

Objects can vibrate at other frequencies besides their natural frequencies. This produces sound waves with more than one frequency. Musical instruments get their quality of sound from the combination of frequencies that each kind of instrument has. ✓

What are overtones?

Even though an instrument can vibrate at many different frequencies at the same time, you hear just one note. All of the frequencies do not vibrate at the same intensity. The main tone that is played and heard is

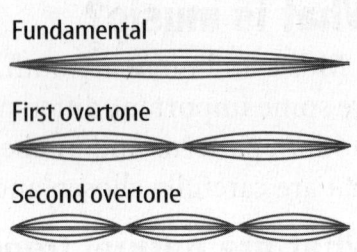

the fundamental frequency. On a guitar, the fundamental frequency is made by the entire string vibrating back and forth. Look at the first example in the figure. It shows how the guitar string vibrates at the fundamental frequency. But the string also vibrates to produce overtones.

An **overtone** is a vibration with a frequency that is a multiple of the fundamental frequency. Suppose a guitar string vibrates at a fundamental frequency of 250 Hz. Multiples of 250 are 500, 750, 1000, and so on. The guitar string will vibrate at its fundamental frequency of 250 Hz. It can also vibrate at overtones of 500 Hz, 750 Hz, 1000 Hz, and so on. The first two guitar-string overtones are shown in the figure. These overtones create the rich sounds of a guitar. Every instrument has a different number and intensity of overtones. These overtones produce each instrument's own sound quality.

✓ Reading Check

1. **Explain** Why do different instruments have different sound qualities?

_____

_____

_____

Picture This

2. **Predict** In the box below, draw what you think the vibration that produces the third overtone would look like.

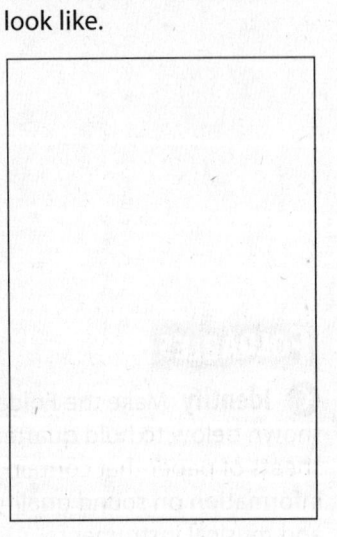

Musical Instruments

A musical instrument is anything that is used to make a musical sound. Violins, cellos, oboes, bassoons, horns, and kettledrums are some of the musical instruments in an orchestra. But people throughout the world play many different kinds of instruments. Australian Aborigines play a woodwind instrument called the didgeridoo (DIH juh ree dew). Caribbean musicians use rubber-tipped mallets to play drums made of steel. An instrument called the nay is played throughout the Arab world. It is similar to a flute.

How do string instruments produce sound?

Violins, electric guitars, and harps are all string instruments with tightly stretched strings. Musicians produce sound by plucking, striking, or moving a bow across the strings. The sound of a vibrating string is soft. A **resonator** (RE suh nay tur) is a hollow space filled with air that makes sound louder when the air inside of it vibrates. As shown in the figure, string instruments usually have a resonator. Imagine a violin string that is stretched tightly between two nails on a board. If you pluck the string, the sound will be very quiet. However, the violin frame and the air inside the body absorb energy from the vibrating string. They begin to vibrate, too. The vibration of the violin body and the air inside makes the sound of the string louder. It also changes the quality of the sound.

Think it Over

3. **Experiment** Close your mouth and gently thump your cheek. Then, open your mouth and thump your cheek again. What causes the difference in sound?

Picture This

4. **Identify** Circle the area where the sound enters the resonator.

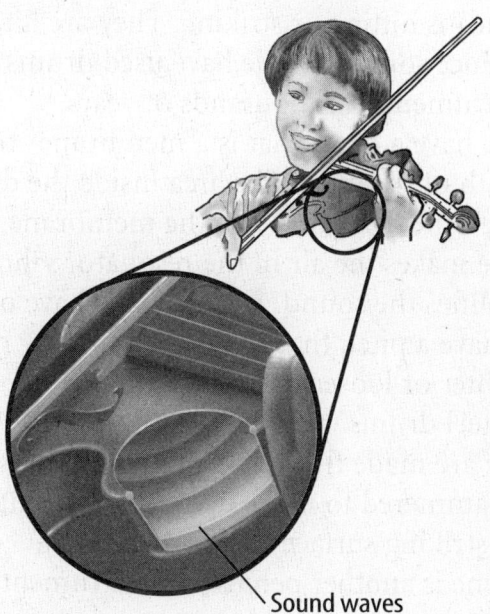

Sound waves

Picture This

5. Label an Illustration In the figures on the previous page and this page, label the source of vibration for each instrument. Then label the resonator of each instrument.

Reading Check

6. Describe Tell how the pitch of a drum can be changed.

_____
_____
_____

What about brass and woodwind instruments?

Vibrating air helps brass and woodwind instruments, like horns, oboes, and flutes, make music. They use different methods to make the air inside vibrate. Brass instruments have cone-shaped mouthpieces, as shown. A mouthpiece is inserted into a metal tube, the resonator. As the musician blows into the instrument, his or her lips vibrate against the mouthpiece. Then air in the resonator starts to vibrate. The resonator produces a pitch.

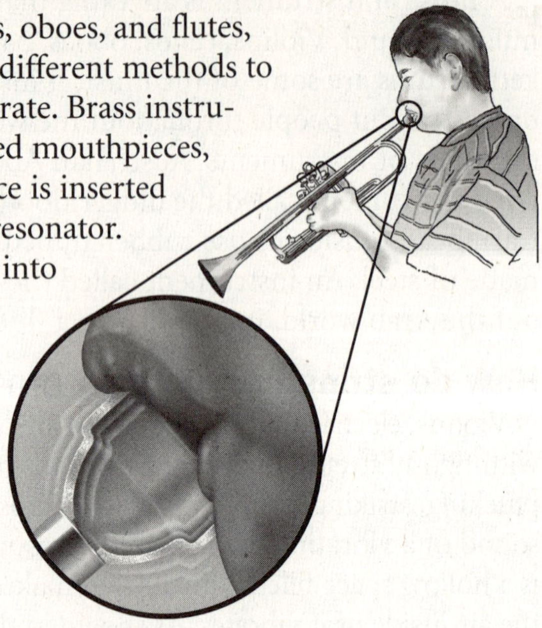

In brass and wind instruments, the length of the vibrating tube of air determines the pitch of the sound. For example, in flutes and trumpets, the musician changes the length of the resonator by opening and closing holes in the tube. In trombones, the musician slides the vibrating tube of air in and out to make the tube shorter or longer.

How do percussion instruments make sound?

Percussion means hitting or striking. They are hit, shaken, or rubbed to produce sound. People have used drums and other percussion instruments for thousands of years.

Some drums have a cover that is a membrane stretched over one or both of the ends. The area inside the drum is the resonator. When the drummer hits the membrane, it vibrates. The membrane makes the air in the resonator vibrate. The resonator amplifies the sound. Some drums have only one pitch. Others have a pitch that can be changed by making the membrane tighter or looser.

Caribbean steel drums were developed in the 1940s in Trinidad. They are made from 55-gallon oil barrels. The end of a barrel is hammered to make up to 32 different striking surfaces. Each striking surface has its own pitch.

The xylophone is another percussion instrument. It has a series of wooden bars. Each bar has its own resonator that is shaped like a tube. The musician hits the bars with mallets. Different kinds of mallets change the sound quality. Other types of percussion instruments are cymbals, rattles, and bells.

188 **CHAPTER 11** Sound

After You Read

Mini Glossary

music: sounds that are carefully chosen and have regular patterns

overtone: a vibration whose frequency is a multiple of the fundamental frequency

resonator: a hollow space filled with air that makes sound louder when the air inside of it vibrates

sound quality: the differences among sounds of the same pitch and loudness

1. Review the vocabulary terms and their definitions in the Mini Glossary above. On the lines below, give examples of a resonator for a string instrument, a wind instrument, and a percussion instrument.

   _____

   _____

2. Complete the graphic organizer by writing the names of some of the instruments mentioned in this section in the correct category.

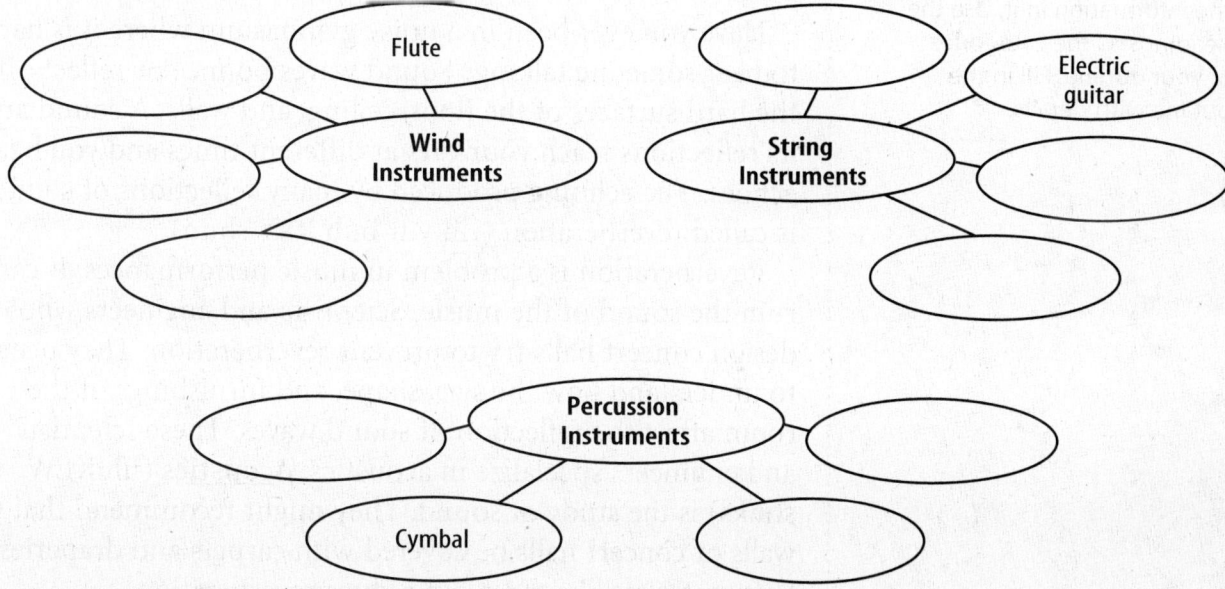

3. You highlighted the vocabulary terms and their definitions. How did these terms help you understand more about music?

   _____

   _____

 Visit gpscience.com to access your textbook, interactive games, and projects to help you learn more about music.

End of Section

Reading Essentials **189**

Sound

section ❹ Using Sound

What You'll Learn
- what affects sound in a concert hall
- how some animals use sound waves
- how sonar is used
- about ultrasound

Study Coach
Outline the Text As you read the section, make an outline of the information in it. Use the headings as the categories of your outline. Fill in the outline with details.

FOLDABLES
ⓓ Compare Make a 4-door Foldable like the one shown to organize and compare ideas about sound.

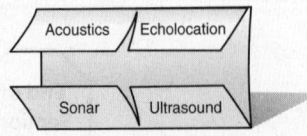

● Before You Read

You use sound when your alarm clock goes off and wakes you up. Write two ways that you have used sound today.

_____

_____

● Read to Learn

Acoustics

Have you ever been in a noisy gymnasium where it is hard to hear someone talking? Sound waves bounce or reflect off the hard surfaces of the floor, ceiling, and walls. A sound and its reflections reach your ears at different times and you hear echoes. The echoing produced by many reflections of sound is called reverberation (rih vur buh RAY shun).

Reverberation is a problem in music performances. It can ruin the sound of the music. Scientists and engineers who design concert halls try to prevent reverberation. They need to understand how the size, shape, and furnishings of the room affect the reflection of sound waves. These scientists and engineers specialize in acoustics. **Acoustics** (uh KEW stihks) is the study of sound. They might recommend that the walls of concert halls be covered with carpets and draperies. Soft materials like these can reduce reverberation.

Echolocation

At night, bats swoop around in darkness without bumping into anything. They even are able to catch insects in the dark. Many species of bats depend on echolocation. **Echolocation** is the process of locating objects by making sounds and interpreting the sound waves that reflect, or bounce back. Bats can get information about an insect, such as how far away it is and how big it is, from the sound waves that bounce back. Dolphins also use echolocation to locate objects.

190 CHAPTER 11 Sound

Sonar

More than 140 years ago, a ship named the *Central America* disappeared in a hurricane. It held 21 tons of gold that would be worth many millions of dollars today. When the ship was wrecked, there was no way to search for it. The *Central America* and its gold lay at the bottom of the ocean under 2,400 m of water. In 1988, crews used sonar to find the wreck. Sonar is a system that uses reflections of underwater sound waves to detect objects.

The figure shows how sonar works. A sound wave is sent toward the ocean floor from a boat on the surface of the water. When the sound wave hits something solid, it is reflected. The sound wave bounces back to the surface. An underwater microphone, called a hydrophone, is attached to the bottom of the boat. The hydrophone picks up the reflected sound. The speed of sound in water is known. So the distance from the boat to the object can be found by measuring how much time it takes the sound to bounce back to the hydrophone.

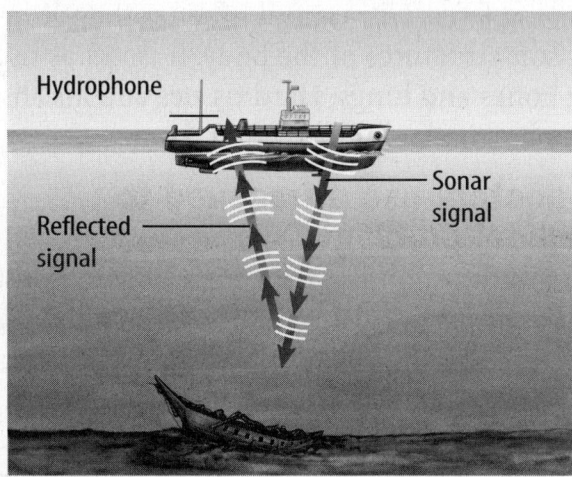

Many uses have been developed for sonar. Navy ships use sonar to locate submarines. Fishing crews use sonar to find schools of fish. Scientists use sonar to map the ocean floor. Most sonar uses ultrasonic or high-frequency waves. High-frequency waves reveal more details.

Other Uses for Ultrasonic Waves

Ultrasonic waves are used in other ways besides echolocation and sonar. They also are used to clean delicate items, such as jewelry. Ultrasonic waves are aimed at a dirt buildup. The ultrasonic waves cause the buildup to vibrate quickly until the dirt breaks up into small pieces. These small pieces can be removed easily.

Picture This

1. **Explain** A fishing crew is using sonar to find schools of fish. The crew's boat is over a rocky area of the ocean floor. What problems might the crew have when using sonar in this area?

_____
_____
_____
_____

Applying Math

2. **Calculate** Sound travels at 1,498 m/s in water. The *Central America* was 2,400 m under the surface of the ocean. How long would it take a sound wave from a ship to bounce off the wrecked ship and return to the surface? Round your answer to the nearest thousandth.

_____

One of the most important uses of ultrasonic waves is in medicine. A process called ultrasound uses ultrasonic waves to gather images of the inside of a person's body. In the past, doctors had to perform surgery to do this. Special instruments are used to send ultrasonic waves into a part of a patient's body. The reflected waves create an image that allows doctors to study many conditions such as pregnancy, heart disease, and cancer.

How does ultrasound imaging work?

Like X rays, ultrasound can be used to make images of body parts inside the body. A technician aims the ultrasound waves toward the target area. The sound waves reflect off the organs or tissues. A computer turns the reflected waves into video pictures, or images. The images are called sonograms. Doctors can use sonograms to find medical problems.

Ultrasound is used to examine many parts of the body. These include the heart, liver, gallbladder, pancreas, spleen, kidney, breast, and eye. Doctors use ultrasound to keep track of a developing fetus. Ultrasound is better than X rays for examining soft structures of the body. It is not as useful for examining bones and lungs. Hard tissues and air absorb the ultrasonic waves instead of reflecting them.

How do doctors use ultrasound to treat medical problems?

Doctors use ultrasound to treat some medical problems. Small, hard deposits of calcium can form in kidneys. The deposits are called kidney stones. In the past, doctors had to operate to remove kidney stones. Now ultrasonic treatments are used to break up the stones. As shown in the figure, many ultrasonic waves create vibrations that break the stones into small pieces. The pieces then pass out of the body with urine. Gallstones can be treated in the same way. Patients who are treated with ultrasound recover more quickly than people who have surgery.

Reading Check

3. Define What is a sonogram?

Picture This

4. Infer Kidney stones sometimes leave the kidney and begin to move down behind the pelvic bone. Kidney stones cannot be treated with ultrasound if they are behind the pelvic bone. Why do you think this is so?

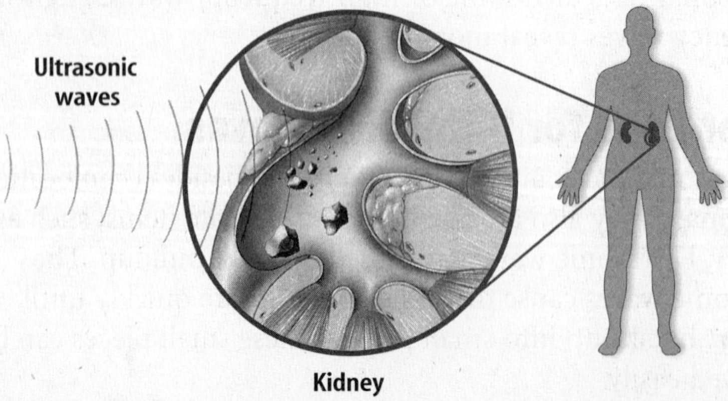

Ultrasonic waves

Kidney

After You Read

Mini Glossary

acoustics: the study of sound

echolocation: the process of sensing things by emitting sounds and interpreting the sound waves that are reflected back

sonar: a way to use the reflection of underwater sound waves to find objects

1. Read the vocabulary terms in the Mini Glossary above. How are echolocation and sonar similar and different?

   _____
   _____
   _____

2. Complete the diagram with ways that people and animals use sound.

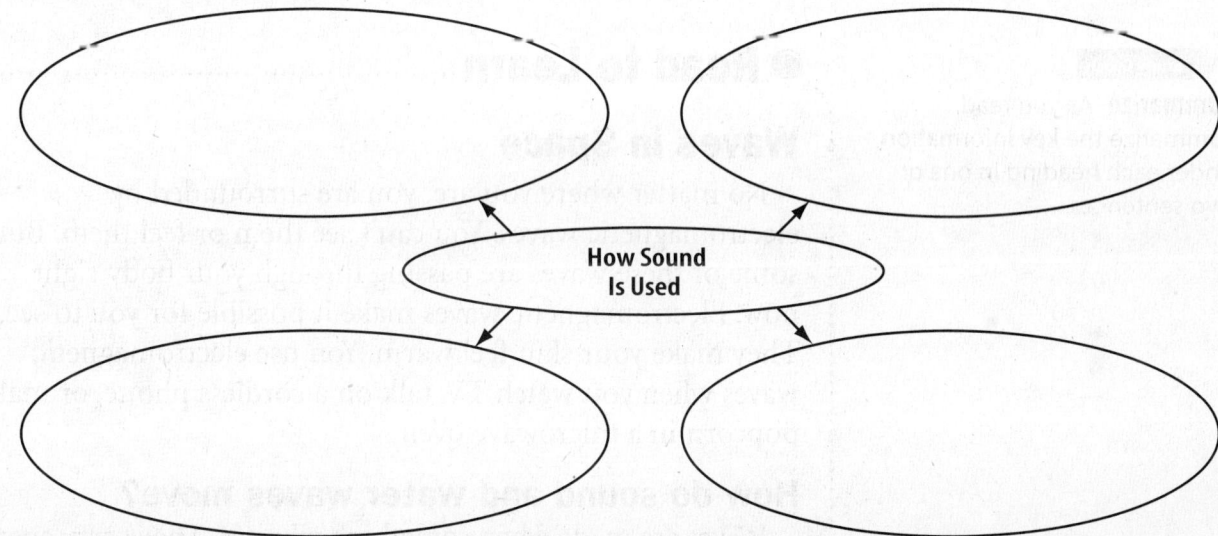

3. You made an outline of the information in this section as you read it. How did this strategy help you learn the material in the section? Would you recommend that a friend use this strategy?

   _____
   _____

 Visit **gpscience.com** to access your textbook, interactive games, and projects to help you learn more about using sound.

Chapter 12 Electromagnetic Waves

section ❶ What are electromagnetic waves?

What You'll Learn
- how electromagnetic waves are formed
- how electric charges produce electromagnetic waves
- properties of electromagnetic waves

Before You Read

Light is transmitted by electromagnetic waves. Without light, you would not be able to see. On the lines below, write three things you could not do without light.

_____

_____

_____

Study Coach

Summarize As you read, summarize the key information under each heading in one or two sentences.

Read to Learn

Waves in Space

No matter where you are, you are surrounded by electromagnetic waves. You can't see them or feel them. But some of these waves are passing through your body right now. Electromagnetic waves make it possible for you to see. They make your skin feel warm. You use electromagnetic waves when you watch TV, talk on a cordless phone, or make popcorn in a microwave oven.

How do sound and water waves move?

Waves are made when something vibrates. They carry energy from one place to another. Both sound waves and water waves move through matter. Water waves move through water, a liquid. Sound waves move through matter that is solid, liquid, or gas. Sound waves and water waves travel because energy is transferred from one particle to another particle.

How are electromagnetic waves made?

Electromagnetic waves do not need matter to transfer energy. **Electromagnetic waves** are made by vibrating electric charges and can travel through space where there is no matter. They do transfer energy from particle to particle. Instead, they travel by transferring energy between vibrating electric and magnetic fields. ✓

Reading Check

1. Apply What vibrates when electromagnetic waves transfer energy?

_____

_____

194 CHAPTER 12 Electromagnetic Waves

Electric and Magnetic Fields

What happens if you move a magnet close to a metal paper clip? The paper clip moves toward the magnet and sticks to it. The magnet moved the paper clip without touching it because every magnet is surrounded by a magnetic field. There is a magnetic field around magnets even if the space around the magnet contains no matter.

Electric charges are surrounded by electric fields in the same way. An electric field surrounds an electric charge even if the space around the charge has no matter. An electric field allows electric charges to exert forces on each other even when they are far apart.

How do moving charges create magnetic fields?

An electric charge is surrounded by an electric field. Electric charges also can be surrounded by magnetic fields. An electric current in a wire is the flow of electrons in one direction. The movement of these electrons creates a magnetic field around the wire as shown in the figure. So, any moving electric charge is surrounded by an electric field and a magnetic field.

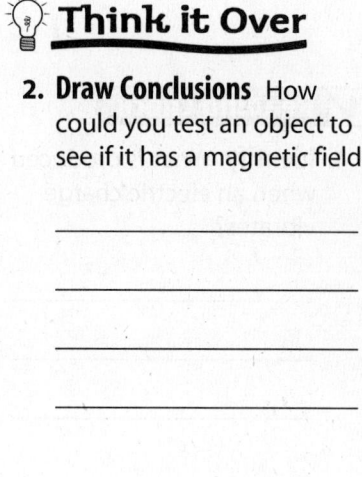

Think it Over

2. **Draw Conclusions** How could you test an object to see if it has a magnetic field?

_____
_____
_____
_____

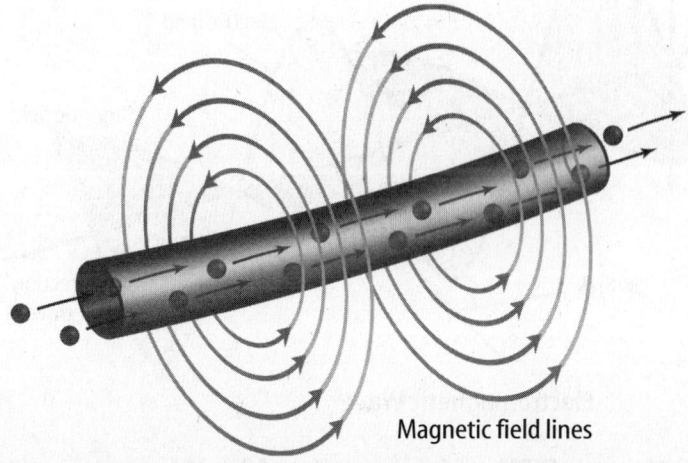

Magnetic field lines

Picture This

3. **Label** the electrons moving through the wire. What is the flow of electrons in one direction in a wire called?

_____

What happens when electric and magnetic fields change?

A changing magnetic field creates a changing electric field. One example of this relationship can be seen in a transformer. A transformer transfers electric energy from one circuit to another circuit. In the main coil of a transformer, changing electric current produces a changing magnetic field. This changing magnetic field then creates a changing electric field in another coil. This electric field produces an electric current in the coil. The reverse is also true. A changing electric field creates a changing magnetic field.

Reading Essentials   **195**

Making Electromagnetic Waves

Waves are made when something vibrates. Electromagnetic waves are made when an electric charge vibrates. When an electric charge vibrates, the electric field around it changes. Remember, a changing electric field creates a changing magnetic field. Then the changing magnetic field creates a changing electric field. How do the changing fields become a wave? Look at the figure below. This process of changing electric and magnetic fields continues. The magnetic and electric fields create each other again and again. ☑

An electromagnetic wave travels in all directions. The figure only shows a wave traveling in one direction. The electric and magnetic fields vibrate at right angles to the direction the wave travels. Remember, in a transverse wave, the matter in the medium moves at a right angle to the direction the wave travels. So, an electromagnetic wave is a transverse wave.

Reading Check

4. **Identify** What is produced when an electric charge vibrates?

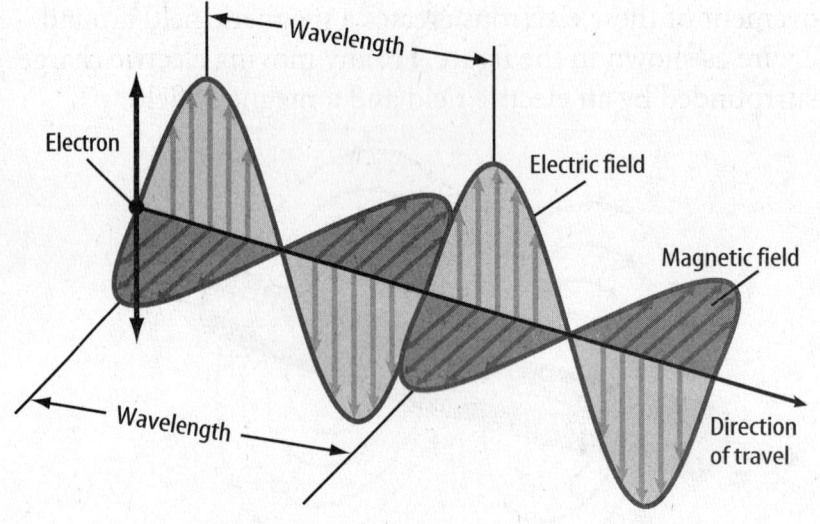

Electromagnetic Wave

Properties of Electromagnetic Waves

All matter contains charged particles that are always moving. Therefore, all objects emit electromagnetic waves. The wavelengths of the waves become shorter as the temperature of the material increases. As an electromagnetic wave moves, its electric and magnetic fields meet objects. These vibrating fields can exert forces on charged particles and magnetic materials. The forces make the charged particles and magnetic materials move. For example, electromagnetic waves from the Sun cause the electrons in your skin to vibrate and gain energy. The energy carried by an electromagnetic wave is called **radiant energy**. Radiant energy makes a fire feel warm. Radiant energy also allows you to see.

What is the speed of electromagnetic waves?

All electromagnetic waves travel at 300,000 km/s in space. Light is an electromagnetic wave. So, the speed of electromagnetic waves in space is usually called the "speed of light." The speed of light is nature's speed limit. Nothing in the universe travels faster than the speed of light. The speed of electromagnetic waves through matter depends on what material the waves travel through. Electromagnetic waves usually travel the slowest in solids and the fastest in gases. The table shows the speed of visible light in some materials.

Speed of Visible Light

Material	Speed (km/s)
Vacuum	300,000
Air	slightly less than 300,000
Water	226,000
Glass	200,000
Diamond	124,000

Applying Math

5. **Calculate** Use the table at left. The average distance from the Sun to Earth is about 150,000,000 km. About how long does it take light from the Sun to reach Earth? Round your answer to the nearest minute.

What is the wavelength and frequency of an electromagnetic wave?

Electromagnetic waves can be described by their wavelength and frequency. Look at the figure on the previous page again. The wavelength is the distance from one crest to the next.

The frequency of any wave is the number of wavelengths that pass a point in 1 s. The frequency of an electromagnetic wave is the same as the frequency of the vibrating charge that makes the wave. This frequency is the number of vibrations of the charge in one second. As the frequency of an electromagnetic wave increases, the wavelength becomes smaller.

Waves and Particles

The difference between a wave and a particle might seem obvious. A wave is a disturbance that carries energy. A particle is a piece of matter. But the difference is really not so clear.

Can a wave be a particle?

In 1887, a scientist named Heinrich Hertz discovered that by shining light on a metal, electrons were ejected. Hertz found that whether or not electrons were ejected depended on the frequency of the light and not the amplitude. This result seemed mysterious because the energy carriedby a wave depends on its amplitude and not its frequency. Years later, Albert Einstein explained Hertz's discovery. Electromagnetic waves can behave as a particle called a **photon**, whose energy depends on the frequency of the waves.

Picture This

6. **Draw and Label** On your own paper, draw two electromagnetic waves. The wavelength of the second wave should be half the wavelength of the first wave. Label the wavelength of each wave.

Reading Check

7. **Identify** Write the correct word to complete the sentence: Electromagnetic waves can behave as particles that are called _____.

Reading Essentials **197**

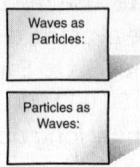

A **Find Main Ideas** Make two quarter-sheets of paper into note cards to organize information about waves and particles.

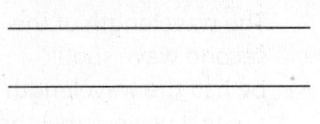

Picture This

8. Describe Look at the second part of the figure showing electrons being sprayed at two slits. Describe the pattern that they form.

_____
_____
_____
_____

Can a particle be a wave?

The discovery that electromagnetic waves could behave as a particle led to other questions. Scientists wondered whether matter could behave as a wave.

Look at the figure below. The first part shows paint particles being sprayed at two narrow openings, or slits. The paint particles cover only the area behind the slits.

Paint particles sprayed at two slits coat only the area behind the slits.

The second part shows a beam of electrons being sprayed at two slits. You might expect the electrons to strike only the area behind the slits. But scientists found that electrons form an interference pattern.

Electrons fired at two slits form an interference pattern, similar to patterns made by waves.

An interference pattern is formed by waves when they pass through two slits and interfere with each other just like the water waves in the third part of the diagram. This experiment shows that electrons can behave like waves. In fact, all particles, not only electrons, can behave like waves.

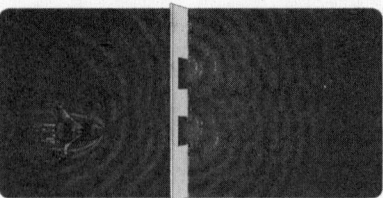

Water waves make an interference pattern after they pass through two slits.

198 CHAPTER 12 Electromagnetic Waves

After You Read

Mini Glossary

electromagnetic waves: waves made by vibrating electric charges that can travel through space where there is no matter

photon: an electromagnetic wave that behaves like a particle and whose energy depends on the frequency of the waves

radiant energy: the energy carried by an electromagnetic wave

1. Review the terms and their definitions in the Mini Glossary. Choose one term and use it in a sentence that shows your understanding of the term.

   _____
   _____
   _____

2. Write a fact about electromagnetic waves on the lines under each heading.

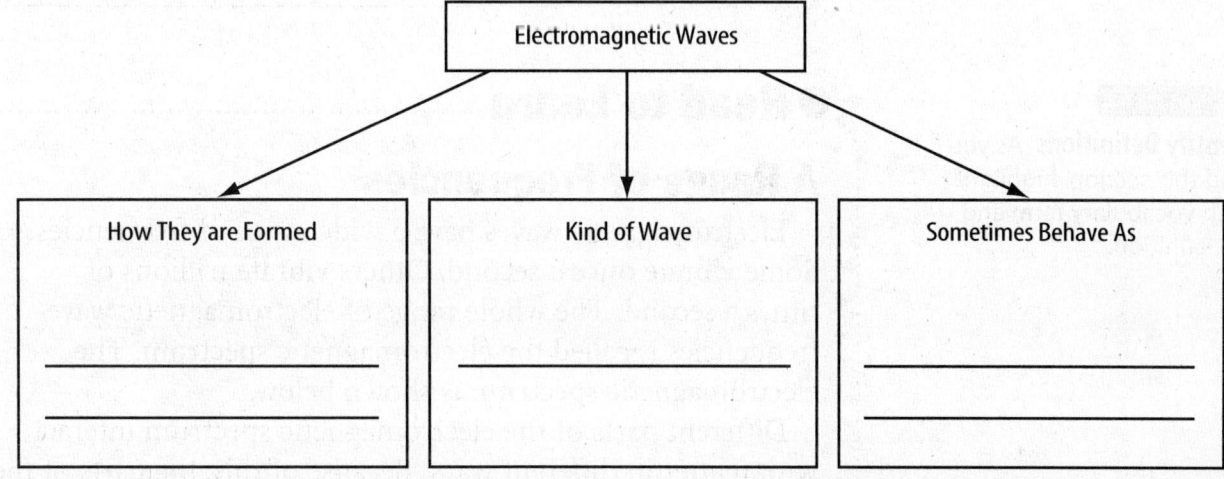

3. As you read this section, you summarized the important information under each heading. How did that strategy help you learn the material in this section?

   _____
   _____
   _____

 Visit **gpscience.com** to access your textbook, interactive games, and projects to help you learn more about what electromagnetic waves are.

Electromagnetic Waves

section ❷ The Electromagnetic Spectrum

What You'll Learn
- the different kinds of electromagnetic waves
- the properties of electromagnetic waves
- how electromagnetic waves are used

● Before You Read

Radio waves, microwaves, and X rays are examples of electromagnetic waves. On the lines below, write a sentence about how you or someone you know has used electromagnetic waves.

_____
_____
_____

Mark the Text

Identify Definitions As you read this section, highlight each vocabulary term and its definition.

● Read to Learn

A Range of Frequencies

Electromagnetic waves have a wide range of frequencies. Some vibrate once a second. Others vibrate trillions of times a second. The whole range of electromagnetic wave frequencies is called the electromagnetic spectrum. The electromagnetic spectrum is shown below.

Different parts of the electromagnetic spectrum interact with matter in different ways. Because of this, the parts of the spectrum have different names. The electromagnetic waves that humans can see are called visible light. They are a small part of the whole electromagnetic spectrum. But a number of devices have been created to detect the other frequencies. For example, the antenna of a radio detects radio waves.

Picture This
1. **Identify** Which electromagnetic waves have the lowest frequency?

_____
_____

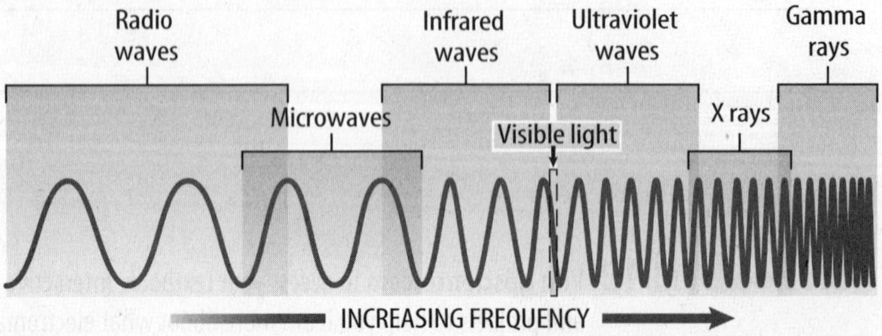

Radio waves | Microwaves | Infrared waves | Visible light | Ultraviolet waves | X rays | Gamma rays

INCREASING FREQUENCY →

200 CHAPTER 12 Electromagnetic Waves

Radio Waves

Even though you can't see them, radio waves are moving everywhere. **Radio waves** are low-frequency electromagnetic waves with wavelengths longer than about 1 mm. Radio stations make use of these waves. They use microphones to change sound waves from voices and music into radio waves. The radio waves carry signals that can be picked up by radios. Radios then change the signals back into sound waves. You cannot hear actual radio waves. Remember, you hear sounds when compressions and rarefactions from a sound wave reach your ears. A radio wave does not produce compressions and rarefactions. It needs to be turned into a sound wave by a radio before you can hear it.

What are microwaves?

Microwaves are radio waves with wavelengths less than 30 cm. They have a higher frequency and shorter wavelength than the waves used by radios. Microwaves with wavelengths between about 1 cm and 20 cm are used for communication. Cell phones and satellites use microwaves of this wavelength.

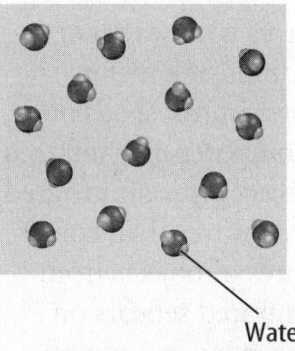

Normally water molecules are randomly arranged.

Water molecules

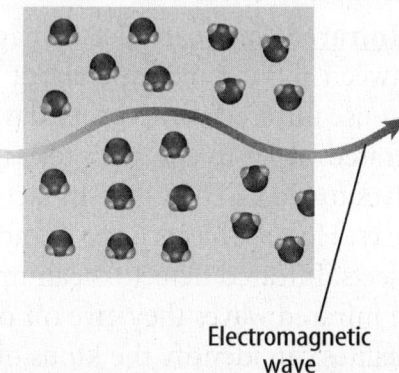

The microwaves cause the water molecules to flip back and forth.

Electromagnetic wave

In microwave ovens, microwaves interact with water molecules in food. The figure shows how microwave ovens work. Each water molecule is positively charged on one side and negatively charged on the other side. There is a vibrating electric field inside a microwave oven. It causes the water molecules in food to rotate billions of times a second. The rotation causes friction. Friction creates thermal energy. Thermal energy made by the interactions between the water molecules heats food. ✓

FOLDABLES

B Find Main Ideas Make a layered book Foldable using four sheets of paper. Write Spectrum, Waves, and Rays on the tabs.

Think it Over

2. Draw Conclusions Suppose a person standing on land is using radar to measure how far away a ship is. One minute later, the measurement is taken again. This time, it takes longer for the radio signal to return. Is the ship moving toward the shore or moving away from the shore? Why?

_____
_____
_____
_____

Reading Check

3. Explain Why is it impossible to heat items that do not contain water in a microwave oven?

_____
_____
_____

How does radar work?

Radar stands for **RA**dio **D**etecting **A**nd **R**anging. Radar is used to find the position and speed of objects. Radio waves are sent toward an object. The waves bounce off the object and return. The time this takes gives the object's position. Radar can show the position of airplanes, boats, and cars. It also can measure the speed of moving vehicles.

What is magnetic resonance imaging?

Researchers developed a technique called magnetic resonance imaging, or MRI, in the 1980s. MRI uses radio waves to help diagnose illnesses. The patient lies inside a large tube surrounded by a strong magnet, a radio wave emitter, and a radio wave detector.

Protons in the hydrogen atoms in bones and soft tissue act like magnets. They line up with the strong magnetic field of the MRI. Energy from radio waves makes some of the protons flip. When the protons flip, they release energy. Different tissues release different amounts of energy. A radio receiver detects the released energy and makes a map of the body's tissues. A picture of the inside of the patient's body is made.

Infrared Waves

<u>Infrared waves</u> are electromagnetic waves with wavelengths between about 1 mm and about 750 billionths of a meter. You use infrared waves every day. A remote control sends out infrared waves to control a television. A computer uses infrared waves to read CD-ROMs. In fact, every object gives off infrared waves. Hotter objects give off more infrared waves than cooler objects. Infrared detectors can make pictures of objects from the infrared waves they give off or emit. Infrared sensors on satellites can identify the kinds of plants growing in a region.

Visible Light

<u>Visible light</u> is the range of electromagnetic waves that people can see. Visible light has wavelengths of about 750 billionths to 400 billionths of a meter. The electromagnetic waves you can see have different wavelengths. You see the different wavelengths as different colors. Blue light has the shortest wavelength. Red light has a longest wavelength. The light looks white if all the colors are present.

💡 Think it Over

4. Predict If your body heat were to show up on an infrared detector, which parts of your body would be giving off the least amount of infrared waves?

_____

_____

_____

Applying Math

5. Write Decimals You can write 10 billionths as a decimal by moving the decimal point in 1.0 ten places to the left. Write 10 billionths as a decimal.

_____

Ultraviolet Waves

Ultraviolet waves are electromagnetic waves with wavelengths from about 400 billionths to 10 billionths of a meter. Ultraviolet, or UV, waves have enough energy to enter skin cells. Being exposed to too many UV rays can cause skin damage and cancer. Sunlight contains ultraviolet waves.

Most of the ultraviolet radiation that reaches Earth's surface are longer-wavelength rays. They are called UVA rays. Shorter-wavelength rays are called UVB rays. UVB rays are the rays that cause sunburn. Both UVA and UVB rays can damage the skin and cause skin cancer.

Can UV radiation be useful?

Some UV rays are useful. A few minutes of UVs from the Sun each day helps your body make vitamin D. Vitamin D is needed for healthy bones and teeth.

UVs are used to sterilize objects like medical supplies and hospital equipment. When ultraviolet light enters a cell, it damages protein and DNA molecules. This can kill some single celled organisms such as bacteria.

Ultraviolet waves make some materials light up, or fluoresce (floor ESS). Fluorescent materials absorb ultraviolet waves. Then they emit the energy as visible light. Police detectives sometimes use fluorescent powder to find fingerprints when solving crimes.

What is the ozone layer?

The ozone layer is an area in Earth's upper atmosphere. It is about 20 km to 50 km above Earth's surface. Ozone is a molecule made up of three oxygen atoms. The ozone layer is necessary to life on Earth because it absorbs most of the Sun's harmful ultraviolet waves. Over the past few decades, the amount of ozone in the ozone layer has decreased.

Ozone has decreased because of the presence of certain chemicals in Earth's atmosphere. The chemicals are called chlorofluorocarbons, or CFCs. CFCs are used in air conditioners, refrigerators, and cleaning fluids. CFC molecules react with ozone molecules. One chlorine atom from a CFC molecule can break apart thousands of ozone molecules. Many countries are using fewer CFCs and other chemicals that destroy ozone. If too much ozone is destroyed, the ozone layer will be damaged or lost altogether. Without the ozone layer, everything on the surface of Earth would become exposed to a much higher level of damaging ultraviolet waves.

✓ Reading Check

6. **Describe** How can ultraviolet waves harm you?

_____

_____

_____

💡 Think it Over

7. **Think Critically** What could happen to humans if the ozone layer were destroyed?

_____

_____

_____

_____

Think it Over

8. Infer Why do you think MRIs might cause less harm than X rays?

_____
_____
_____

Reading Check

9. Identify Which statement is *not* true. (Circle your answer.)

a. Gamma rays are low-frequency waves.

b. X rays are high-energy waves.

c. Gamma rays are used to treat diseases.

Picture This

10. Identify Highlight the parts on the X ray that you can see clearly.

_____

X Rays and Gamma Rays

X rays and gamma rays are the electromagnetic waves with the shortest wavelengths. They also have the highest frequencies. Both X rays and gamma rays are high energy electromagnetic waves. <u>X rays</u> have wavelengths between 10 billionths and 10 trillionths of a meter. They have enough energy to go through skin and muscle. Doctors and dentists use low levels of X rays to take pictures of internal organs, bones, and teeth. High levels of X rays are dangerous and can cause cancer. X rays are projected only at very specific areas of the body, as shown in the figure. Lead aprons or shielding are used to protect other areas from exposure. X rays cannot travel through lead.

<u>Gamma rays</u> are electromagnetic waves with wavelengths shorter than 10 trillionths of a meter. They are the highest-energy electromagnetic waves. They can travel through several centimeters of lead. Gamma rays are produced in the nuclei of atoms. Both X rays and gamma rays are used in radiation therapy. Radiation therapy is used to kill diseased cells in the human body. X rays and gamma rays can kill both healthy and diseased cells. Doctors carefully control the amount of X-ray or gamma ray radiation the diseased area receives. This reduces the damage to healthy cells.

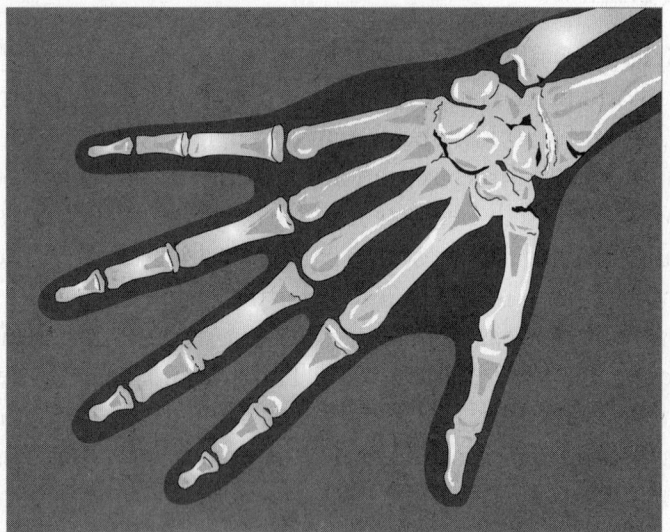

X Ray of Bone

After You Read

Mini Glossary

gamma rays: electromagnetic waves with wavelengths shorter than 10 trillionths of a meter

infrared waves: electromagnetic waves with wavelengths between about 1 mm and about 750 billionths of a meter

microwave: a kind of radio wave with wavelengths of less than 30 cm

radio waves: low-frequency electromagnetic waves with wavelengths longer than about 1 mm

ultraviolet waves: electromagnetic waves with wavelengths from about 400 billionths to 10 billionths of a meter

visible light: the range of electromagnetic waves that people can see, with wavelengths of about 750 billionths to 400 billionths of a meter

X ray: an electromagnetic wave with a wavelength of between 10 billionths of a meter and 10 trillionths of a meter

1. Draw a line to match the name of each ray in the first column with its wavelength in the second column.

Column 1	Column 2
Visible light	less than 30 cm
X ray	400 billionths to 10 billionths of a meter
Microwave	longer than 1 mm
Gamma ray	1 mm to 750 billionths of a meter
Radio wave	10 billionths to 10 trillionths of a meter
Ultraviolet wave	10 billionths to 400 billionths of a meter
Infrared wave	shorter than 10 trillionths of a meter

2. Write the names of the waves listed in Question 2 in the electromagnetic spectrum pyramid.

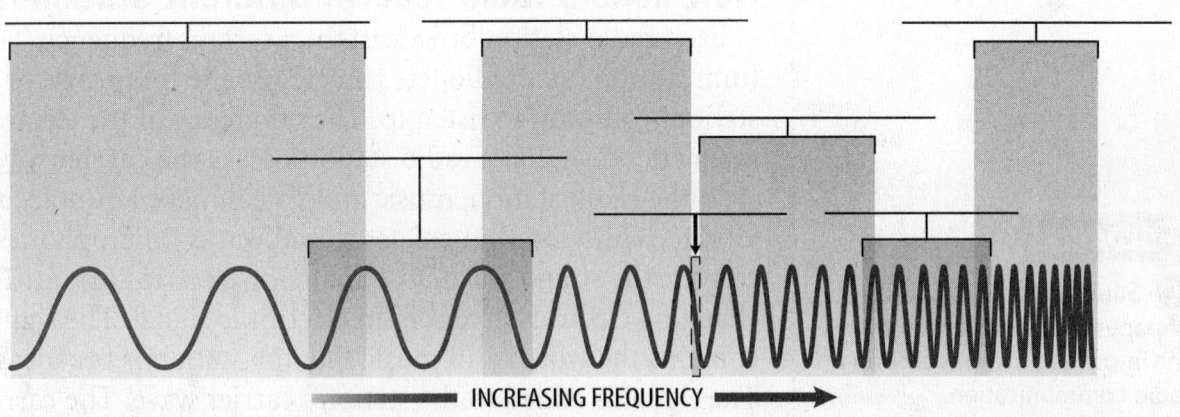

 Visit gpscience.com to access your textbook, interactive games, and projects to help you learn more about the electromagnetic spectrum.

chapter 12 Electromagnetic Waves

section ● Radio Communication

What You'll Learn
- how radio waves transmit information
- the difference between amplitude modulation and frequency modulation
- how people communicate with radio waves

Study Coach
Create a Quiz After you have read this section, create a quiz based on what you have learned. After you have written the quiz questions, be sure to answer them.

FOLDABLES
C Summarize Fold one sheet of paper into sixths to summarize the information in this section on radio communication.

| Radio Communication |
| Radio |
| Television |
| Telephones |
| Satellites |
| GPS |

● Before You Read

You learned that all electromagnetic waves have frequencies. Every radio station broadcasts at a certain frequency. What is the frequency of your favorite radio station? What do you think this number means?

_____

_____

● Read to Learn

Radio Transmission

When you listen to the radio, you hear music and words that were created far away. Radio stations need to send sounds over great distances. They do this by changing sound waves to electromagnetic waves. Your radio then changes the electromagnetic waves back to sound waves again.

How does a radio receive different stations?

Each radio station broadcasts at a certain frequency. The tuning knob on a radio lets you choose the frequency, or station, you want to listen to. The frequency of the electromagnetic wave that a radio station uses is the **carrier wave**.

At the radio station, music and voices make air molecules vibrate. This vibration creates sound waves. Microphones convert the sound waves to a changing electric current. The changing electric current is an electronic signal. The signal contains the words and music that the station is broadcasting. The signal is added to the station's carrier wave. The carrier wave has been changed, so it is a modified carrier wave. The modified carrier wave vibrates electrons in the station's antenna. These vibrating electrons create a radio wave. The radio wave travels out in all directions at the speed of light. The radio wave makes electrons in your radio's antenna vibrate.

206 CHAPTER 12 Electromagnetic Waves

These vibrating electrons produce a changing electric current. The current contains the carrier wave and the signal. If your radio is tuned to the station's frequency, the radio removes the carrier wave from the electronic signal. The signal then makes the radio's speakers vibrate and creates sound waves. The sound waves travel to your ears. Your brain interprets the sound waves as music and words.

What is AM radio?

The figure below shows the two ways the carrier wave can be modified. The first way is amplitude modulation, or AM. AM radio stations broadcast information by changing the amplitude of the carrier wave. Look at the wave labeled *amplitude modulation*. Notice that the amplitude changes, but the frequency does not. The original sound is changed into an electrical signal that changes the amplitude of the carrier wave. You tune an AM radio to the frequency of the carrier wave. AM frequencies range from 540,000 Hz to 1,600,000 Hz. AM radio stations give their frequencies in kilohertz. A station that gives its frequency as 810 AM means that it is broadcasting at a frequency of 810,000 Hz.

What is FM radio?

The second way a carrier wave can be modified is called frequency modulation, or FM. FM radio stations transmit broadcast information by changing the frequency of the carrier wave. Look at the wave labeled *frequency modulation*. Notice that the frequency of the wave changes, but not the amplitude. The strength of the FM waves is always the same. Because of this, FM signals are usually clearer than AM signals. FM frequencies range from 88 million to 108 million Hz. This is much higher than AM frequencies. FM radio stations give their frequencies in megahertz. *Mega-* means "million," so a station that gives its frequency as 89.9 FM is broadcasting at a frequency of 89,900,000 Hz.

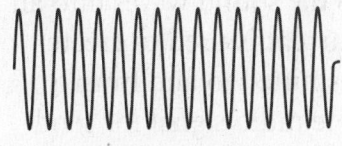
Carrier wave

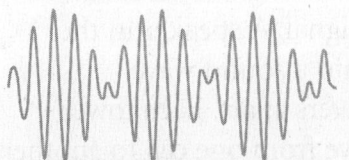
Amplitude modulation

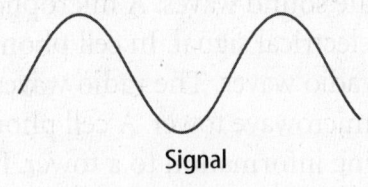
Signal

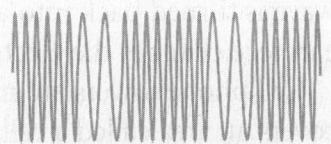
Frequency modulation

Applying Math

1. **Use Decimals** Suppose one radio station broadcasts at 1250 AM. Another station broadcasts at 91.7 FM. What is the frequency of each station, in hertz?

   _____

   _____

Picture This

2. **Interpret a Scientific Illustration** Look at the figure that shows frequency modulation. How does the wavelength of the wave change as its frequency increases?

   _____

   _____

   _____

Applying Math

3. **Calculate** A standard television has about 165,000 rectangular spots. The screen of a 19-in. TV is about 141 square inches. About how many rectangular spots are there in one square inch of a TV screen? Round your answer to the nearest whole number.

_____

Reading Check

4. **Identify** What focuses the electronic beams inside a CRT?

_____

Think it Over

5. **Think Critically** What would happen if you drove into an area where there was no cell phone tower while you were talking on a cell phone?

_____

Television

Television might seem like magic. It is not if you know how it works. Television and radio work in similar ways, using radio waves. At a television station, sounds and images are changed into electronic signals. The signals are broadcast by carrier waves. Television sound is sent by FM radio waves. Information about the color and brightness is sent at the same time by AM signals.

What is a cathode-ray tube?

Many television sets and computer monitors display images on a cathode-ray tube, or CRT. A **cathode-ray tube** is a sealed vacuum tube in which beams of electrons are produced. The CRT in a color TV produces three electron beams. A magnetic field focuses the three beams inside the CRT. They strike the inside of the screen. The screen is covered with rectangular spots. There are more than 100,000 of these spots on a television screen.

There are three types of spots. One type glows red when electrons hit it. Another type glows green. The third type glows blue. The spots are grouped together. There is a red spot, a green spot, and a blue spot in each group. The three electron beams of the cathode-ray tube move back and forth across the screen. One beam controls the brightness of the red spots. The other two beams control the brightness of the blue spot and the green spot. The information in the signal from the TV station controls how bright each spot is. The three spots together can form any color. You see a full-color image on the television.

Telephones

Just 20 years ago, you never would have seen someone walking down the street talking on a telephone. Today, cell phones are seen everywhere. When you talk into a telephone, you create sound waves. A microphone turns the sound waves into an electrical signal. In cell phones, this current is used to create radio waves. The radio waves are transmitted to and from a microwave tower. A cell phone uses one radio signal for sending information to a tower. It uses another signal for receiving information from the tower. At the receiving end, the radio wave is turned back into an electric signal. A speaker in the earpiece changes the electric signal into a sound wave.

Cell phone towers are many kilometers apart. Each tower covers an area called a cell. If you move from one cell to another cell, a control station moves your signal to the new cell.

208 CHAPTER 12 Electromagnetic Waves

✓ **Reading Check**

6. Explain Why do cordless phones use two different frequencies?

_____

_____

_____

Picture This

7. Explain What do the circles in the figure represent?

_____

How do cordless telephones work?

Cell phones and cordless telephones are transceivers. A <u>transceiver</u> is a device that transmits one radio signal and receives another radio signal from a base unit. You can talk and listen at the same time using a transceiver because the two signals are at different frequencies. Cordless telephones work much like cell phones. With a cordless phone, you must be close to the base unit. Many kinds of cordless phones use the same frequency. Sometimes, when someone nearby is using a cordless telephone, you can hear the conversation on your phone. Many cordless phones have a channel button. You can use the channel button to switch your call to another frequency. ✓

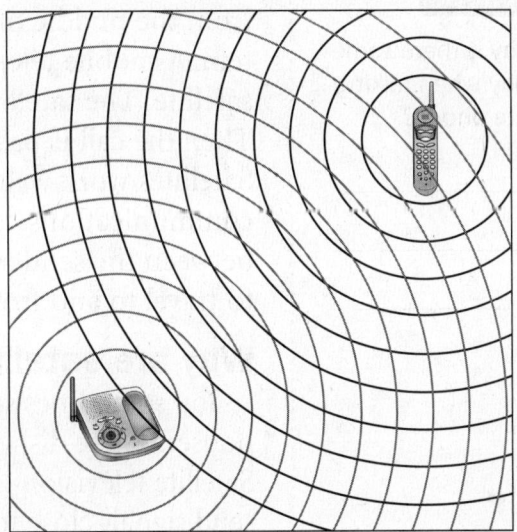

How do pagers work?

Pagers are another way to send signals. A pager is a small radio receiver. Each pager has its own phone number. A caller can dial the pager number and leave a message at a central terminal. The caller enters a call-back number using a telephone keypad or by entering a text message from a computer. At the terminal, the message is changed into an electronic signal. The signal is transmitted by radio waves. Many pagers use the same frequencies. Each pager in the area receives all the messages that are transmitted in its frequency. However, each pager has a special identification number. The identification number is sent along with each message. A pager only picks up messages with its identification number. Newer pagers can send messages as well as receive them.

Communications Satellites

How can you send information to the other side of the world? Radio waves cannot be sent through Earth. Instead, radio signals are sent to satellites. The satellites can communicate with other satellites or with stations on the ground.

Thousands of satellites have been launched and orbit Earth. Many of these satellites are used for communication. A radio or television station sends a high-frequency microwave signal to the satellite. The satellite amplifies the signal and sends it back to a different place on Earth. The satellite avoids interference by using a different frequency to receive and to send messages.

What is a satellite telephone system?

If you have a mobile telephone, you can make a phone call from the middle of the ocean using a satellite telephone system. A mobile telephone transmits radio waves directly to a satellite. The satellite sends the signal to a ground station. Then the call is passed on to the regular telephone network. Satellites work well for one-way transmissions. But two-way communications can be difficult. There can be a delay between the sender and the receiver because of the time used to travel to and from the satellite.

Why are satellites used for television signals?

You sometimes see satellite-reception dishes attached to houses. These dishes are receivers for television satellite signals. Satellite television is sometimes used instead of systems that send signals close to the ground. Communications satellites use shorter microwaves rather than the longer-wavelength radio waves used for normal television broadcasts. Microwaves travel more easily through the atmosphere.

The Global Positioning System

People sometimes get lost while hiking. If they are carrying a Global Positioning System receiver, it is much less likely to happen. The **Global Positioning System (GPS)** is a system of satellites, ground monitoring stations, and portable receivers that determine your exact location on Earth. The system consists of 24 satellites. To determine the location of an object, a GPS receiver measures the time it takes for radio waves to travel from four different satellites to the receiver. It gives information around the world 24 hours a day. GPS satellites are owned and operated by the U.S. Department of Defense, but the microwaves they send out can be used by anyone. GPS receivers are used in airplanes, ships, cars, and even by hikers.

Reading Check

8. **Explain** Why is there sometimes a delay when talking on a satellite phone?

_____
_____
_____

Think it Over

9. **Apply** Some cars have GPS systems installed in them. Why would drivers want to have such a system in their car?

_____
_____
_____

After You Read

Mini Glossary

carrier wave: the frequency of the electromagnetic wave that a radio station uses

cathode-ray tube: a sealed vacuum tube in which beams of electrons are produced

Global Positioning System (GPS): a system of satellites, ground monitoring stations, and portable receivers that determines your exact location on Earth

transceiver: a device that transmits one radio signal and receives another radio signal from a base unit

1. Read the vocabulary terms and their definitions in the Mini Glossary. In the space below, write the name of a device that uses a cathode-ray tube.

   _____

2. Complete the list to organize the information from the section about radio transmission.

1.	Sounds make air molecules vibrate and create sound waves.
2.	
3.	
4.	The carrier wave vibrates electrons in station's antenna and creates a radio wave.
5.	
6.	
7.	The vibrating electrons produce a changing electric current.
8.	
9.	
10.	The sound waves travel to your ears and your brain interprets them as music and voices.

 Visit gpscience.com to access your textbook, interactive games, and projects to help you learn more about radio communication.

End of Section

Light

section ❶ The Behavior of Light

What You'll Learn
- how light waves interact with matter
- regular and diffuse reflection
- index of refraction
- why prisms bend light

Study Coach

Create a Quiz After you read each paragraph, write a question that you think your teacher might ask on a quiz. When you finish reading the section, try to answer all of your quiz questions.

✓ Reading Check

1. **Determine** Which of the following materials is opaque?

   a. plastic food wrap
   b. poster board
   c. waxed paper
   d. window glass

● Before You Read

Rainbows can appear when the Sun is shining during a rain shower. What do you think causes rainbows in the sky?

_____

_____

● Read to Learn

Light and Matter

Imagine looking around a dark room. Your eyes slowly get used to the darkness. You begin to see things that look familiar. The objects look different than they do with the light on. Everything is a shade of gray or even black. When you turn on the light, the objects are colorful again. What you see depends on the amount of light in the room. It also depends on the color of the objects in the room. For you to see an object, it must reflect light back to your eyes.

What are opaque, transparent, and translucent objects?

Objects can absorb light, reflect light, or transmit light. Transmitting light means that light can pass through the object. Whether light is absorbed, reflected, or transmitted depends on what the object is made of. Think of a coffee mug. Most likely, you can't see through it. This is because no light passes through it. It only absorbs or reflects light. A material that only absorbs or reflects light is **opaque** (oh PAYK). ☑

Have you ever seen a drinking glass made of dark-colored glass? You can see light through the glass, but you can't really see what's behind it. **Translucent** (trans-LEW-sunt) materials allow some light to pass through. You can't see clearly through translucent materials.

Matter that is clear, such as glass or water, is transparent. **Transparent** materials allow almost all the light to pass through. You can see clearly through transparent materials.

212 CHAPTER 13 Light

Reflection of Light

Did you look in a mirror this morning before you left for school? When you saw your reflection, you actually saw light that reflected off of your body. The light then reflected off of the mirror and traveled to your eye. Reflection happens when light waves bounce off objects.

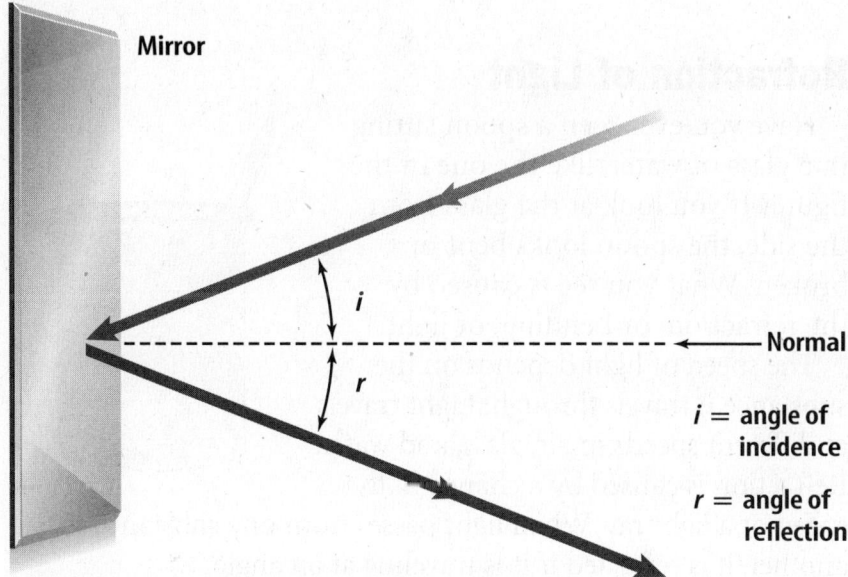

Mirror
Normal
i = angle of incidence
r = angle of reflection

Applying Math

2. **Observe** Suppose the light from the flashlight hits the mirror at a 60° angle. What will be the angle of the reflected beam of light?

_____

What is the law of reflection?

Have you ever shone a flashlight at a mirror in a dark room? The beam of light does not always shine straight back at you. Depending on how you hold the flashlight, the beam might hit the ceiling or the floor. When light is reflected, it follows the law of reflection. According to the law of reflection, the angle at which a light wave strikes a surface is the same as the angle at which it is reflected. The figure above shows the law of reflection. Light follows this law when it is reflected from any surface, whether it is a mirror or a piece of paper.

How is regular reflection different from diffuse reflection?

You can see your reflection in a window. But you cannot see your reflection in a brick wall. Why not? How well you can see your reflection depends on the smoothness of the surfaces. A glass window has a very smooth surface and a brick wall has a very rough surface. Smooth surfaces like the glass window usually reflect light in an even pattern. They produce sharp images. Reflection of light waves from a smooth surface is called regular reflection. Rough or uneven surfaces like the brick wall reflect light in many directions. Reflection of light from a rough surface is called diffuse reflection.

FOLDABLES

A **Find Main Ideas** Make a Foldable like the one shown to write down the main ideas in this section about reflection of light and refraction of light.

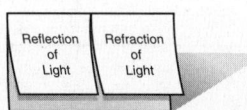

Reading Essentials **213**

Think it Over

3. Explain Car side mirrors do not work when they are covered with frost. What does frost do to the surface of a mirror to cause a diffuse reflection?

_____
_____
_____

Picture This

4. Identify How do you know that light waves are being refracted when they pass from the air to the water in the glass?

_____
_____
_____

How smooth must a surface be for regular reflection?

Some surfaces seem smooth, but they still cause diffuse reflection. For example, the surface of a metal pot might look smooth. But if you looked at it under a microscope, you would see that it is not smooth. The rough surface causes diffuse reflection.

Refraction of Light

Have you ever seen a spoon sitting in a glass of water, like the one in the figure? If you look at the glass from the side, the spoon looks bent or broken. What you see is caused by the refraction, or bending, of light.

The speed of light depends on the substance it travels through. Light travels at different speeds in air, glass, and water. Refraction is caused by a change in the speed of a light ray. When light passes from one substance to another, it is refracted if it is traveling at an angle.

What is the index of refraction?

The amount that light rays bend when they are refracted depends on the speed that light travels in each material. The greater the difference in speeds, the more the light will be bent. Every material has an index of refraction. The **index of refraction** is a property of a material that indicates how much light slows down when it travels in the material.

The greater the index of refraction, the more light slows down in the material. For example, glass has a greater index of refraction than air. This means light moves more slowly in glass than in air. Glass lenses in eyeglasses, microscopes, and cameras use refraction to focus light.

What are prisms?

A prism refracts sunlight into colorful patterns. How does bending light create colors? White light, such as sunlight, is made up of different wavelengths. Red waves have the longest wavelengths of visible light and violet waves have the shortest. The different wavelengths are refracted, or bent, different amounts by a prism. The longer red wavelengths are bent the least. The shorter violet wavelengths are bent the most. The light that leaves the prism is separated into different colors because of the different amounts that each wavelength is bent.

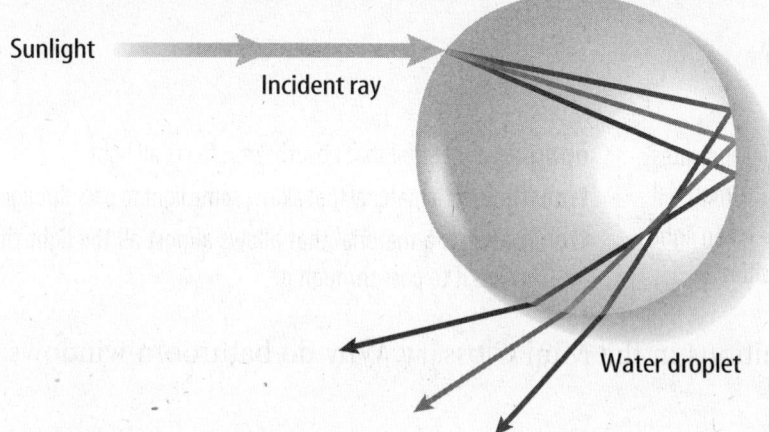

Picture This

5. **Observe** Circle the two areas in the figure that show where sunlight is refracted by the water droplet.

How do rainbows form?

Does the refracted light from a prism remind you of a rainbow? Like prisms, raindrops also can refract light. Look at the figure above. Sunlight enters water droplets in the atmosphere. The droplets separate the light into individual colors, depending on wavelength.

Humans can usually see only seven colors in a rainbow or in the refracted light of a prism. The colors are red, orange, yellow, green, blue, indigo, and violet, in order from longest wavelength to shortest wavelength.

What causes mirages?

Have you ever been riding in a car on a hot, sunny day? When you looked ahead, you may have seen what looked like a pool of water on the road. As you got closer, the pool of water seemed to disappear. What you saw was a mirage. A **mirage** is an image of a faraway object. The image is made when light is refracted through air layers of different densities.

Mirages result when the air at ground level is much warmer or much cooler than the air above ground level. As the temperature of air changes, the density of air changes also. These changes in density cause light waves to refract as they pass through air layers with different temperatures.

Reading Check

6. **Explain** What happens to light waves as they pass through air layers with different temperatures?

_____

After You Read

Mini Glossary

index of refraction: a property of a material that indicates how much light slows down when it travels in the material

mirage: an image of a far away object that is made when light is refracted through air layers of different densities

opaque: a material that absorbs or reflects all light

translucent: a material that allows some light to pass through it

transparent: a material that allows almost all the light that strikes it to pass through it

1. Review the terms and their definitions in the Mini Glossary. Why do bathroom windows often have translucent glass?

   _____

   _____

2. Complete the graphic organizer to organize the information you learned in this section about the behavior of light.

   Reflection
   - Happens when light waves _____ off objects.
   - The angle at which light strikes a surface and the angle at which it is reflected are the _____.

   Regular reflection
   - _____ surfaces
   - _____ images

   Diffuse reflection
   - _____ surfaces
   - _____ light

   Refraction
   - The _____ of light.
   - A property of a material that indicates how much light slows down when traveling in the material is the _____.

   Prisms
   - _____ light
   - _____ wavelengths

   Rainbows
   - formed by light refracting in _____

   Mirages
   - formed when light is refracted through _____

 Visit gpscience.com to access your textbook, interactive games, and projects to help you learn more about the behavior of light.

Chapter 13: Light

section 2 Light and Color

Before You Read

Colors are important in everyday life. Name a time when seeing color is important.

_____

_____

What You'll Learn
- how you see color
- the difference between light color and pigment color
- what happens when colors are mixed

Read to Learn

Colors

Why do some apples appear red, while others look green or yellow? You learned that white light is made up of light of different wavelengths. Each wavelength is a different color. The color of an object, like an apple, depends on which wavelengths of light it reflects. For example, when white light hits a red apple, only the red wavelengths are reflected. This is why the apple appears to be red. The apple absorbs all other wavelengths of light.

Some objects reflect all wavelengths of visible light. Visible light is light that we can detect with our eyes. Objects that reflect all wavelengths of visible light appear to be white. What about black objects? Black is not a color of visible light. Black objects absorb all wavelengths of light. Since almost no light is reflected from these objects, they appear to be black.

How can colors be filtered?

Have you ever worn tinted glasses? Maybe you noticed that tinted glasses change the color of almost everything you see. Yellow glasses make everything look yellow. Red glasses make the world look red. If you put a see-through green plastic sheet over this page, the paper would look green. Tinted glasses and see-through plastic sheets are filters. A filter is a transparent material that lets one or more colors pass through, but absorbs the rest. The color of a filter is the color of light that it lets through.

Mark the Text

Identify the Main Idea After you read each paragraph, highlight the topic sentence, or the main idea, of the paragraph.

What happens when you look through a colored filter?

When you look at a baseball field, you see green grass. But what would happen if you looked at it through a green filter? Since a green filter lets green light pass, the grass would still look green. Suppose you looked at the same field with a red filter. Since a red filter lets only red light pass, the grass would look black. The grass absorbs the red light so no color would be reflected through the filter and back to your eyes.

Seeing Color

As you approach an intersection the traffic light changes from green to yellow to red. What could happen at the intersection if you couldn't see the color changes? Your safety could depend on seeing colors. How do you see colors?

How does your eye detect light?

Look at the figure below. Light enters your eye through the lens. The lens focuses light onto the retina. The retina is an area on the inside of your eyeball. The retina has two types of cells that absorb light. When retina cells absorb light energy, chemical reactions happen inside the cells. The chemical reactions change light energy into nerve signals that go to your brain.

The two types of retina cells are cones and rods.

What are cones and rods?

Your retina has three types of cones. Cones are sensitive to color and bright light. Each type detects different wavelengths, or colors of light. Red cones detect mostly red and yellow light. Green cones detect mostly yellow and green light. Blue cones detect mostly blue and violet light.

The second type of retina cell is a rod. Rods are sensitive to dim light and aid night vision.

✓ Reading Check

1. **Infer** What color would green grass appear to be if you looked at it through a red filter?

_____

💡 Think it Over

2. **Infer** Nocturnal animals are awake mostly at night. Which type of retina cells would nocturnal animals probably have the most of?

_____

Picture This

3. **Use a Scientific Illustration** In the figure, underline the name of the kind of cell that allows you to see colors. Circle the name of the kind of cell that allows you to see in dim light.

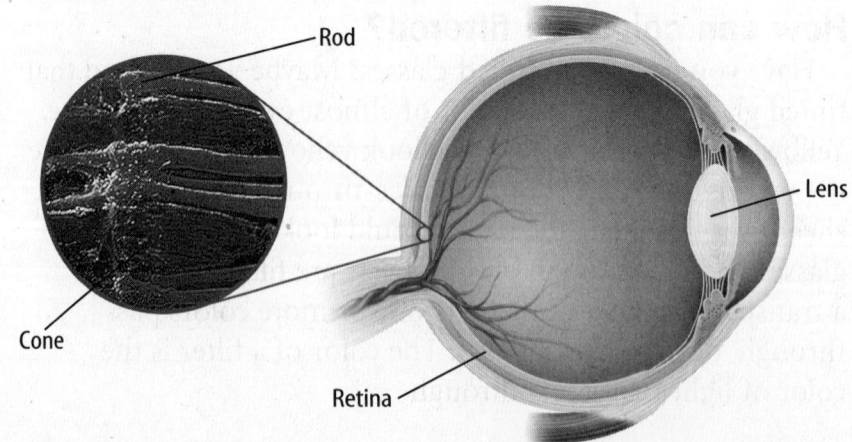

How do you interpret color?

When you see a stop sign, how do you know it is red? Red light reflected by the stop sign enters your eye and is focused on the retina. Red cone cells send a signal to your brain. Your brain then understands the signal to mean "red."

What is color blindness?

If one or more sets of cones in your retinas did not work correctly, you would not be able to detect certain colors. This condition is called color blindness. About eight percent of men and one-half percent of women have a form of color blindness. Most people who have color blindness are not truly blind to colors. They have trouble telling the difference between some colors, usually between red and green. People learn to deal with color blindness in many ways. For example, the color of a traffic light can be identified by its position.

Applying Math

4. **Convert** What does "one-half percent" mean?

_____

Mixing Colors

Have you ever noticed the hundreds of different paint colors customers can choose in a hardware store? You may even have mixed paints to make new colors in art class. It is possible to create different paint colors by mixing pigments. A **pigment** is a colored material that is used to change the color of other substances. A pigment's color depends on the wavelengths of light it reflects.

What happens when you mix colored light?

All the colors you see are made by mixtures of three colors of light—red, green, and blue, known as the primary colors of light. They correspond to the three different types of cones in the retina of your eye. Mixing them in different proportions produces all visible colors. As the figure shows, if red, green, and blue light are mixed equally, the result is white light.

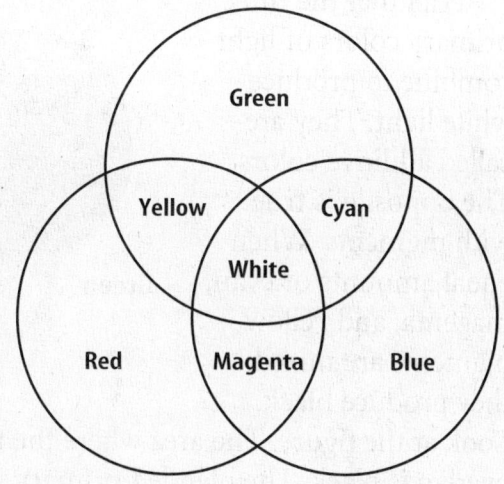

Picture This

5. **Observe** What color appears when red and blue light are mixed?

_____
_____

FOLDABLES

B Classify Make the following Foldable to help you organize information about light color and pigment color.

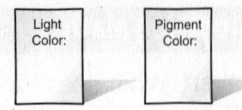

Think it Over

6. Explain A hardware store clerk uses green pigment to make green paint. Is green pigment a primary pigment, or is it made from a mixture of primary pigments? Why or why not?

_____
_____
_____
_____

Picture This

7. Observe What color appears when magenta and cyan pigments are mixed?

_____

What are paint pigments?

Paints are made with pigments. Paint pigments are made from chemical compounds. Titanium dioxide is a bright white paint pigment. Lead chromate is a yellow pigment that is used to paint the yellow stripes on highways. If you mixed equal amounts of red, green, and blue paint, would you get white paint? No, because mixing paint is different from mixing colored light.

How is mixing pigments different from mixing light?

There are three primary colors of pigments, just as there are three primary colors of light. But the primary pigment colors are different. They are cyan, magenta, and yellow. Cyan is a greenish blue. Magenta is a bluish red. You may have seen these color names if you have ever put ink cartridges in a color printer. You can make any pigment color by mixing different amounts of the three primary pigment colors.

A primary pigment's color depends on the color of light it reflects. Pigments both absorb and reflect many colors. Your eyes and brain, however, see the range of light wavelengths as one color. For example, a yellow pigment appears yellow in white light because it reflects red, orange, yellow, and green light. It absorbs blue and violet light. So, the color of a mixture of primary pigments depends on the primary colors of light that the pigment reflects.

How are black pigments made?

Recall that the three primary colors of light combine to produce white light. They are called additive colors. The opposite is true with pigments. When equal amounts of cyan, magenta, and yellow pigments are mixed, they produce black.

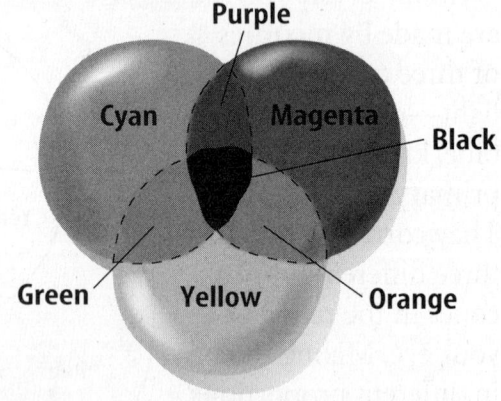

Look at the figure. The area where the three primary pigments overlap is black. The blended primary pigments absorb all colors of light. Because black is the result of no light being reflected, primary pigments are called subtractive colors.

● After You Read
Mini Glossary

pigment: a colored material that is used to change the color of other substances

1. Review the term and its definition in the Mini Glossary. Write a sentence describing a product you have used that contains pigments.

   _____

   _____

2. Complete the Venn diagram to organize the information you learned in this section about color.

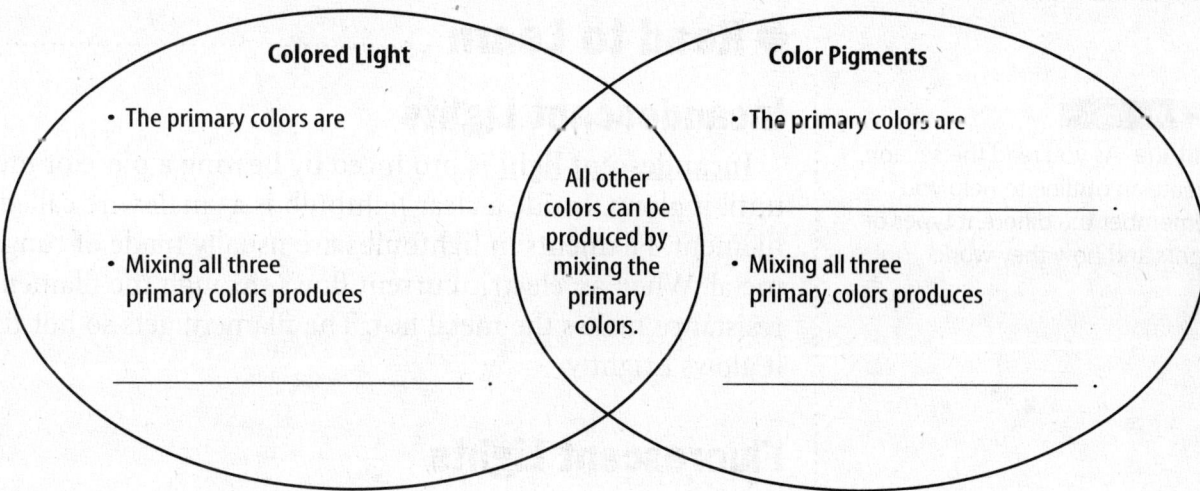

3. Think about what you have learned. How did highlighting the main idea of each paragraph help you as you read the section?

   _____

   _____

   _____

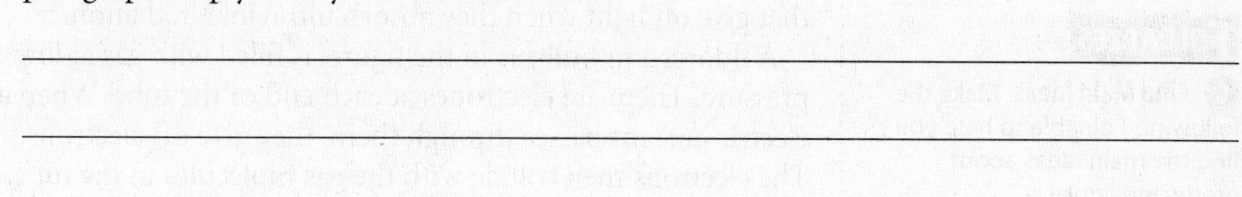 Visit **gpscience.com** to access your textbook, interactive games, and projects to help you learn more about light and color.

Reading Essentials **221**

Light

section ❸ Producing Light

What You'll Learn
- how incandescent and fluorescent lightbulbs work
- about different lighting devices
- about coherent light
- how lasers are used

Study Coach

Outline As you read the section, create an outline to help you remember the different types of lights and how they work.

FOLDABLES

C Find Main Ideas Make the following Foldable to help you find the main ideas about producing light.

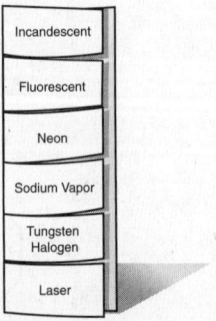

● Before You Read

Lasers have many uses. You have probably seen light that was produced by a laser. Describe what this light looked like.

_____

_____

● Read to Learn

Incandescent Lights

<u>Incandescent light</u> is produced by heating a piece of metal until it glows. Inside a clear lightbulb is a small wire called a filament. Filaments in lightbulbs are usually made of tungsten metal. When an electric current flows through the filament, resistance makes the metal hot. The filament gets so hot that it glows brightly.

Fluorescent Lights

You may have fluorescent (floo-RE-sunt) lights in your home or school. A <u>fluorescent light</u> uses phosphors to change ultraviolet (UV) radiation to visible light. Phosphors are substances that give off light when they absorb ultraviolet radiation.

A fluorescent bulb, as in the figure, is filled with gas at low pressure. There are electrodes at each end of the tube. When an electric current passes through them, they give off electrons. The electrons then collide with the gas molecules in the tube. The collisions make the gas molecules give off UV radiation. The tube is coated on the inside with phosphors. The phosphors give off visible light when they absorb the UV radiation.

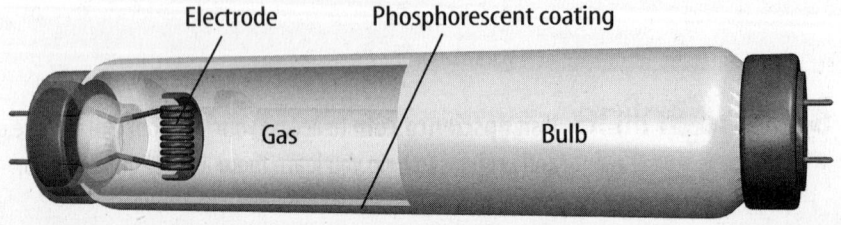

222 CHAPTER 13 Light

Which type of lighting is more efficient?

Fluorescent lights are more efficient than incandescent lights. A fluorescent bulb uses as little as one-fifth the energy that an incandescent bulb uses to make the same amount of light. Fluorescent bulbs also last much longer than incandescent bulbs. Because fluorescent bulbs are more efficient, they cost less to use over the life of the bulb. They also help reduce energy usage. By reducing energy usage, they could reduce the amount of fossil fuels needed to generate electricity. This would reduce the amount of carbon dioxide and pollutants that are released into Earth's atmosphere.

Neon Lights

Neon lights are glass tubes filled with gas. They work much the same way as fluorescent bulbs. An electric current flows through the tube. The electrons of the electric current collide with gas molecules, as shown in the figure. The collisions produce visible light. If the tube contains only neon gas, the light is bright red. Different colors are produced by adding other gases to the tube.

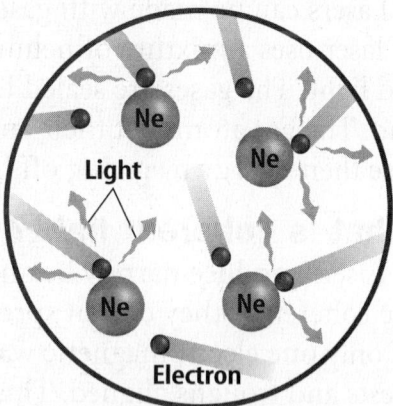

Sodium-Vapor Lights

Sodium-vapor lights are often used for outdoor lighting. A sodium vapor lamp has a tube filled with neon gas, a small amount of argon gas, and a small amount of sodium metal. When the lamp is turned on, the gas mixture in the tube becomes hot. The sodium metal turns to vapor. The vapor gives off a yellow-orange glow.

Tungsten-Halogen Lights

Tungsten-halogen lights give off bright light. These lights have a tungsten filament, like incandescent bulbs. The filament is inside a quartz tube or bulb. The tube is filled with a halogen gas, such as fluorine or chlorine. The gas allows the filament to get hotter than the filament in an incandescent lightbulb. As a result, the light produced is much brighter. The bulb also lasts longer.

Applying Math

1. **Calculate** Fluorescent lights use one-fifth the energy of incandescent lights. What percent is one-fifth?

   _____

Think it Over

2. **Compare** How do the collisions in a fluorescent tube differ from the collisions in a neon tube?

   _____
   _____
   _____

Lasers

Lasers produce light waves that have the same wavelength. A laser's light begins when a number of light waves are given off at the same time. To produce these light waves, a number of atoms are given the same amount of energy. The atoms then release their energy and send off identical light waves. These light waves bounce off mirrors at opposite ends of the laser. One of the mirrors is only partly reflective. It allows some light to escape. This escaped light forms the beam that you see. Some of the light waves do not escape. These waves continue to bounce between the mirrors. This causes other atoms in the laser to produce more identical waves. As this process continues, a steady stream of laser light is produced. ☑

Lasers can be made with gases, liquids, or solids. One kind of laser uses a mixture of helium and neon gases to produce red light. The gases are sealed in a tube with a mirror at each end. The gas atoms get their energy from a flashtube. They lose their energy by giving off light waves.

What is coherent light?

Lasers produce narrow beams of light. Waves of laser light are coherent—they do not spread out. **Coherent light** is light of only one electromagnetic wavelength that travels with its crests and troughs aligned. This means the crests and troughs are always the same distance from each other. The first figure shows coherent light. Coherent light does not spread out because all the waves travel in the same direction. Because the beam does not spread out, the light energy stays concentrated in a small area.

What is incoherent light?

Incoherent light can contain more than one wavelength. The electromagnetic waves are not aligned. The second figure shows incoherent light. Notice that the waves have different wavelengths and they do not travel with their crests and troughs aligned. Light from a lightbulb is incoherent. Beams of incoherent light spread out. The energy of the light waves spreads out with the beam. This makes the intensity of incoherent light much less than the intensity of a laser beam. Incoherent light is better for lighting a room.

Reading Check

3. **Describe** Why is one mirror in a laser only partly reflective?

Picture This

4. **Identify** Circle the incoherent light and put a box around coherent light.

Using Lasers

Lasers have properties that make them important tools. They are used in CD players and even in surgery. A laser beam is narrow and does not spread out as it travels. So, lasers can carry energy to small areas. In industry, powerful lasers are used for cutting and welding. Surveyors and builders use lasers for measuring. Scientists use laser light reflected from mirrors on the Moon to measure the Moon's orbit with great accuracy. Information can be sent in pulses of light from lasers. This makes lasers useful in communication. In some telephone systems, pulses of laser light send conversations through long glass fibers called optical fibers.

How do compact discs work?

Compact discs (CDs) are plastic discs with reflective surfaces. They are used to store sound, images, and text in digital form. When a CD is made, information is burned into the surface of the disc with a laser. The laser creates millions of tiny pits, where information is stored.

A CD player also uses a laser to read the disc. Look at the figure below. The figure on the right shows the pits on the bottom surface of a CD. The laser shines on the spinning disc. As the beam hits the pits, different amounts of light are reflected to a light sensor. The reflected light is then converted to an electric signal. The signal creates sound in the speakers.

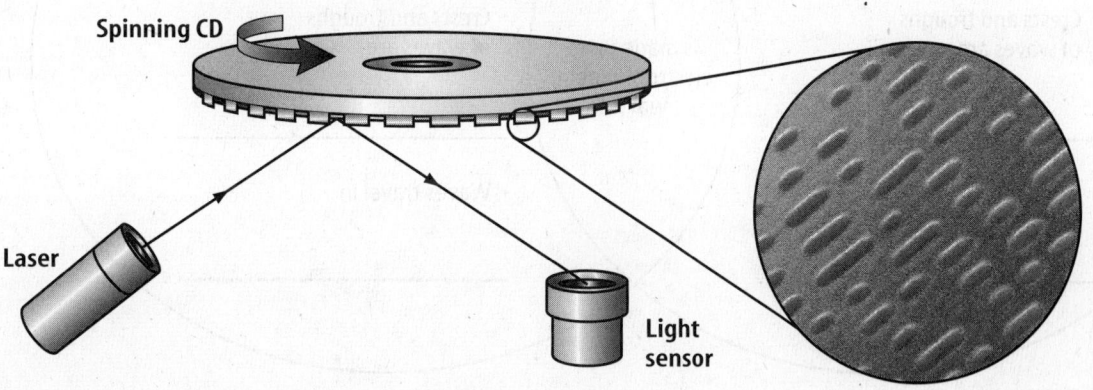

How are lasers used in medicine?

Lasers are often used in eye surgery. They can be used to remove cataracts, reshape the cornea, and repair the retina. Surgeons can use lasers in place of scalpels to cut body tissues. Laser energy seals off blood vessels as it cuts. This reduces bleeding during surgery. Most lasers do not cut deeply through the skin. This makes them useful in removing tumors or birthmarks on the surface of the skin.

Think it Over

5. Explain Why is a laser a good tool to make sure something is level?

_____

_____

_____

Picture This

6. Identify Circle the part of the figure that shows where information is stored on a CD.

After You Read
Mini Glossary

coherent light: light of only one wavelength that travels with its crests and troughs aligned

fluorescent light: light that uses phosphors to convert ultraviolet light to visible light

incandescent light: light that is produced by heating a piece of metal until it glows

incoherent light: light that can contain more than one wavelength and travels in many directions with its electromagnetic waves not aligned

1. Review the terms and their definitions in the Mini Glossary. Write a sentence using a term for a device that produces incoherent light.

   _____

   _____

2. Complete the Venn diagram to organize the information you learned in this section about producing light.

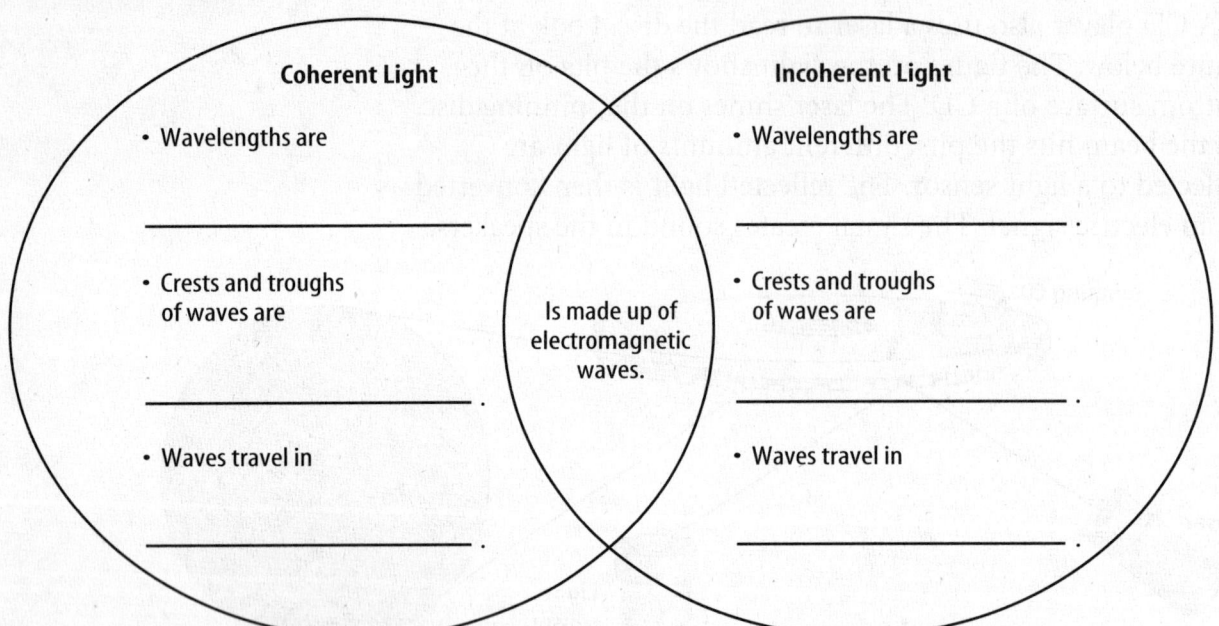

3. Think about what you have learned. How did outlining as you read help you learn about the different type of lights and lighting devices?

   _____

   _____

End of Section

 Visit gpscience.com to access your textbook, interactive games, and projects to help you learn more about producing light.

Light

section ❹ Using Light

Before You Read

Have you ever worn sunglasses on a very bright day? Describe why you think sunglasses allow you to see more clearly.

___

What You'll Learn
- difference between polarized and unpolarized light
- how a hologram is made
- when total internal reflection occurs
- about optical fibers

Read to Learn

Polarized Light

Do you know what makes "polarized" sunglasses different from other sunglasses? The difference is in the vibrations of the light waves that pass through the lenses. Transverse waves have repeating high points, or crests, and repeating low points, or troughs. If you hold a rope at one can end and shake it, you create a transverse wave in any direction. Light is a transverse wave that can vibrate in any direction. In **polarized light**, the waves vibrate in only one direction.

What are polarizing filters?

Light that passes through a polarizing filter becomes polarized. A polarizing filter acts like a group of parallel slits. Only light waves vibrating in the same direction as the slits can get through. It is similar to transverse waves on a rope passing through a picket fence, as in the figure.

Study Coach

Make Flash Cards Each time you read a heading on the page, make a flash card with that heading. Then write the main idea of the information under the heading on the card.

FOLDABLES

ⓓ Organize Information Make the following Foldable to help you organize information about polarized light, holography, optical fibers, and optical scanners.

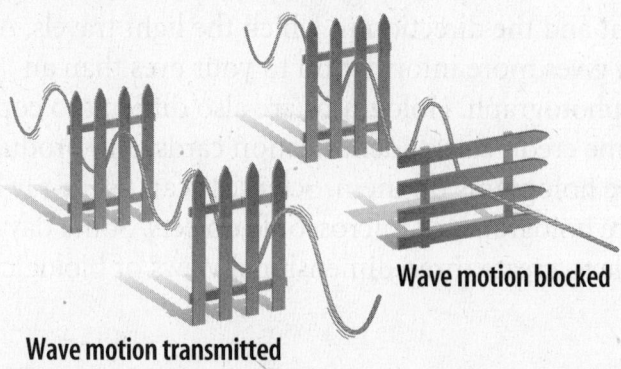

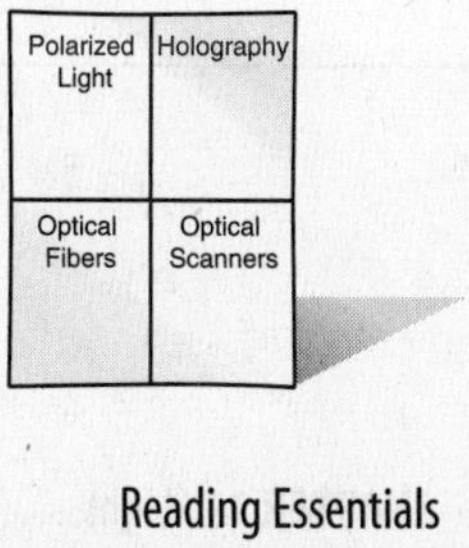

Wave motion transmitted

Wave motion blocked

Reading Essentials **227**

Light passes through one polarizing filter. Then it hits another polarizing filter with slits at right angles to the first filter. The first filter only lets light through that was polarized in one way. None of this light is polarized in the same direction as the slits on the second filter. No light gets through the second filter.

Polarized lenses are useful for reducing glare. They do this without affecting your ability to see. Some of the light reflected from a flat surface, such as a lake or car hood, becomes horizontally polarized. The lenses of polarized sunglasses have vertical polarized filters. They block out the reflected light that is polarized horizontally.

Holography

Have you ever seen a three-dimensional image that seems to float in the air? If you walk around the image, you can see it from different angles. It is as if you are looking at a real object. Such images are produced by holography. **Holography** is a method that produces a hologram, a complete, three-dimensional photographic image of an object. ✓

How is a hologram made?

It takes a laser to make a hologram. First, a laser is shined onto an object. The laser light reflects from the object onto photo film. At the same time, a second beam split from the laser is shined directly on the film. The light from the two beams creates a pattern on the film. When the film is developed, the pattern looks nothing like the original object. But laser light shines onto the film, it produces a holographic image.

How do holograms contain more information than ordinary photos?

A photograph records only the brightness of light that is reflected by an object. A hologram records both the brightness of the light and the direction in which the light travels. A hologram gives more information to your eyes than an ordinary photograph. Holograms are also difficult to copy. This is why some credit cards, identification cards, and product labels have holograms on them. Scientists can use X-ray lasers to produce holograms of microscopic objects. Some day it may be possible to create three-dimensional views of biological cells.

✓ Reading Check

1. **Analyze** What kind of photographic image is a hologram?

_____

_____

💡 Think it Over

2. **Draw Conclusions** Why is a holographic image of a microscopic object better than a two-dimensional image of the object?

_____

_____

_____

Optical Fibers

Optical fibers are transparent glass fibers that carry light from one place to another. They can transmit light because of a process called total internal reflection.

What is total internal reflection?

Remember that as light passes from one substance to another, it is refracted. Refraction happens because the speed of the light is different in the two substances, such as water and air. Light rays also can be reflected.

Critical Angle Look at the figure. A line that is perpendicular to a surface is called the normal. When light travels from water to air, the direction of the light ray is bent away from the normal. If the underwater light ray makes a larger angle, the light ray in the air bends closer to the surface of the water. At a certain angle, the refracted ray is traveling along the surface of the water. This is called the critical angle. For a light ray traveling from water into air, this is about 49°.

Total Internal Reflection If the angle of the original beam is increased past 49°, refraction stops. Instead of being refracted, the light ray is reflected off the surface of the water. It is reflected back into the water, as if there were a mirror on the surface. No light escapes the surface. This is called total internal reflection. **Total internal reflection** happens when light traveling from one medium to another is completely reflected at the boundary between the two materials. For total internal reflection to happen, light must travel slower in the first material than in the second material. It also must hit the boundary at an angle greater than the critical angle.

Applying Math

3. **Use Geometry** An acute angle is an angle less than 90°. An obtuse angle is an angle greater than 90° and less than 180°. Is the critical angle an acute angle or an obtuse angle? Explain.

Picture This

4. **Conclude** Use a protractor to measure the angle labeled "Total internal reflection." Is this an acute angle or an obtuse angle? Explain.

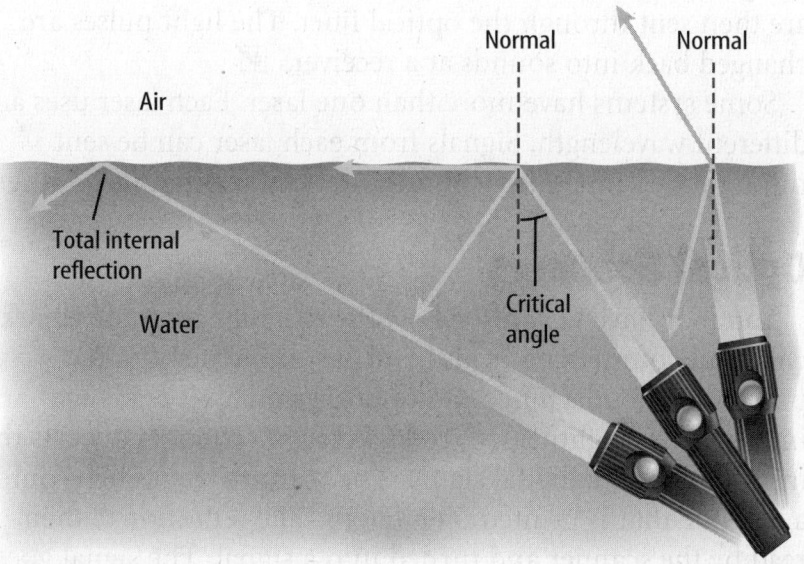

 **Think it Over**

5. Apply What allows light to reflect off the surface of an optical fiber without leaking out?

_____
_____
_____

Picture This

6. Identify Highlight the boundary where total internal reflection takes place.

Reading Check

7. Determine What are sounds converted into before they can be sent through optical fibers?

_____
_____

How does light travel through an optical fiber?

Total internal reflection allows light to be carried through optical fibers. Look at the figure below. Light enters one end of the fiber. It is reflected continuously until it comes out the other end. Almost no light is lost or absorbed. An optical fiber is almost like a water pipe, only it carries light instead of water.

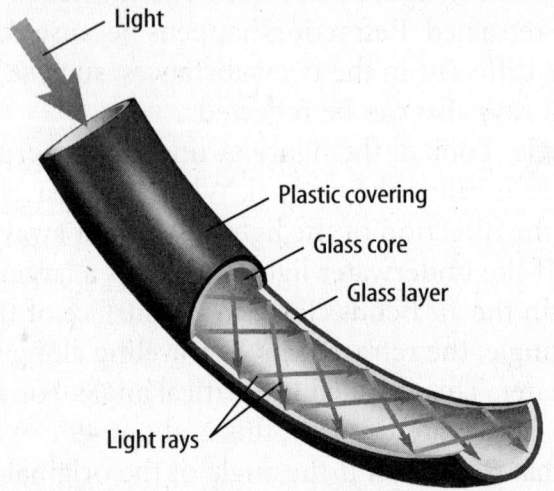

How are optical fibers used?

Optical fibers are most often used to communicate information. They are used as telephone lines, television cable lines, and computer cables. Signals can't leak from optical fibers, so they don't interfere with other nearby fibers. This allows the messages to be transmitted very clearly.

To send telephone conversations through optical fibers, sounds are changed to digital signals. Those signals are pulses of light from a light-emitting diode or a laser. The light pulses are then sent through the optical fiber. The light pulses are changed back into sounds at a receiver.

Some systems have more than one laser. Each laser uses a different wavelength. Signals from each laser can be sent through the same fiber without interfering with one another.

Optical Scanners

You may have wondered how the machine at a supermarket checkout counter can read the prices of the items you purchase. The machine is an optical scanner. It reads the intensity, or brightness, of reflected light. Then it converts this information to a digital signal. The scanner shines light onto a barcode that is printed on an item. The reflection is then read by the scanner and turned into a signal. The signal goes to a computer, which finds the price of the item.

After You Read

Mini Glossary

polarized light: light in which the waves vibrate in only one direction

holography: a method that produces a hologram, which is a complete, three-dimensional photographic image of an object

total internal reflection: when light traveling from one medium to another is completely reflected at the boundary between the two materials

1. Review the terms and their definitions in the Mini Glossary. What is polarized light?

   _____
   _____

2. On the lines below, list three things that must happen for total internal reflection to take place.

   | Total Internal Reflection |

   1. Light traveling from one medium to another is completely reflected at the _____. between two materials.

   2. Light must travel _____ in the first material than in the second material.

   3. Light must hit the boundary at an angle _____ than the _____ angle.

3. Think about what you have learned. How did making flash cards help you learn about how light is used?

   _____
   _____
   _____

Science Online Visit **gpscience.com** to access your textbook, interactive games, and projects to help you learn more about using light.

Reading Essentials 231

chapter 14 Mirrors and Lenses

section ❶ Mirrors

What You'll Learn
- how three kinds of mirrors work
- the difference between real and virtual images
- uses of plane, concave, and convex mirrors

Mark the Text

Locate Information Many headings in this section are questions. Underline the answers to these questions.

FOLDABLES

Ⓐ Find Main Ideas Make a Foldable like the one shown below to write down main ideas about light and vision.

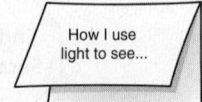

● Before You Read

Turn off the lights and try to read this page. Turn the lights back on and try again. On the lines below, explain why you think it is easier to read a printed page with the lights on.

_____

_____

_____

● Read to Learn

How do you use light to see?

Light comes from sources like lightbulbs, candles, and the Sun. When light reflects, or bounces, off of an object and travels to your eye, you see the object. For example, light from the lightbulbs in your classroom reflects off this book. The light travels from the book to your eyes. As a result, you see the words on the page. The more light there is to reflect off of objects, the easier it is to see them. When there is no light to reflect off objects, you cannot see anything. This is why it is hard to read a book in the dark.

What is a light ray?

First, think about light as waves. A light source like the Sun or a lightbulb sends out waves of light in all directions. Now, think about light as being many straight lines of light coming out from a source. Each line is called a light ray. Light rays refract and reflect just like light waves. The figure shows how a candle puts out rays of light in all directions. You can change the direction of light rays by reflecting them off a shiny object, like a mirror.

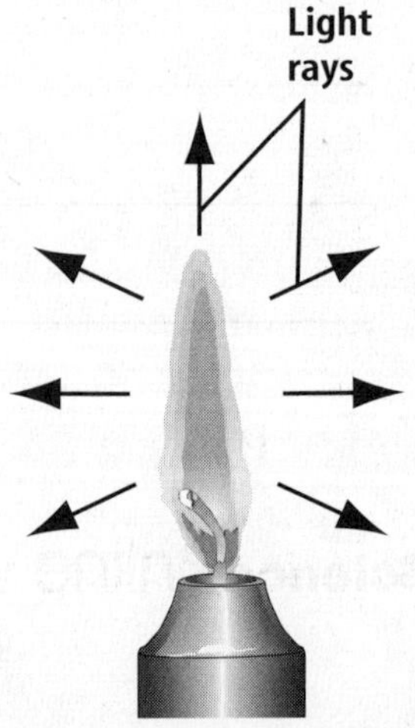

Light rays

232 CHAPTER 14 Mirrors and Lenses

Seeing Reflections with Plane Mirrors

Imagine a lake with trees growing along the shore. If you look at the surface of the water, you will see an image of the trees. An image is the picture you see when light reflects off of a smooth surface. You might see an image in a store window or a shiny car hood.

A very common place to see an image is in a flat, smooth mirror called a **plane mirror**. This is the kind of mirror you see in most bathrooms and dressing rooms. ✓

How does a plane mirror work?

Suppose you stand in front of a mirror and turn on a light as shown below. The light bulb puts out rays of light. Some of the light rays hit you. Then the rays reflect off of you in all directions. Some of these rays hit the mirror. The mirror reflects these rays in all directions. Some of the reflected light rays hit your eyes. You see your image in the mirror. If there are no light rays to reflect, there is no image for you to see.

Reading Check

1. **Define** What is a plane mirror?

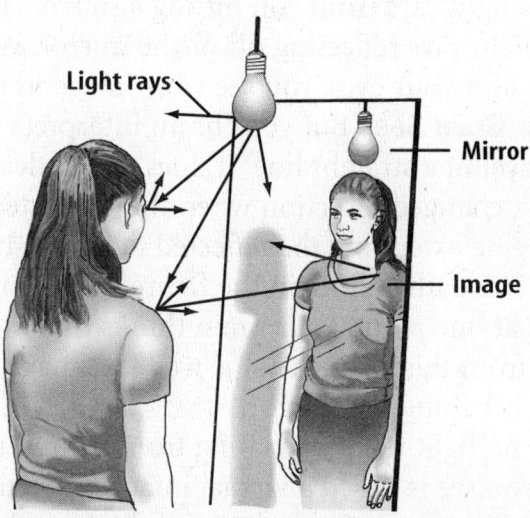

Picture This

2. **Infer** What will happen to the image in a plane mirror if the lights are turned out? Why?

What does an image in a plane mirror look like?

Your image in a plane mirror appears right side up. Your head is at the top of the image and your feet are at the bottom. But the image that faces you in the mirror is reversed, or opposite. Your left side appears on the right side of the image. Your right side appears on the left side of the image. Suppose you stand 1 m in front of a plane mirror. Your image will look like it is standing 1 m behind the mirror. To you, your image appears to be 2 m away.

Reading Essentials **233**

Picture This

3. **Apply** Look at the figure. Which action would make the image look like it is farther behind the mirror?

   a. Turn out the lights.

   b. Turn on more lights.

   c. Move closer to the mirror.

   d. Move away from the mirror.

What is a virtual image?

The figure below shows why an image in a plane mirror looks like it is behind the mirror.

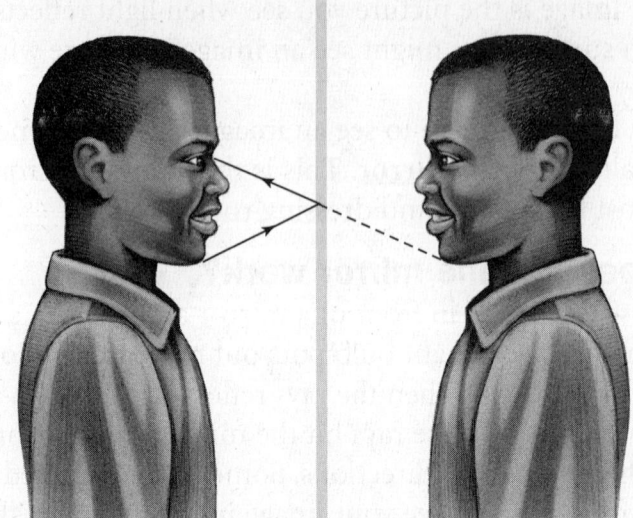

Suppose you see your reflection in a plane mirror. The arrows show light rays from you hitting a mirror. The arrows show these light rays reflecting off of the mirror. When these reflected rays hit your eyes, you see your image in the mirror.

What Your Brain Sees But your brain interprets light rays as if they travel in a straight line. It does not understand that the light rays changed direction when they reflected off of the mirror. Imagine extending the reflected light rays back behind the mirror. The dashed lines in the figure show that the rays would meet at one point. Your brain thinks that the light rays are coming from this point. That is why the image of the box looks like it is behind the mirror.

However, no light really is coming from behind the mirror. The image you see is called a virtual image. A virtual image is an image that is not real, even though it looks real. No light rays actually meet to create the image. The virtual image always appears to be as far behind the mirror as the object is in front of it.

4. **Describe** What is a virtual image?

_____

_____

_____

Concave Mirrors

A plane mirror is flat. It produces a right side up, virtual image. However, not all mirrors are flat. Some mirrors have curved surfaces. Mirrors with curved surfaces form different kinds of images. One kind of mirror with a curved surface is a concave mirror.

What is a concave mirror?

A <u>concave mirror</u> has a surface that curves in. The edges of a concave mirror are closer to you than the center of the mirror. Many shaving mirrors and makeup mirrors are concave. Have you ever looked at your image inside the bowl of a shiny spoon? The inside of the bowl of the spoon is a concave mirror, too.

How does a concave mirror work?

Every concave mirror has an optical axis. The <u>optical axis</u> is an imaginary line perpendicular to the surface of the mirror. It also passes through the center of the mirror. The figure below shows a side view of a concave mirror and the location of its optical axis.

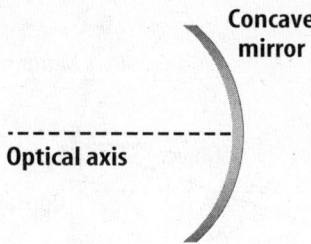

Picture This

5. **Draw** Connect the top and bottom of the concave mirror with a line. Is this line parallel or perpendicular to the optical axis?

_____

Parallel Light Rays Some light rays travel parallel to the optical axis on their way to a concave mirror. These rays reflect off the mirror. As the figure below shows, the reflected rays all cross each other at the same point. This point is called the focal point. The distance from the center of the mirror to the focal point is called the focal length. ✓

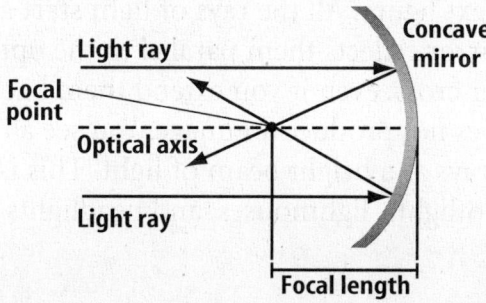

Reading Check

6. **Identify** What is the focal point of a concave mirror?

_____
_____
_____

Intersecting Light Rays As shown below, some light rays travel through the focal point on their way to a concave mirror. The mirror reflects these rays parallel to the optical axis.

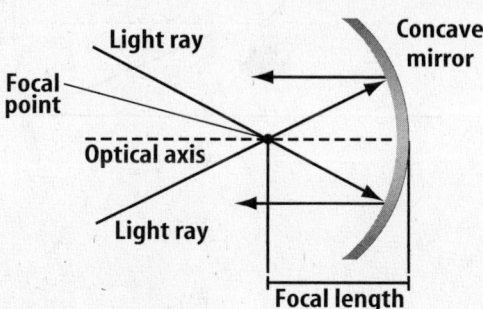

Picture This

7. **Identify** In the last two figures, use a highlighter to trace the light rays and the paths of their reflection.

Reading Essentials **235**

What does an image in a concave mirror look like?

The image of an object in a concave mirror does not always look the same. It depends on how close the object is to the mirror.

Suppose you put a candle less than one focal length from the mirror. Look at the first figure. Like an image in a plane mirror, the image of the candle in the concave mirror is a virtual image. The image is right side up. But, the image is larger than the candle. Have you ever seen a concave shaving mirror or makeup mirror? When you stand less than one focal length away, the image of your face looks larger.

Picture This

8. Infer What will happen to the image if the candle is moved closer to the mirror?

_____

_____

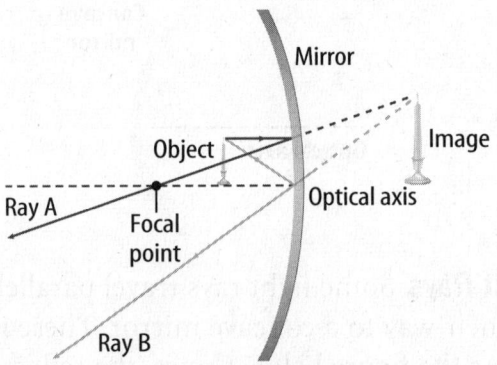

How are light beams created?

What happens if you put an object right at the focal point? Look at the next figure. All the rays of light start at the focal point. The mirror reflects them parallel to the optical axis. The rays never cross, even if you extend them backward. So, the mirror does not produce an image. You see all those parallel light rays as a bright beam of light. This is how flashlights, spotlights, lighthouses, and headlights create light beams.

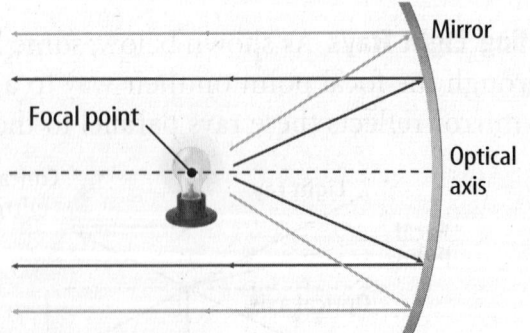

Picture This

9. Highlight the part of the figure that shows why the mirror does not produce an image.

What are real images?

What happens when the distance from the candle to the mirror is between one and two focal lengths? This time the mirror will produce a real image. A **real image** is formed when rays of light converge to form the image. Rays of light really pass through the location of the image. The image is larger than the candle. And, the image is upside down.

How do mirrors decrease size?

If the distance between the candle and the mirror is more than two focal lengths, you get another kind of image. Look at the figure below. The mirror produces a real image. The image is also upside down. However, the image is smaller than the candle.

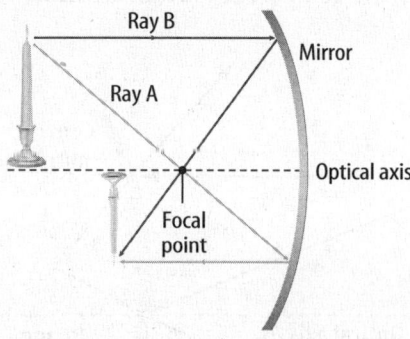

Picture This
10. **Determine** where 2 focal lengths are and place a mark at that point on the optical axis.

Convex Mirrors

Concave mirrors are only one kind of curved mirror. Another kind of mirror with a curved surface is a convex mirror. Convex mirrors have some special uses in everyday life.

What is a convex mirror?

Have you seen a side-view mirror on a car that says "Objects in mirror are closer than they appear?" A convex mirror has a surface that curves out, like the back of a spoon. The center of a convex mirror is closer to you than the edges of the mirror. The security mirrors in banks and stores are convex mirrors. Some rear-view mirrors and side-view mirrors in cars are convex, too. ✓

Reading Check
11. **Define** What is a convex mirror?

_____
_____
_____

Think it Over

12. Infer Why might a store-owner use convex mirrors, instead of plane or concave mirrors, to watch the store?

_____
_____
_____

Picture This

13. Compare and Contrast How are the images in a plane mirror and a convex mirror alike? How are they different?

_____
_____
_____
_____

How does a convex mirror work?

When a convex mirror reflects light rays, it spreads the light rays apart from each other. Because of this, a convex mirror can show you the image of a large area. This is called a wide field of view. For example, the wide field of view of the convex mirrors in cars lets drivers see more of the traffic around them.

What does an image in a convex mirror look like?

So why do side-view mirrors say that the objects "are closer than they appear?" The light rays that reflect off of a convex mirror do not cross each other. Like a plane mirror, a convex mirror produces a virtual image. The image of an object in a convex mirror is right side up. But, the image is smaller than the object. If an image appears smaller, then we think it is farther away.

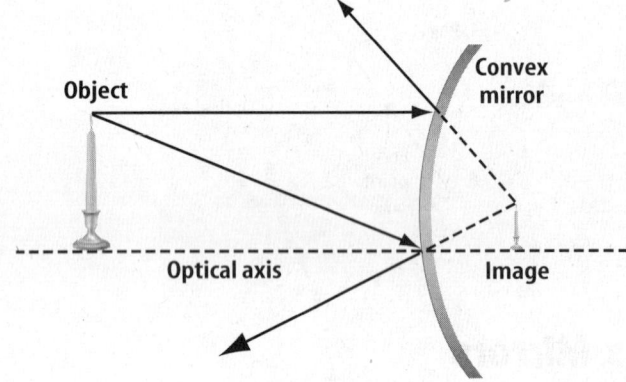

The table below summarizes the images formed by plane, concave, and convex mirrors.

Images Formed by Mirrors				
Mirror Shape	Position of Object	Virtual/Real	Image Created Upright/Upside Down	Size
Plane		virtual	upright	same as object
Concave	Object more than two focal lengths from mirror	real	upside down	smaller than object
	Object between one and two focal lengths	real	upside down	larger than object
	Object at focal point	none	none	none
	Object within focal length	virtual	upright	larger than object
Convex		virtual	upright	smaller than object

After You Read

Mini Glossary

concave mirror: a mirror with a curved surface so that the edges are closer to you than the center of the mirror is

convex mirror: a mirror with a curved surface so that the center is closer to you than the edges of the mirror are

focal length: the distance from the center of a concave mirror to the mirror's focal point

focal point: if light rays parallel to the optical axis hit a concave mirror, the reflected rays cross at this point

optical axis: an imaginary line perpendicular to the center of a concave mirror

plane mirror: a flat, smooth mirror

real image: an image formed when rays of light converge to form the image

virtual image: an image that looks real, but no light rays actually pass through it

1. Review the terms and their definitions in the Mini Glossary. Write a sentence that uses the terms *concave mirror*, *optical axis*, and *focal point*.

   _____

   _____

2. Complete the chart below to organize the information from this section.

Kind of Mirror	Position of Object	Is image right side up or upside down?	Is image larger, smaller or the same size as the object?
Plane	✗		
Concave	object more than two focal lengths from mirror		
Concave	object between one and two focal lengths		
Concave	object at focal point	✗	✗
Concave	object within one focal length		
Convex	✗		

3. How did underlining the answers to questions asked in this section's headings help you learn about mirrors and images?

   _____

   _____

 Visit **gpscience.com** to access your textbook, interactive games, and projects to help you learn more about mirrors and images.

chapter 14 Mirrors and Lenses

section ❷ Lenses

What You'll Learn
- what convex and concave lenses look like
- how convex and concave lenses form images
- how lenses can help people see better

Study Coach

Make a Chart Make a two-column chart. As you read, list the main ideas in the left column. In the right column, list the details that support each main idea.

Picture This

1. **Predict** A light ray travels along the optical axis to a convex lens. Draw a line on the figure to show what will happen to the ray as it passes through the lens.

● Before You Read

Tools used to look at objects contain at least one lens. On the lines below, make a list of tools that have lenses in them.

_____

_____

● Read to Learn

What is a lens?

What do your eyes have in common with cameras, eyeglasses, and microscopes? They all contain at least one lens. A lens is made of transparent material. Remember, transparent means that almost all the light rays that hit the material pass through it. A lens has at least one curved surface. The curved surface makes light rays refract, or bend, as they pass through the lens.

Like mirrors, lenses produce images. What an image looks like depends on the shape of the lens. A lens can be convex or concave.

Convex Lenses

A <u>convex lens</u> is thicker in the middle than at the edges. The optical axis of a convex lens is an imaginary line perpendicular to the thickest point on the lens. The figure shows a convex lens.

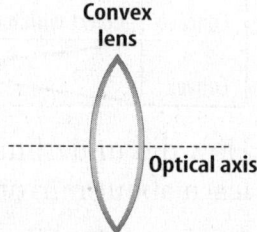

How does a convex lens work?

Some light rays travel parallel to the optical axis on their way to a convex lens. The lens refracts these light rays in a special way. The refracted rays all cross each other at the same point, called the focal point. The focal length of the lens is the distance from the center of the lens to the focal point.

240 CHAPTER 14 Mirrors and Lenses

The focal length of a convex lens depends on the shape of the lens. Look at the figures below. If the curved sides of the lens are very rounded, the light rays bend sharply. The focal point is close to the lens. So, the focal length is short. If the sides of the lens are flatter, the light rays bend less. The focal point is farther from the lens. So, the focal length is longer.

B Organize Take notes about lenses in a 2-tab Foldable such as the one shown below.

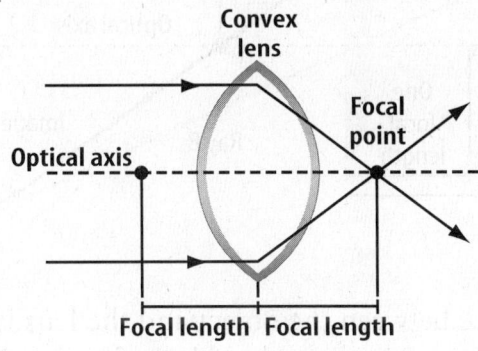

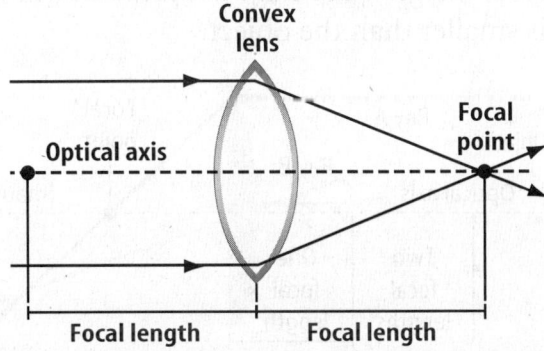

What does an image from a convex lens look like?

The image of an object from a convex lens does not always look the same. It depends on how close the object is to the lens.

Suppose you put an object less than one focal length from a convex lens. The lens will produce a virtual image of the object. The image is right side up. But, the image is larger than the object. Have you ever used a magnifying glass? This tool is a convex lens.

Picture This

2. **Compare** Extend the top horizontal line in each figure. Compare the angles formed by these lines and the bent rays. Which angle is greater, the one in the figure of more curved lens or the one on the figure with the flatter lens?

_____
_____

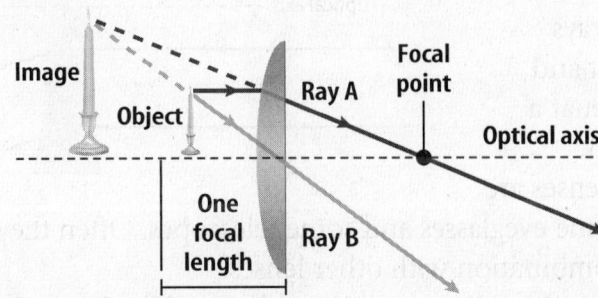

Picture This

3. **Describe** Suppose you need to create an upside down image of an object. You have a convex lens. Where should you put the object?

_____
_____

Reading Essentials **241**

Picture This

4. Apply Rob has a convex lens with a focal length of 5 in. He wants to produce a larger, right-side-up image of a pencil. Which figure shows where Rob should put the pencil? Circle your answer.

a. top figure

b. middle figure

c. bottom figure

What would happen if you put the object between one and two focal lengths from the lens? This time the lens will produce a real image. The image is larger than the object. And, the image is upside down.

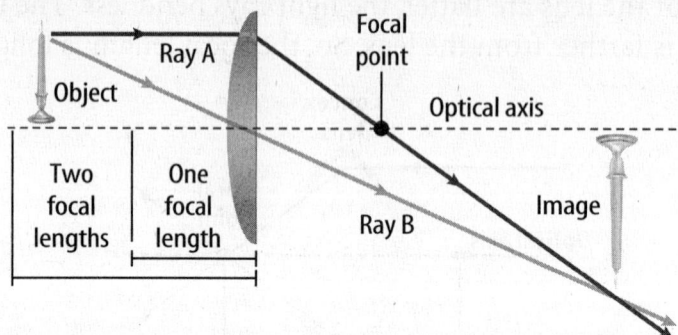

If the distance between the object and the lens is more than two focal lengths, you get another kind of image. The lens produces a real image. The image is upside down. However, the image is smaller than the object.

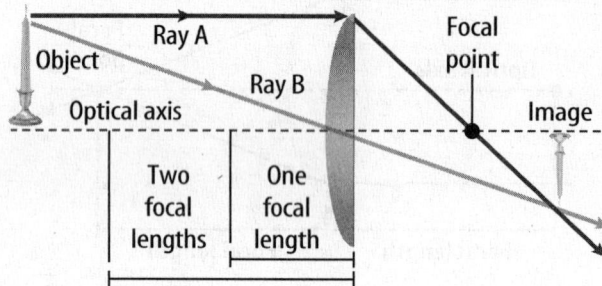

Concave Lenses

A **concave lens** is thinner in the middle and thicker at the edges, as shown below. The optical axis of a concave lens is an imaginary line perpendicular to the thinnest point on the lens.

When a concave lens refracts light rays, it bends the rays outward, away from the optical axis. The rays spread out and never meet at a focal point.

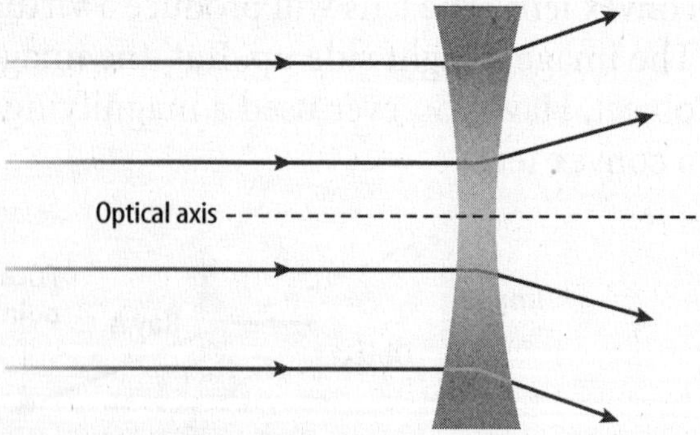

Concave lenses are used in some eyeglasses and some telescopes. Often they are used in combination with other lenses.

A concave lens creates a virtual image. The image is right side up. But, the image is smaller than the object.

Think it Over

5. Classify Is a concave lens most like a concave mirror or a convex mirror? Explain your choice.

_____

_____

Lenses and Eyesight

What determines how well you can see the words on this page? If you do not need eyeglasses, it is because the parts of your eyes work together to let you see clearly.

How do your eyes work?

Light enters your eye through the cornea. The **cornea** (KOR nee uh) is a transparent covering over your eyeball. The cornea bends light rays to bring them together. Then the light goes through an opening in your eye called the pupil. Behind the pupil is a convex lens. The lens also brings the light rays together. Then the rays form an image on your retina. The **retina** is the inner lining of your eye. It has cells on it that can change the image into electrical signals. The optic nerve sends the signals to your brain.

Reading Check

6. Explain On what part of the eye does the image form?

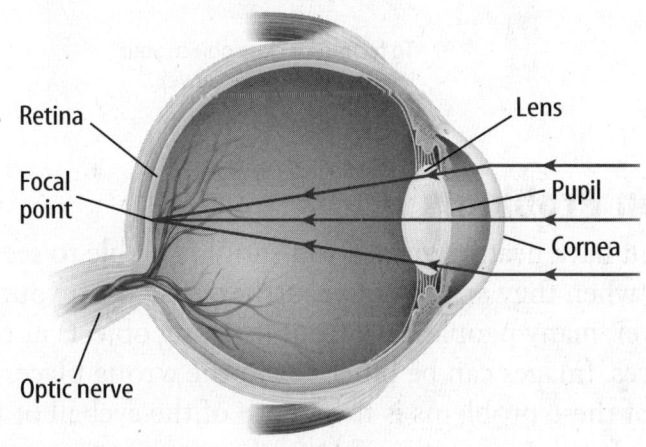

How do you clearly see both near and far objects?

You can see the watch on your wrist clearly. Then you can look at a clock across the room and see it clearly, too. How is this possible? For you to see an object clearly, its image must form on your retina. If the image forms in front of the retina or behind the retina, the object will look blurry.

Remember that the location of an image from a convex lens depends on the focal length and the location of the object. For an image to be formed on the retina, the focal length of the lens must change as the object's distance changes.

Picture This

7. Find Highlight the two parts of the eye that bring light rays together. What are these two parts?

Picture This

8. Identify Circle the figure that shows what your retina does to let you clearly see a nearby object.

How does a lens change shape?

Fortunately, the lens is made of material that can change shape. Muscles are attached to the lens of an eye. They can pull the lens into rounder or flatter shapes. This changes the lens' focal length.

If the object is far away, the muscles make the lens flatter. This increases the focal length to keep the image on the retina. If the object moves closer to you, muscles make the lens rounder. This shortens the focal length to keep the image on the retina.

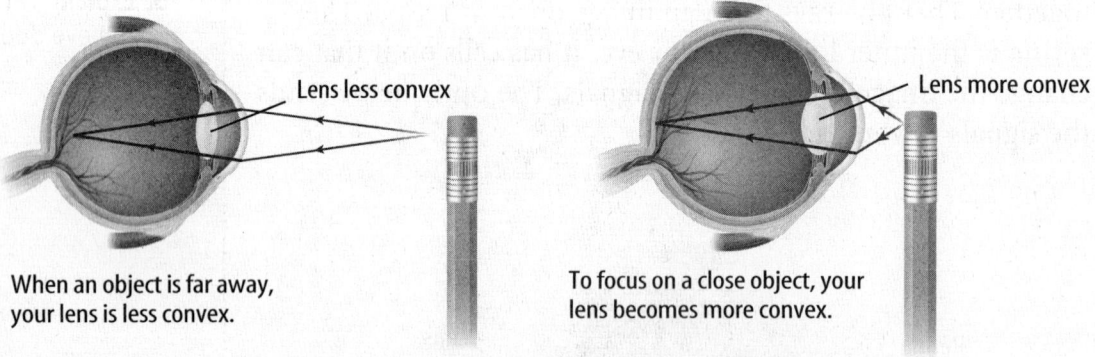

When an object is far away, your lens is less convex.

To focus on a close object, your lens becomes more convex.

Vision Problems

If you have healthy vision, you should be able to see objects clearly when they are 25 cm or farther away from your eyes. However, many people have trouble seeing objects at certain distances. Images can be blurry or in the wrong place. One cause of these problems is the length of the eyeball or the fact that the lenses may not be able to change shape properly. Diseases of the retina can cause vision problems. But, eyeglasses, contact lenses, and surgery often can correct vision problems and help people see more clearly.

What does it mean to be nearsighted?

If you have nearsighted friends, you know that they can see clearly only when objects are nearby. Their eyeballs may be too long which means their lenses may not be able to flatten enough. This means that they cannot make the focal length of their lenses long enough to form an image on the retina. Instead, the image forms in front of the retina. Eyeglasses with concave lenses can help increase the focal length. Nearsighted people see objects that are near better than objects that are far away.

FOLDABLES

C Compare Make three note cards out of quarter sheets of paper to take notes and compare different vision problems.

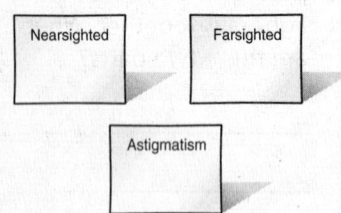

What does it mean to be farsighted?

Some people can see objects clearly that are far away. But, nearby objects look blurry to them. These people are farsighted. Their eyeballs may be too short which means their lenses may not be able to curve enough. This means that they cannot make the focal length of their lenses short enough to form an image on the retina. To correct this problem, they may wear glasses with convex lenses. As you can see in the figure, the convex lenses help shorten the focal length. Farsighted people see objects that are far away better than objects that are near.

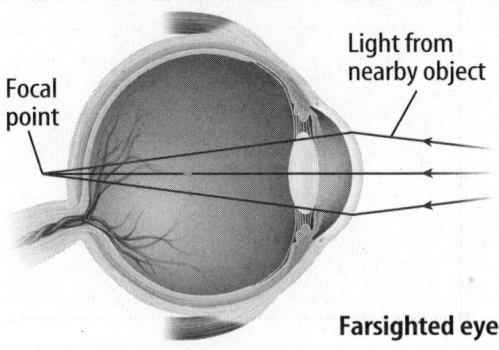

Farsighted eye

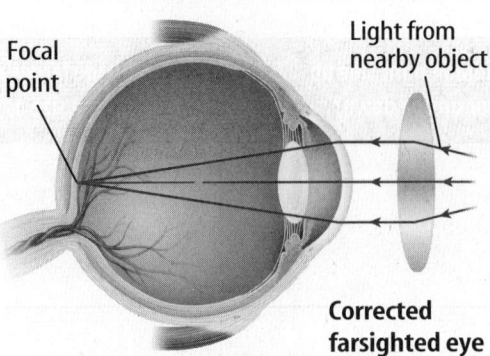

Corrected farsighted eye

Farsightedness is common as people get older. The lenses in the eyes of older people often become less flexible. The muscles cannot make the lens curve enough to form an image on the retina. ☑

What is astigmatism?

Another vision problem, called astigmatism (uh STIHG muh tih zum), occurs when the surface of the cornea is curved unevenly. When people have astigmatism, their corneas are more oval than round in shape. Astigmatism causes blurry vision at all distances. Eyeglasses with unevenly curved lenses can help people with this condition. The lenses cancel out the effects of the uneven cornea.

💡 Think it Over

9. Compare and Contrast What are the differences between farsightedness and nearsightedness?

_____
_____
_____
_____

Picture This

10. Measure Use a metric ruler to measure the distance from the outer edge of the cornea to the focal point in each figure. By how much did this length change?

_____

✓ Reading Check

11. List What are the causes of farsightedness?

_____
_____
_____

Reading Essentials **245**

After You Read
Mini Glossary

concave lens: a lens that is thinner in the middle than at the edges
convex lens: a lens that is thicker in the middle than at the edges
cornea: the transparent covering over the eyeball
retina: the inner lining of the eye

1. Review the terms and their definitions in the Mini Glossary. In the first box, sketch a convex lens. In the second box, sketch a concave lens. Add light rays to show how the lenses affect them.

2. Complete the chart below to organize the information from this section.

Kind of Lens	Position of Object	Is image right side up or upside down?	Is image larger, smaller, or the same size as the object?
Convex			
Convex			
Convex			
Concave	✕		

Science Online Visit **gpscience.com** to access your textbook, interactive games, and projects to help you learn more about mirrors and images.

 Mirrors and Lenses

section ❸ Optical Instruments

⦿ Before You Read

Astronomers study stars that are too far away to be seen by using just their eyes. Chemists study molecules that are too small to be seen by using just their eyes. On the lines below, describe how you think these scientists look at these objects.

_____

_____

What You'll Learn
- about refracting and reflecting telescopes
- why a telescope in space would be useful
- how a microscope uses lenses to magnify small objects
- how a camera creates an image

⦿ Read to Learn

Telescopes

Have you ever had a hard time seeing an object that is far away? When you see an object, only some of the light reflected from its surface enters your eye. The figure shows that as an object gets farther away, less light from the object enters your eye. So, the farther away an object gets, the dimmer and less detailed it looks.

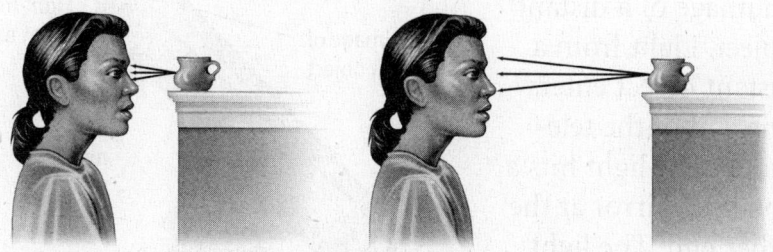

A telescope uses a lens or a combination of lenses and mirrors to gather light from distant objects. A telescope can gather more light than your eye can. This is because the lens or mirror in a telescope is much larger than your eye. In fact, the largest telescopes can gather more than a million times more light than the human eye. So, when you look at a distant galaxy through a telescope, more light enters your eye. The galaxy appears brighter and more detailed.

Study Coach

Summarize Read the section. Close your book and make a list of the section's main ideas. Look at the section again to check your list. Then, write a summary of the section's main ideas.

FOLDABLES

D Compare and Contrast Make a 3-tab Foldable to organize and compare information about optical instruments.

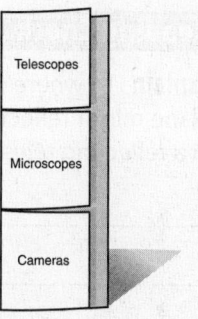

How does a refracting telescope work?

One common type of telescope is the refracting telescope. A simple **refracting telescope** uses two convex lenses to gather light and form an image of a distant object. Light from a distant object, like a planet, passes through the first lens. This is the objective lens. Since the light rays are from very far away, they enter the lens nearly parallel to the optical axis. This first convex lens refracts these rays to form a real image. The second convex lens is the eyepiece. When you look through the eyepiece lens, you see a larger, upside down, virtual image of the real image.

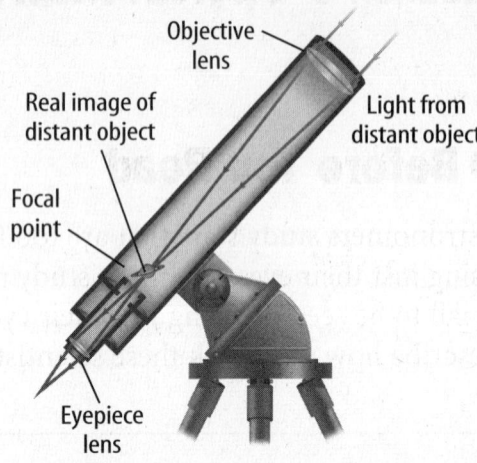

There are disadvantages to refracting telescopes. To see distant objects like planets, the objective lens must be as large as possible. The larger the lens is, the heavier it is. The weight of a large lens can make it sag in the middle. This deforms the image. Also, heavy glass lenses are expensive and difficult to make.

Reflecting telescopes

Most large telescopes today are reflecting telescopes. A **reflecting telescope** uses a concave mirror, a plane mirror, and a convex lens to collect light and form an image of a distant object. Light from a distant object enters one end of the telescope. The light hits a concave mirror at the other end. The light reflects off of this mirror. The rays start to come together. But, before the rays reach the focal point, they hit a plane mirror. The plane mirror reflects the light toward the telescope's eyepiece. A real image of the object forms. Just as in a refracting telescope, a convex lens in the eyepiece then magnifies this image.

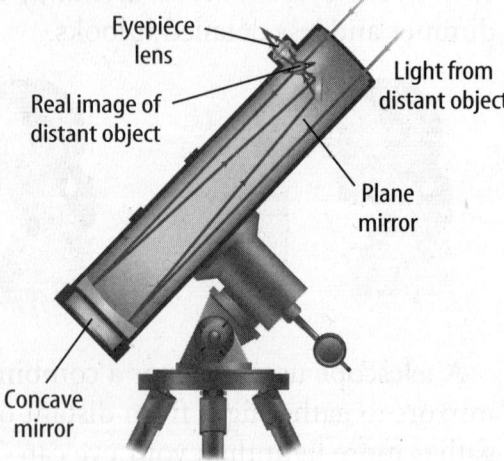

Picture This

1. **Describe** What type of image do you see when using both types of telescopes shown on this page?

   _____

   _____

✔ Reading Check

2. **Explain** To where does the plane mirror reflect the light in a reflecting telescope?

   _____

   _____

Telescopes in Space

The light from objects in space has to pass through the atmosphere to get to a telescope on Earth. However, the atmosphere blurs your view of any object in outer space. To get around this problem, the National Aeronautics and Space Administration (NASA) built the *Hubble Space Telescope*. On April 25, 1990, NASA used the space shuttle *Discovery* to put this telescope in orbit above Earth's atmosphere. Images from the *Hubble Space Telescope* are much sharper and more detailed than any telescope on Earth can produce.

Microscopes

A telescope works well for looking at distant objects. But, if you need to look at tiny objects, like the cells in a butterfly wing, you will need a microscope. A **microscope** uses two convex lenses with short focal lengths to magnify objects. These lenses are the objective lens and the eyepiece lens.

The figure shows a simple microscope. You set the object you want to look at on a transparent slide under the objective lens. The microscope shines light from under the object. The light passes by or through the object. Then it travels through the objective lens. The objective lens is a convex lens. It forms a larger, real image of the object because the distance from the object to the lens is between one and two focal lengths. The eyepiece lens is also a convex lens. The eyepiece lens creates a larger, virtual image of the real image. This final image can be hundreds of times larger than the original object.

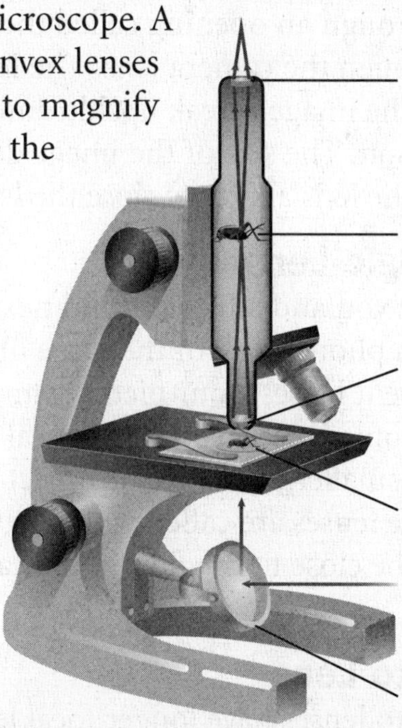

Cameras

How does a camera make an image of a person on a small piece of film? A camera works by gathering and bending light with a lens. Then the lens forms an image on film. The film responds to light and permanently records the image.

Reading Check

3. Explain Why is the *Hubble Space Telescope* able to produce clearer images than telescopes on Earth?

Picture This

4. Predict How does light reach the objective lens in this microscope?

Picture This

5. Circle the part of the figure that has the same function as the retina in an eye.

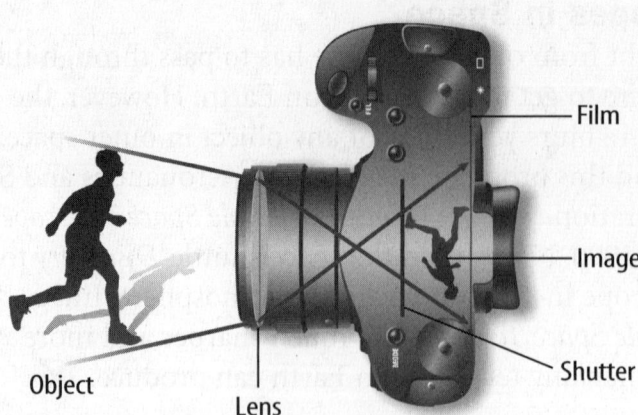

When you take a picture with a camera, a shutter opens and closes. The shutter stays open long enough to let light enter the camera. The light reflecting off the person enters the camera through an opening called the aperture. The light passes through the camera lens. The lens forms the image on the film. The image is real, upside down, and smaller than the actual person. The size of the image depends on the focal length of the lens and how close the lens is to the film.

Wide-Angle Lenses

Suppose you and a friend stand next to each other. You each take a photograph of the same object. If your cameras have different lenses, your pictures might look different. For example, some lenses have short focal lengths. These lenses produce a smaller image of the object but have a wide field of view. These lenses are called wide-angle lenses. A wide-angle lens must be close to the film in the camera to form a clear image.

Telephoto Lens

Telephoto lenses have longer focal lengths. They produce larger images. But, they have a narrower field of view than wide-angle lenses do. A camera with a telephoto lens is easy to recognize. The telephoto lens usually sticks out from the front of the camera. This increases the distance between the lens and the film in the camera.

Reading Check

6. Describe How does a camera put an image on film?

_____

_____

_____

After You Read
Mini Glossary

microscope: an instrument that uses two convex lenses with fairly short focal lengths to magnify extremely small objects

reflecting telescope: a telescope that uses a concave mirror, a plane mirror, and a convex lens to collect light and form an image of a distant object

refracting telescope: a telescope that uses two convex lenses to gather light and form an image of a distant object

1. Review the terms and their definitions in the Mini Glossary. Choose one term. In the space below, draw a picture to represent the term. Label each part of the drawing.

2. Place a check mark in each box to show what is found in each item.

	Plane Mirror	Convex Lens	Concave Mirror
Refracting Telescope			
Reflecting Telescope			
Camera			

3. Think about what you have learned. How did summarizing your reading help you understand optical instruments?

 Visit **gpscience.com** to access your textbook, interactive games, and projects to help you learn more about optical instruments.

Reading Essentials 251

chapter 15 Classification of Matter

section ❶ Composition of Matter

What You'll Learn
- what are substances and mixtures
- how to identify elements and compounds
- the difference between solutions, colloids, and suspensions

Before You Read

Matter is all around you. You breathe matter, sit on it, and drink it every day. What words would you use to describe different kinds of matter?

_____

_____

Mark the Text

Underlining Look for different descriptions of matter as you read each paragraph. Underline these descriptions. Read the underlined descriptions again after you've finished reading the section.

Read to Learn

Pure Substances

Have you ever seen a print that looked like a real painting? Did you have to touch it to find out? The smooth or rough surface told you whether it was a painting or a print. Each material has its own properties. The properties of materials can be used to classify them into categories.

Materials are made of a pure substance or of a mixture of substances. A **substance** is a type of matter that is always made of the same material or materials. A substance can be either an element or a compound. Some substances you might recognize are helium, aluminum, water, and salt.

What are elements?

All substances are made of atoms. A substance is an **element** if all the atoms in the substance are the same. The graphite in your pencil is an element. The copper coating on most pennies is an element, too. In graphite, all atoms are carbon atoms. In copper, all atoms are copper atoms. The metal under the copper coating of a penny is another element, zinc. There are about 90 elements found on Earth. More than 20 other elements have been made in laboratories. Most of the 20 human-made elements are unstable. They exist for only a short time in laboratories. You may recognize the elements that are shown in the figure on the next page.

Atoms in Elements

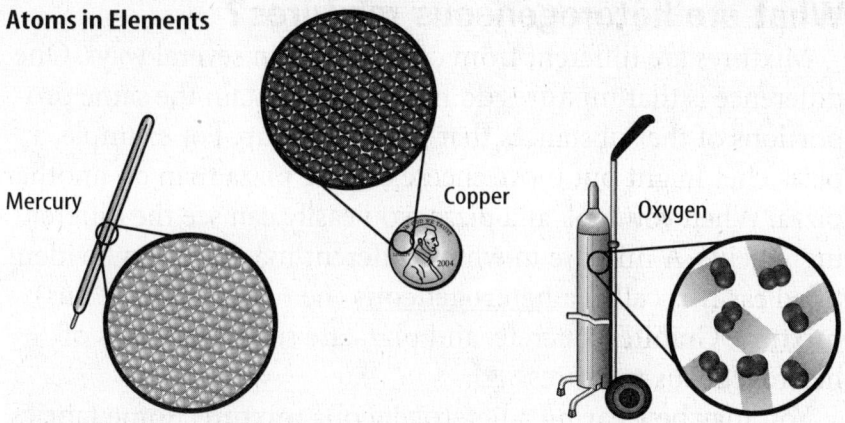

What are compounds?

A <u>compound</u> is a substance with two or more elements that are combined in a fixed proportion. A common compound is water. Water is made up of the elements hydrogen and oxygen. Each water molecule contains two hydrogen atoms and one oxygen atom. A molecule is the smallest particle of a compound that has all the properties of the compound. Chalk is another compound. It contains calcium, carbon, and oxygen. Each chalk molecule contains one calcium atom, one carbon atom, and three oxygen atoms.

Can you imagine putting a silvery metal and a greenish-yellow poisonous gas on your food? Table salt is a compound that fits this description. Another name for table salt is sodium chloride. It is made up of sodium, a silvery metal, and chlorine, a greenish-yellow poisonous gas. Many compounds look different from the elements in them. The table lists some common compounds and the elements in them.

Some Common Compounds	
Compound	**Elements in the Compound**
Water (clear liquid)	Hydrogen (clear gas) Oxygen (clear gas)
Table salt (white solid)	Sodium (silvery metal) Chlorine (green-yellow gas)
Carbon dioxide (clear gas)	Carbon (black solid) Oxygen (clear gas)
Table sugar (white solid)	Carbon (black solid) Hydrogen (clear gas) Oxygen (clear gas)

Mixtures

Is pizza one of your favorite foods? Do you like soft drinks? If so, you like two foods that are mixtures. A mixture is a material made up of two or more substances that can easily be separated.

Picture This

1. **Determine** What are the elements in each object shown in the figure?

   _____
   _____
   _____

Think it Over

2. **Infer** How do you know that the water molecules from a faucet and the water molecules from a river are the same?

   _____
   _____
   _____

Picture This

3. **Infer** Which of the following combine to make water? Circle the correct answer.

   a. two liquids

   b. two gases

   c. a liquid and a gas

Reading Check

4. Explain How can you tell that a pizza is a heterogeneous mixture?

_____
_____
_____

FOLDABLES

A Classify Make the following Foldable to help you classify heterogeneous and homogeneous mixtures, colloids, and suspensions.

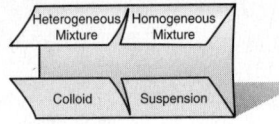

Reading Check

5. Explain Why is soda in an unopened bottle a homogeneous mixture?

_____
_____
_____

What are heterogeneous mixtures?

Mixtures are different from compounds in several ways. One difference is that mixtures do not always contain the same proportions of the substances that make them up. For example, a pizza chef might put more cheese on one pizza than on another pizza. When you look at a pizza, you easily can see the different ingredients. A mixture in which different materials can be identified easily is called a **heterogeneous** (he tuh ruh JEE nee us) **mixture**. Granite, concrete, and pizza are some examples of heterogeneous mixtures.

You may be wearing a heterogeneous mixture. Some fabrics are labeled as permanent-press. These fabrics resist wrinkles. Permanent-press fabric contains fibers of two materials. The materials are polyester and cotton. The amounts of polyester and cotton can change from one piece of fabric to another. Look at the labels on some of your clothes. See if some contain different amounts of polyester and cotton.

You probably cannot tell that permanent-press fabric is a heterogeneous mixture by looking at it. It looks like it is made up of only one material. However, you might be able to see the mixture with a microscope. Under a microscope, the polyester fibers probably would look different from the cotton fibers.

Many of the substances around you are heterogeneous mixtures. Some have materials that are easy to see, such as those in pizza. Others have materials that are not easy to tell apart, such as the fibers in permanent-press fabrics. In fact, some of the parts of heterogeneous mixtures can be mixtures themselves. For example, the cheese in pizza is a mixture, but you cannot see the materials. Cheese contains many compounds, such as milk, proteins, butterfat, and food coloring.

What are homogeneous mixtures?

Remember the soft drink you had with your pizza? Soft drinks are mixtures. They contain water, sugar, flavorings, coloring, and carbon dioxide gas.

Soft drinks in sealed bottles are homogeneous mixtures. A **homogeneous** (hoh muh JEE nee us) **mixture** contains two or more substances blended evenly throughout. You cannot see the different substances in a homogeneous mixture. However, when a soft drink is poured into a glass, the carbon dioxide gas forms bubbles. You then can see that the gas is separate from the other ingredients. When this happens, the soda becomes a heterogeneous mixture.

254 CHAPTER 15 Classification of Matter

What is a solution?

Vinegar is another homogeneous mixture. It looks clear, but it contains particles of acetic acid mixed with water. Homogenous mixtures, such as soft drinks and vinegar, also are known as solutions. A **solution** is a homogeneous mixture of particles so small that they cannot be seen with a microscope and will never settle to the bottom of their container. Solutions stay constantly and evenly mixed. Look at the diagram. It summarizes the differences between substances and mixtures.

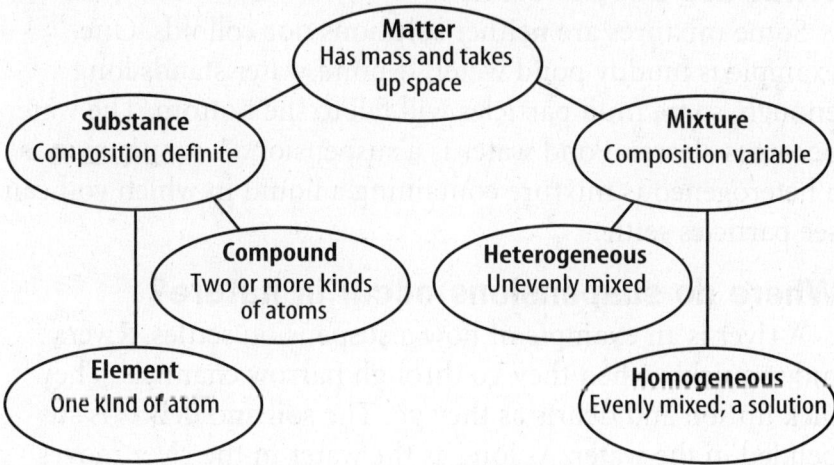

Picture This

6. **Identify** In the diagram, highlight the words that explain the difference between a substance and a mixture.

What are colloids?

Milk is an example of a special mixture called a colloid. A **colloid** (KAH loyd) is a type of mixture with particles that are larger than those in solutions, but not heavy enough to settle to the bottom of their container. Remember how a pizza was still a pizza even if the proportion of its ingredients were changed? Milk is similar to a pizza in that way. Milk contains water, fat, and proteins, but like any mixture, these substances can be in different proportions. What makes milk a colloid is that these ingredients form large particles, but they are not heavy enough to settle.

Paint is an example of a liquid colloid. Gases and solids can also be colloids. For example, fog and smoke are colloids. Fog is made up of liquid water particles suspended in air. Smoke contains solids suspended in air.

Do colloids and solutions look the same?

One way to tell the difference between a colloid and a solution is by how each looks. Fog looks white because its particles are large enough to scatter light. Sometimes it is not easy to tell that a liquid is a colloid. For example, some shampoos and gelatins are colloids called gels that look almost clear.

Think it Over

7. **Give an Example** Name a colloid that you can see through.

Reading Essentials   **255**

How do you identify colloids?

You can tell if a liquid is a colloid by shining a beam of light through it. You cannot see a light beam as it passes through a solution. But you easily can see a light beam in a colloid because its large particles scatter light. Small particles in solutions do not scatter light. Have you ever noticed how fog scatters the light from a car's headlights? The scattering of light by particles in a colloid is known as the **Tyndall effect**.

What are suspensions?

Some mixtures are neither solutions nor colloids. One example is muddy pond water. If pond water stands long enough, some mud particles will fall to the bottom. The water becomes clearer. Pond water is a suspension. A **suspension** is a heterogeneous mixture containing a liquid in which you can see particles settle.

Where do suspensions occur in nature?

A river is an example of how a suspension settles. Rivers move quickly when they go through narrow channels. They pick up soil and debris as they go. The soil and debris is suspended in the water. As long as the water in the river moves fast enough, the suspended soil does not settle. When the river slows, the particles fall out of the suspension and settle on the bottom of the river. This also happens when rivers flow into large bodies of water such as lakes or oceans. After many years, the soil builds up and forms a delta made up of mud and debris. Dams also can cause soil suspended in river water to settle. Look at the table to compare the properties of different types of mixtures.

Think it Over

8. Infer Why would a dam cause suspended particles to settle out of river water?

_____
_____
_____
_____

Applying Math

9. Comparison Which type of mixture has the largest particles? Explain how you know.

_____
_____
_____
_____

Comparing Solutions, Colloids, and Suspensions			
Description	Solutions	Colloids	Suspensions
Settle upon standing?	no	no	yes
Separate using filter paper?	no	no	yes
Particle size	0.1–1 nm	1–100 nm	>100 nm
Scatter light?	no	yes	yes

After You Read
Mini Glossary

colloid: a type of mixture with particles that are larger than those in solutions, but not heavy enough to settle to the bottom of their container

compound: a substance with two or more elements that are combined in the same proportion

element: a substance in which all the atoms are the same

heterogeneous mixture: a mixture in which different materials can be identified easily

homogeneous mixture: a mixture that contains two or more substances blended evenly throughout

solution: a homogeneous mixture of particles so small that they cannot be seen with a microscope and that will never settle to the bottom of their container

substance: a type of matter that is always made of the same material or materials

suspension: a heterogeneous mixture containing a liquid in which you can see particles settle

Tyndall effect: the scattering of light by particles in a colloid

1. Review the terms and their definitions in the Mini Glossary. The oxygen that you breathe is made up of tiny particles that are actually two atoms of oxygen bonded together. Would you say that oxygen is an element or a compound? Explain.

   _____
   _____

2. Fill in the blanks with an example of each type of matter.

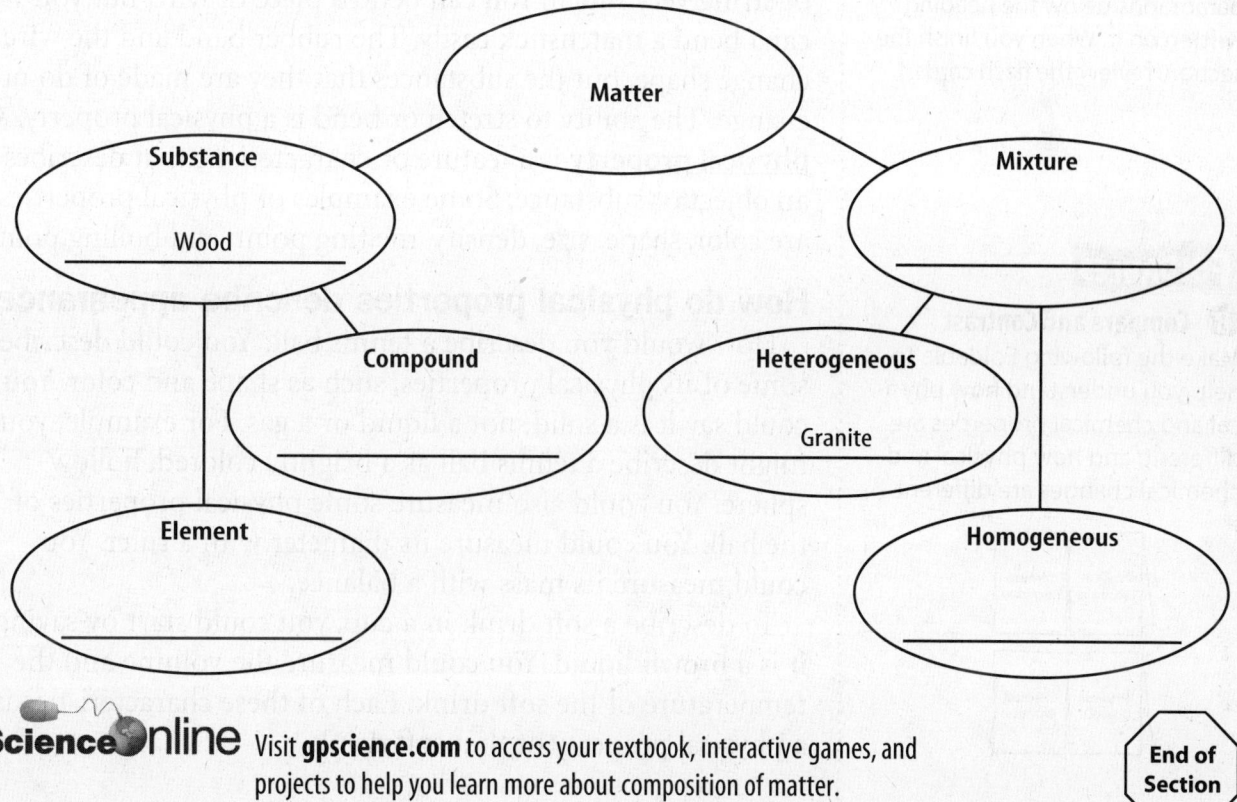

Science Online Visit gpscience.com to access your textbook, interactive games, and projects to help you learn more about composition of matter.

End of Section

Reading Essentials **257**

Classification of Matter

section ❷ Properties of Matter

What You'll Learn
- to identify substances using physical properties
- differences between physical and chemical changes
- how to identify chemical changes
- the law of conservation of mass

Study Coach

Make Flash Cards For each bold heading in this section, make a flash card. The flash card should contain the main point of the paragraphs below the heading written on it. When you finish the section, review the flash cards.

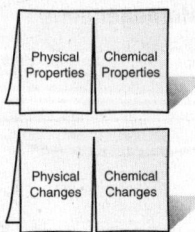

Ⓑ Compare and Contrast Make the following Foldable to help you understand how physical and chemical properties are different, and how physical and chemical changes are different.

● Before You Read

When you see someone, how do you identify that person as a friend or a stranger? How do you identify a friend on the phone? What are some things about people that help you recognize them? On the lines below, list some things you use to identify people.

_____

_____

● Read to Learn

Physical Properties

You can stretch a rubber band, but you can't stretch a piece of string very much. You can bend a piece of wire, but you can't bend a matchstick easily. The rubber band and the wire change shape, but the substances that they are made of do not change. The ability to stretch or bend is a physical property. A **physical property** is a feature or characteristic that describes an object or substance. Some examples of physical properties are color, shape, size, density, melting point, and boiling point.

How do physical properties describe appearance?

How would you describe a tennis ball? You could describe some of its physical properties, such as shape and color. You could say it is a solid, not a liquid or a gas. For example, you might describe a tennis ball as a brightly colored, hollow sphere. You could also measure some physical properties of the ball. You could measure its diameter with a ruler. You could measure its mass with a balance.

To describe a soft drink in a cup, you could start by saying it is a brown liquid. You could measure the volume and the temperature of the soft drink. Each of these characteristics is a physical property of that soft drink.

258 CHAPTER 15 Classification of Matter

How do physical properties describe behavior?

Some physical properties describe the behavior of a material or substance. You know that a magnet attracts objects that contain iron, such as a safety pin. Attraction to a magnet is a physical property of iron. Every substance has physical properties that make it useful for certain tasks.

Some metals, such as copper, are useful because they bend easily and can be drawn out into wires. Other metals, such as gold, are useful because they easily can be pounded into sheets as thin as 0.1 micrometers. That's about four-millionths of an inch. This property of gold makes it useful for decorating objects. Gold that has been flattened this way is called gold leaf.

Think again about the soft drink. If you knock over the cup, the drink will spill. If you knock over a jar of honey, however, it does not flow as quickly. The ability to flow is a physical property of liquids.

How are physical properties used to separate materials?

You can use the physical properties of size and hardness to separate some substances. Removing the seeds from a watermelon is easy. The seeds are small and hard, and the flesh of the large watermelon is soft and juicy. You may have used a screen or sifter to remove pebbles from soil. The small pieces of soil quickly fall through the sifter, leaving the larger pebbles behind.

Look at the figure. The dish contains sand mixed with iron filings. You probably would not be able to sift out the iron filings. They are about the same size as the sand particles. What you could do is pass a magnet through the mixture. The magnet attracts only the iron filings. A magnet does not attract sand. This is an example using the physical property of magnetism to separate substances in a mixture. Recycling companies use magnetism to separate scrap iron from other metals.

Using Magnetism to Separate Materials

Reading Check

1. **Explain** Why does honey flow more slowly than water?

Think it Over

2. **Apply** Sifting probably works best for separating what type of materials?

Picture This

3. **Explain** Why would it be difficult to sift the iron filings from the sand?

Physical Change

When you break a stick of chewing gum, you change its size and shape. You do not change the identity of the materials that make up the gum.

Why does the identity of a substance stay the same?

When a substance freezes, boils, evaporates, or condenses, it undergoes a physical change. A **physical change** is a change in size, shape, or state of matter. Heat might be added or removed during a physical change. Changes in energy do not change the identity of the substance being heated or cooled. All substances have distinct properties that are constant, or never change. The properties of density, specific heat, boiling point, and melting point are constant for substances. These properties can be used to identify unknown substances in a mixture.

Iron is a substance that changes states when it absorbs or releases energy. At high temperatures, iron melts. However, iron has the same physical properties that identify it as iron, whether it is in the liquid or solid state.

What is distillation?

Distillation is the process of separating substances in a mixture by evaporating a liquid and condensing its vapor. A laboratory distillation process is shown below. The liquid is heated until it vaporizes. Then, the vapors are cooled until they condense. A solid material is left behind. Distillation is used to make drinking water out of salt water.

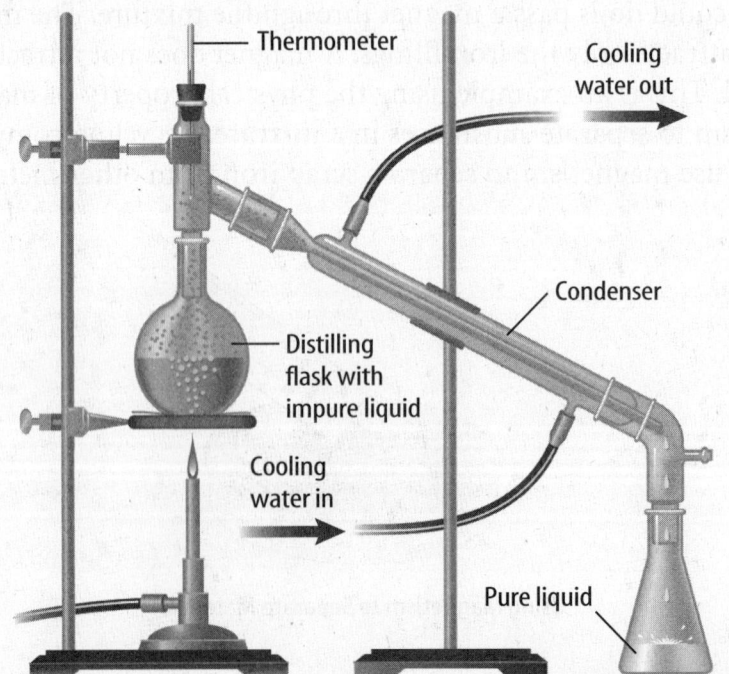

Reading Check

4. Explain What affect does an energy change have on the identity of a substance?

_____

_____

_____

Picture This

5. Observe Where is the solid material left behind in the distillation process?

_____

_____

Liquids with different boiling points also can be distilled. The mixture is heated slowly until it begins to boil. Vapors of the liquid with the lowest boiling point form first. They are condensed and collected. As temperature increases the second liquid boils. Its vapors are condensed and collected. Distillation is used often in industry. For instance, natural oils such as mint are distilled.

Chemical Properties and Changes

A <u>chemical property</u> is a characteristic of a substance that indicates whether it can undergo a change that results in a new substance. A warning on a can of liquid paint thinner or lighter fluid states that the liquid is flammable (FLA muh buhl). If a substance is flammable, it can burn. When a substance burns, it produces new substances during a chemical change. Therefore, whether or not a substance is flammable is a chemical property. Knowing which substances are flammable helps you to use them safely.

Detecting Chemical Change

Your senses tell you when a chemical change has happened. Leave a pot of chili cooking on the stove too long and you will smell it burning. The smell tells you a new substance formed.

How does the identity of a substance change?

You smell a rotten egg or see rust on a bike. These are signs that a chemical change has taken place. A <u>chemical change</u> is a change of one substance to another. When fireworks explode, they produce heat, light, and sound. Heat, light, and sound are often signs of a rapid energy release and of a chemical change.

The only proof that a chemical change has taken place is that a new substance is formed. For example, when hydrogen and oxygen combine in a rocket engine, they use chemical changes to produce heat, light, and sound. But there are no such clues when iron combines with oxygen to form rust. Rust forms slowly. The only clue that iron has changed to a new substance is the presence of rust. Burning and rusting are chemical changes because new substances are formed.

How can a chemical change be used to separate substances?

You can separate substances using a chemical change. One example is cleaning silver. Silver chemically reacts with sulfur compounds in the air to form silver sulfide, or tarnish. A different chemical reaction changes the tarnish back to silver using warm water, baking soda, and aluminum foil.

Think it Over

6. Infer How can knowing a chemical property such as flammability help you to use a product safely?

_____
_____
_____

Reading Check

7. Explain What is the only proof that a chemical change has taken place?

_____
_____

Weathering—Chemical or Physical Change?

The forces of nature continuously shape Earth's surface. Rocks split, rivers carve deep canyons, sand dunes shift, and interesting formations develop in caves. These changes are known as weathering. Do you think weathering changes are physical or chemical? Weathering changes are physical and chemical changes.

What is physical weathering?

As a stream cuts through rock to form a canyon, small particles of rock are carried downstream. The large rocks and the small particles of rock have the same characteristics. Because the particles are not changed, this type of weathering is physical.

What is chemical weathering?

Limestone is made up mostly of a chemical called calcium carbonate. Calcium carbonate does not dissolve easily in water. But if the water is even slightly acidic, calcium carbonate reacts. A new substance, calcium hydrogen carbonate, is formed. This substance dissolves in water. This change in limestone is a chemical change. The calcium carbonate changes to calcium hydrogen carbonate in the chemical reaction. Rainwater can dissolve limestone because of this reaction. This chemical change leads to weathering. Chemical changes like this one create caves and the rock formations that are found in them.

The Conservation of Mass

A chemical property of wood is that it burns. After you burn a log, all that is left is a small pile of ash. Because you saw smoke and light, felt heat, and saw a change in appearance as the log was burning, you know a chemical change took place.

If you look closely at the ashes, you will see that the ashes have less mass than the log. Did the rest of the matter that made up the log disappear? No, it did not. Suppose you could burn a log in a sealed container. The container would have to hold enough oxygen for the fire and contain all the smoke the fire produced. If you measured the mass in the container before and after the fire, you would find that both are the same.

No mass is lost during a fire because no mass is gained or lost in a chemical change. According to the **law of conservation of mass**, the mass of all substances that are present before a chemical change equals the mass of all the substances that remain after the change.

Think it Over

8. Apply How are shifting sand dunes an example of physical weathering?

Applying Math

9. Calculate If a 2-kg log is burned, what is the mass of the ash, smoke, and carbon dioxide produced by the chemical change?

After You Read
Mini Glossary

chemical change: a change of one substance to another
chemical property: a characteristic of a substance that indicates whether it can undergo a certain chemical change
distillation: the process of separating substances in a mixture by evaporating a liquid and condensing its vapor

law of conservation of mass: the mass of all substances that are present before a chemical change equals the mass of all the substances that remain after the change
physical change: a change in size, shape, or state of matter
physical property: a feature or characteristic that describes an object or substance

1. Review the terms and their definitions in the Mini Glossary. What is the main difference between a physical change and a chemical change?

   _____
   _____

2. Complete the table below by giving an example of the property or change.

Physical property	Example:
Chemical property	Example:
Physical change	Example:
Chemical change	Example:
Separation using physical change	Example:
Separation using chemical change	Example:

3. Imagine explaining physical and chemical changes to a group of elementary school students. Describe some items around your house to use as examples of physical and chemical changes.

   _____
   _____

 Visit **gpscience.com** to access your textbook, interactive games, and projects to help you learn more about the properties of matter.

Reading Essentials 263

chapter 16 Solids, Liquids, and Gases

section ❶ Kinetic Theory

What You'll Learn
- the kinetic theory of matter
- how particles move in the four states of matter
- how particles behave at the melting and boiling points

● Before You Read

Everywhere you go, you are surrounded by the three states of matter—solids, liquids, and gases. Look around you and find a substance in each of the three states of matter. List these substances on the lines below.

_____

_____

Mark the Text

Identify Concepts Highlight each heading in this section that asks a question. Then use a different color to highlight the answers to the questions.

● Read to Learn

States of Matter

An everyday activity such as drinking tea may include solids, liquids, and gases. Think about heating water on the stove for a cup of hot tea. The boiling water is in the liquid state. The steam you can see above the kettle is also in the liquid state. You might drop an ice cube into the tea to cool it. Ice is in the solid state. How are these states alike and different?

What is the kinetic theory of matter?

The **kinetic theory** is an explanation of how particles in matter behave. To explain the behavior of particles, you have to make some basic assumptions. An assumption is an idea that most people accept as true. The three assumptions of the kinetic theory are:

1. All matter is composed of small particles—atoms, molecules, and ions.

2. These particles are in constant motion. The motion has no pattern.

3. These particles are colliding or crashing into each other and into the walls of their container.

Particles lose some energy when they bump into other particles. The amount of energy the particles lose is very small and can be ignored.

FOLDABLES

A Build Vocabulary Make the following Foldable to help you learn the vocabulary terms in this section.

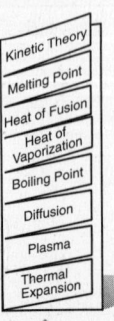

264 **CHAPTER 16** Solids, Liquids, and Gases

A model that you picture in your mind can help you understand the kinetic theory and how particles in matter move. Think of each particle as a tiny ball. These balls constantly are bouncing and bumping into one another.

What is thermal energy?

One assumption of the kinetic theory is that particles are in constant motion. Think about an ice cube. Does an ice cube seem to move? How can a frozen, solid ice cube have motion? Remember that the kinetic theory is describing particles. Atoms in solids are held in place by the attraction between the particles. This attraction gives solids a definite shape and volume. However, the thermal energy in the particles causes them to vibrate, even when the object stays in one place.

Thermal energy is the total energy of the particles in a material. Thermal energy includes kinetic energy and potential energy. Even though the ice cube does not appear to be moving, its particles are moving because of thermal energy. When the temperature of a substance is lowered, its particles will have less thermal energy and will vibrate more slowly.

What is average kinetic energy?

The word *temperature* is used to explain how hot or cold something is. In science, temperature means the average of how fast the particles in a substance are moving. Temperature is the average kinetic energy of particles in a substance. On average, molecules of frozen water at 0°C will move slower than molecules at 100°C. Therefore, water molecules at 0°C have lower average kinetic energy than the molecules at 100°C. Molecules have kinetic energy at all temperatures, even at absolute zero. Absolute zero is about −273°C. Scientists theorize that at absolute zero, particles move so slowly that no more thermal energy can be removed from a substance.

How are particles in a solid arranged?

The figure shows how the particles of a solid are packed together. The particles vibrate constantly. The particles of most materials always will be in the same arrangement in the solid. The type of ordered arrangement formed by a solid is important. It gives the solid its chemical and physical properties.

Solid

Reading Check

1. **Determine** What two kinds of energy are part of thermal energy?

Picture This

2. **Use a Model** Describe materials you might use to make a model of the particles in a solid.

How do solids become liquids?

Think about the ice cube you dropped into the hot tea. The moving particles in the tea bump into the vibrating particles in the ice. These collisions transfer energy from the tea to the ice cube. The particles on the surface of the ice cube vibrate faster. These particles bump into and transfer energy to other ice particles. Soon the particles have enough kinetic energy to overcome the attractive forces. They slip out of their ordered arrangement and the ice melts.

The **melting point** is the temperature at which a solid begins to turn into a liquid. Energy is needed for the particles in a solid to slip out of their ordered arrangement. The **heat of fusion** is the amount of energy needed to change a substance from a solid to a liquid at its melting point.

Reading Check

3. **Explain** What is the temperature at which a solid begins to turn into a liquid called?

_____

Why do liquids flow?

Particles in a liquid have more kinetic energy than particles in a solid. This extra kinetic energy lets particles overcome some of their attraction to other particles. The particles can slide past each other. This is why liquids flow and take the shape of their container. But the particles in a liquid have not overcome all of the attractive forces among them. The particles still cling together. Because of the attractive forces, liquids have a definite volume. A certain amount of liquid will always take up the same amount of space regardless of the shape of the container.

What is a gas state?

Gas particles have enough kinetic energy to overcome the attractions among them. Gases do not have a fixed volume or shape. They can spread far apart or move close together to fill a container. How does a liquid become a gas? The particles in a liquid are moving constantly. Some particles are moving faster than other particles. They have more kinetic energy. Particles that move fast enough can escape the attractive forces of other particles. They enter the gas state. This process is called vaporization. Vaporization can occur in two ways—evaporation and boiling.

Think it Over

4. **Infer** Which particles have the least kinetic energy? Circle the correct answer.

   a. solid particles

   b. liquid particles

   c. gas particles

How do liquids evaporate?

Vaporization occurs when liquids evaporate. Evaporation is vaporization that occurs at the surface of a liquid. It can occur at temperatures below the liquid's boiling point. To evaporate, particles must have enough kinetic energy to escape the attractive forces of the liquid. The particles must be at the surface of the liquid. They also must be traveling away from the liquid.

How does boiling vaporize liquids?

Another way that a liquid vaporizes is by boiling. Boiling occurs at a specific temperature. The temperature depends on the pressure on the surface of the liquid. Air exerts pressure on the surface of a liquid. This external pressure keeps particles from escaping from the liquid. The **boiling point** of a liquid is the temperature at which the pressure of the vapor in the liquid is equal to the external pressure on the surface of the liquid. Particles need energy to overcome the force of pressure. **Heat of vaporization** is the amount of energy needed for the liquid at its boiling point to become a gas.

Why do gases fill their containers?

Gas particles move so quickly and are so far apart that they overcome the attractive forces among them. Therefore, gases do not have a definite shape or a definite volume. The movement of particles and the collisions among them cause gases to diffuse. **Diffusion** is the spreading of particles throughout a given volume until they are evenly distributed. Diffusion occurs in solids, liquids, and gases, but it happens most quickly in gases. Imagine that you spray air freshener in one corner of a room. It is not long before you smell the scent all over the room. Gases will fill the container that they are in even if the container is rather large. The particles continue to move and collide in a random motion within their container.

What is the heating curve of a liquid?

The heating curve graph shows water being heated from −20°C to 100°C. It shows the temperature change as heat is added. In areas **a** and **c** the graph slopes upward because the water's kinetic energy is increasing. In areas **b** and **d** the graph is a horizontal line. The temperature of the water does not change.

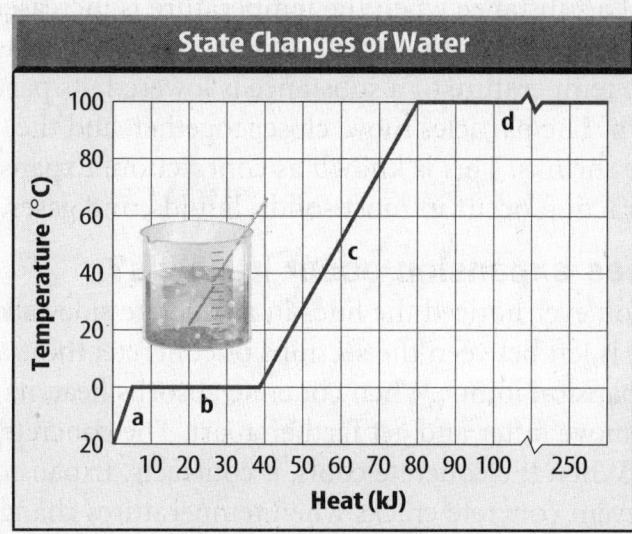

Reading Check

5. **Apply** What do particles need in order to overcome the force of pressure and become a gas?

Think it Over

6. **Sequence** List the following substances in the order in which their particles diffuse, slowest to fastest: liquid water, water vapor, ice.

a. _____

b. _____

c. _____

Applying Math

7. **Interpret a Graph** At what temperature is the water turning into gas?

FOLDABLES

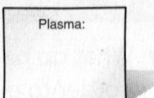

B **Find Information** Make the following Foldable from a quarter sheet of paper to help you take notes about plasma.

Melting Point At 0°C, ice is melting. All of the energy put into the ice at this temperature is used to overcome the attractive forces among the particles in the solid. The temperature stays the same during melting. After the attractive forces are overcome, the particles move more freely. Their temperature increases.

Boiling Point At 100°C, water is boiling or vaporizing. The temperature stays the same again. All of the energy that is put into the water goes to overcoming the remaining attractive forces among the water particles. When all the attractive forces in the water have been overcome, the energy returns to increasing the temperature of the particles.

What is the plasma state?

Solids, liquids, and gases are three familiar states of matter. However, scientists estimate that much of the matter in the universe is plasma. <u>Plasma</u> is matter made up of positively and negatively charged particles. The overall charge of plasma is neutral. Plasma is neutral because it contains equal numbers of positive and negative particles.

The faster the particles move, the greater the force when they collide. The forces produced from high-energy collisions are so large that electrons from the atoms are ripped off. This state of matter is called plasma. Plasma is found in stars, lightning bolts, neon and fluorescent tubes, and auroras.

Thermal Expansion

The kinetic theory explains other characteristics of matter. Recall that particles move faster and separate as temperature rises. The separation of the particles in a substance increases the size of the substance. <u>Thermal expansion</u> is an increase in the size of a substance when the temperature is increased.

The kinetic theory also explains contraction in objects. When the temperature of a substance is lowered, its particles slow down. The particles move closer together, and the substance shrinks. This is known as contraction. Expansion and contraction occur in most solids, liquids, and gases.

How does expansion occur in solids?

Have you ever noticed the lines in a concrete sidewalk? A gap often is left between the sections of concrete; these are called expansion joints. When concrete absorbs heat, its particles move faster and get farther apart. The concrete expands. When the concrete cools, it contracts. Expansion joints prevent concrete cracks when temperatures change.

Think it Over

8. **Explain** How do expansion joints in a sidewalk keep the concrete from cracking?

_____

_____

_____

_____

268 CHAPTER 16 Solids, Liquids, and Gases

How does expansion occur in liquids?

An example of expansion in liquids occurs in a thermometer. One kind of thermometer, as shown below, is a narrow glass tube with a liquid in it. The addition of energy makes the particles of the liquid in the thermometer move faster. The particles in the liquid move farther apart. Even if the liquid expands only a little, it will show a large change on the thermometer's temperature scale.

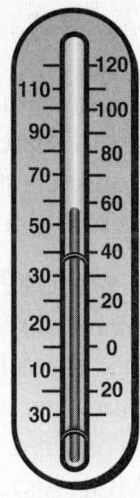

Picture This
9. **Infer** What will happen to the liquid in the thermometer as the temperature cools?

_____

_____

How does expansion occur in gases?

A hot-air balloon is an example of thermal expansion in gases. Hot-air balloons are able to rise because of the thermal expansion of air. When the air in the balloon is heated, the distance between the particles in the air increases. This makes the balloon expand. As the balloon expands, the number of particles per cubic centimeter inside the balloon decreases. The density of the hot air inside the balloon decreases also. The density of the air in the hot-air balloon is lower than the density of the cooler air outside the balloon. This makes the balloon rise. ☑

✓ Reading Check
10. **Infer** What happens to the density of the air inside a hot-air balloon as the balloon expands?

_____

Why does water behave in a different way?

Why does ice float in water? Most substances contract as the temperature is lowered because the particles move closer together. However, water expands when it freezes. Water molecules are unusual. They have some areas that have a highly positive charge and other areas that have a highly negative charge. The diagram shows these regions. The charged regions affect the behavior of water.

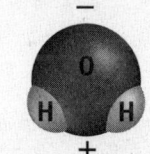

Partial negative charge

Partial positive charge

Picture This
11. **Identify** Draw lines to and label the oxygen atom and the two hydrogen atoms.

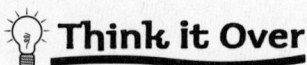

12. Infer What will happen if you completely fill an ice tray with water?

_____

_____

_____

Unlike charges attract each other. As the temperature of water drops, the particles move closer together. The unlike charges attract each other. They line up so that only positive and negative areas are near each other. Because the water molecules line up according to charge, there are empty spaces in the structure. These empty spaces are larger in ice than in liquid water. So water expands when it goes from a liquid to a solid state. Solid ice is less dense than liquid water.

Solid or a Liquid?

Some other substances also have unusual behavior when they change states. Two kinds of materials that react unusually are amorphous solids and liquid crystals. It can be difficult to tell if these materials are in a solid state or a liquid state.

What are amorphous solids?

Ice melts at 0°C. Not all materials have a definite temperature when they change from solid to liquid. Some solids just get softer and slowly turn to liquid over a range of temperatures. These solids do not have the ordered structure of crystals. They are called amorphous (uh MOR fus) solids. Amorphous comes from the Greek word for "without form."

You are familiar with two amorphous solids—glass and plastics. The particles in amorphous solids are like long chains. They can get jumbled and twisted instead of having an ordered structure. As a result, amorphous solids have different properties than crystalline solids.

13. Describe What do the particles in amorphous solids resemble?

_____

What are liquid crystals?

Liquid crystals are groups of materials that do not change states in the usual way. When a substance changes from a solid to a liquid, it usually loses its ordered structure. Liquid crystals start to flow in the melting phase, like most liquids. But liquid crystals keep their ordered structure. Liquid crystals respond to temperature changes and electric fields. Scientists use these properties to make liquid crystal displays. LCDs are used in watches and calculators.

After You Read

Mini Glossary

boiling point: the temperature at which the pressure of the vapor in the liquid is equal to the external pressure acting on the surface of the liquid

diffusion: the spreading of particles throughout a given volume until they are evenly distributed

heat of fusion: the amount of energy needed to change a substance from a solid to a liquid at its melting point

heat of vaporization: the amount of energy needed for the liquid at its boiling point to become a gas

kinetic theory: an explanation of how particles in matter behave

melting point: the temperature at which a solid begins to turn into a liquid

plasma: matter consisting of positively and negatively charged particles

thermal expansion: an increase in the size of a substance when the temperature is increased

1. Read the vocabulary terms and their definitions in the Mini Glossary. How are the heat of fusion and the heat of vaporization similar? How are they different?

_____

_____

_____

2. Complete the graphic organizer about the arrangement and behavior of particles in solids, liquids, and gases.

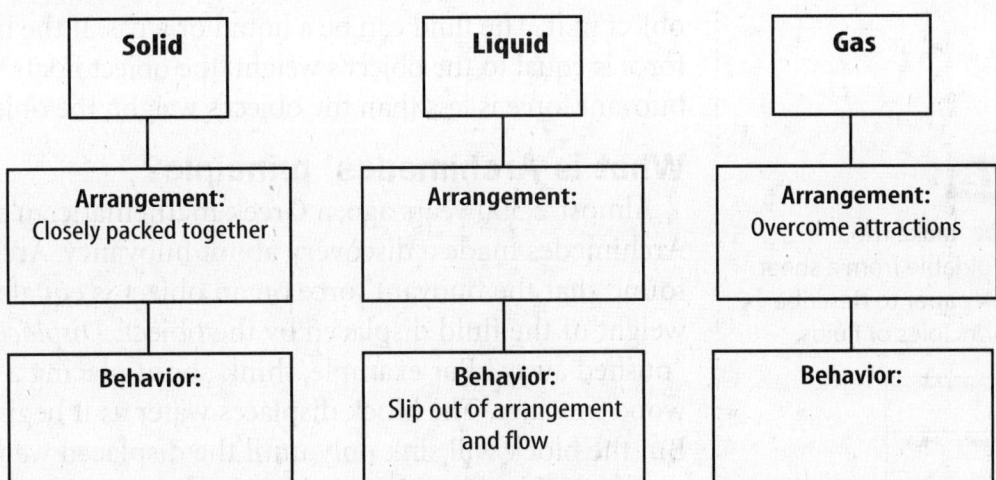

Science Online Visit gpscience.com to access your textbook, interactive games, and projects to help you learn more about kinetic theory.

End of Section

chapter 16 Solids, Liquids, and Gases

section 2 Properties of Fluids

What You'll Learn
- Archimedes' principle
- Pascal's principle
- Bernoulli's principle

Before You Read

Suppose that you have a piece of aluminum foil and have rolled it into a ball. You take another piece of foil the same size and make it into a flat, open box. Then you put both of these objects in a tub of water. What do you think would happen?

_____

_____

Mark the Text

Identify Key People Underline the name of each person discussed in this section. Say the name out loud. Then highlight the sentences that show how that person helped explain the properties of fluids.

Read to Learn

How do ships float?

Some ships are so huge that they are like floating cities. Even though ships are heavy, they are able to float. Ships float because the force pushing up on the ship is greater than the force of the ship pushing down. This force is called buoyancy. **Buoyancy** (BOY un see) is a fluid's ability to exert an upward force on an object in it. The fluid can be a liquid or a gas. If the buoyant force is equal to the object's weight, the object floats. If the buoyant force is less than the object's weight, the object sinks.

What is Archimedes' principle?

Almost 2,300 years ago, a Greek mathematician named Archimedes made a discovery about buoyancy. Archimedes found that the buoyant force on an object is equal to the weight of the fluid displaced by the object. *Displaced* means "pushed away". For example, think about placing a block of wood in water. The block displaces water as it begins to sink. But the block will sink only until the displaced water's weight equals the block's weight. When this happens, the block floats.

Do equal-sized objects float?

Would a steel block the same size as the wood block float in water? They both displace the same volume and weight of water. So, the buoyant force on the blocks is the same. But the steel block sinks. What is different?

FOLDABLES

C Describe Make the following Foldable from a sheet of notebook paper to describe the three principles of fluids.

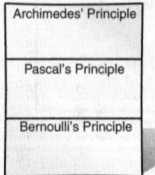

What does density have to do with buoyancy?

The volume of the two blocks and the volume of the water displaced is equal. But each volume has a different mass. Recall that density is mass per unit volume. If the three equal volumes have different masses, they must have different densities. The density of the steel block is greater than the density of water. The density of the wood block is less than the density of water. An object will float if its density is less than the density of the fluid it is placed in.

What would happen if you formed the steel block into the shape of a boat filled with air? Now the same mass takes up a larger volume. The density of the steel boat together with the air inside it is less than the density of water. The boat will float.

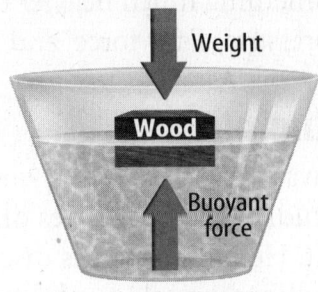

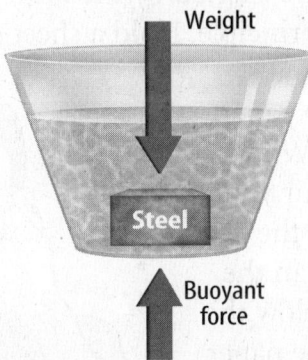

Picture This

1. **Compare** the lengths of the two "weight" arrows. Then compare the lengths of the two "buoyant force" arrows.

_____
_____
_____
_____

Pascal's Principle

When you are under water, you can feel the pressure of the water all around you. **Pressure** is force exerted per unit area or $P = E/A$. Did you know that Earth's atmosphere is a fluid? Earth's atmosphere exerts pressure all around you. Because you are used to the pressure of air, you do not feel the pressure.

Blaise Pascal was a French scientist who lived from 1623 to 1662. Pascal discovered a useful property of fluids. According to Pascal's principle, pressure applied to a fluid is transmitted or sent throughout the fluid. For example, when you squeeze a toothpaste tube, pressure is transmitted through the fluid toothpaste all around the tube. ✓

Reading Check

2. **Explain** According to Pascal's principle, what is transmitted throughout fluid?

_____
_____

Reading Essentials **273**

Picture This

3. Determine What does each letter in the figure stand for?

P → _____

F → _____

A → _____

How is Pascal's principle used?

Hydraulic machines move heavy loads using Pascal's principle. The figure shows how a hydraulic lift works. A pipe filled with fluid connects a small cylinder and a large cylinder. Pressure is applied to the small cylinder. The pressure is transferred through the fluid to the large cylinder. Pressure stays the same throughout the fluid. The greater surface area in the large cylinder provides more force. With a hydraulic machine, you can use your weight to lift something much heavier than you are. In the figure, P stands for pressure, F for force, and A for surface area.

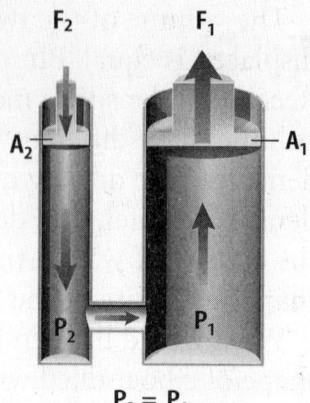

Bernoulli's Principle

Daniel Bernoulli was a Swiss scientist who lived from 1700 to 1782. Bernoulli studied the properties of moving fluids such as water and air. He published his discovery in 1738. According to Bernoulli's principle, as the velocity of a fluid increases, the pressure exerted by the fluid decreases.

You can test this principle. Hold a sheet of paper near your mouth as shown. Blow across the top of the paper. The paper will rise. Why? The velocity of the air you blew over the top of the paper was greater than the velocity of the air below the paper. As a result, the paper moved upward as air pressure above decreased. Engineers use this principle in designing aircraft wings and piping systems.

Picture This

4. Identify Highlight the area that has the greatest air pressure.

_____

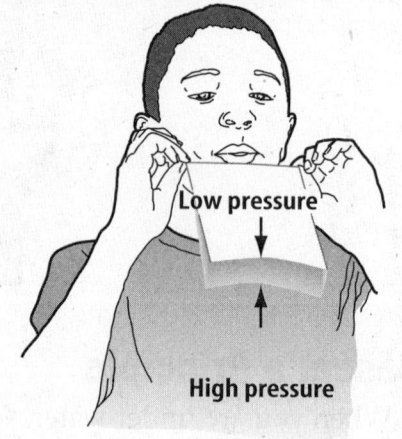

Fluid Flow

Another property of fluid is its tendency to flow. The resistance to flow by a fluid is called <u>viscosity</u>. Viscosity measures how much force is needed for one layer of fluid to flow over another layer. Some fluids tend to flow more easily than others. Water has a low viscosity because it flows easily. Syrup has a high viscosity because it flows slowly.

After You Read

Mini Glossary

buoyancy: the ability of a fluid to exert an upward force on an object that is placed in the fluid

pressure: force exerted per unit area

viscosity: the resistance by a fluid to flow

1. Read the terms and definitions in the Mini Glossary. Circle the term that is explained by Archimedes' principle. On the lines below, write a sentence using the term.

   _____

   _____

2. Match the concepts in Column 1 to the definitions and explanations in Column 2.

   Column 1

   _____ 1. Viscosity

   _____ 2. Bernoulli's principle

   _____ 3. Pressure

   _____ 4. Archimedes' principle

   _____ 5. Buoyancy

   _____ 6. Pascal's principle

   Column 2

   a. As the velocity of a fluid increases, the pressure exerted by the fluid decreases

   b. The ability of a fluid to exert an upward force on an object

   c. Pressure applied to a fluid is transmitted throughout the fluid

   d. The resistance by a fluid to flow

   e. Force exerted per unit area

   f. The buoyant force on an object is equal to the weight of the fluid displaced by the object

3. You underlined the name of each person discussed in this section. You also highlighted how those people helped you understand the properties of fluids. How did this strategy help you learn the material in this section?

   _____

   _____

 Visit **gpscience.com** to access your textbook, interactive games, and projects to help you learn more about properties of fluids.

Reading Essentials 275

chapter 16 Solids, Liquids, and Gases

section ❸ Behavior of Gases

What You'll Learn
- how a gas exerts pressure on its container
- how changing pressure, temperature, or volume affect a gas

● Before You Read

The air that you breathe is a gas. List other gases that you have heard about in a science class.

_____

_____

_____

Study Coach

Writing for Understanding As you read this section, write down each question heading. After each heading, write the answer to the question.

● Read to Learn

Pressure

You learned from the kinetic theory that gas particles are moving constantly and bumping into anything in their path. The collisions of these particles in the air result in pressure. Pressure is the amount of force exerted per unit of area. Pressure can be expressed by the formula $P = F/A$.

Often, gases are placed containers. A balloon and a bicycle tire are gas containers. The balloon and the tire stay inflated because of collisions the air particles have with the walls of the container. The collisions of the particles cause a group of forces to push the walls of the container outward in every direction. If more air is pumped into the balloon, the number of air particles inside the balloon increases. This causes more collisions with the walls of the container and the balloon expands even more. Bicycle tires cannot expand as much as balloons. When more air is pumped into a bicycle tire, the tire's pressure increases.

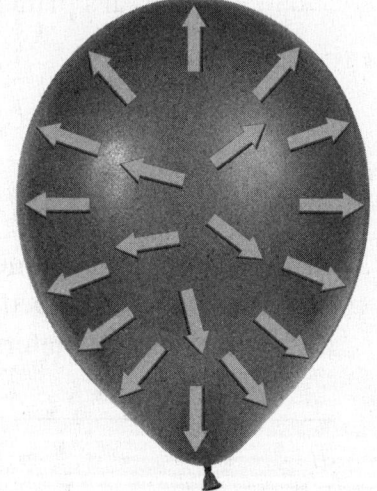

FOLDABLES

D Understanding Cause and Effect Make the following Foldable to help you understand the cause and effect relationship of gases.

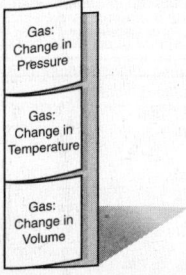

276 CHAPTER 16 Solids, Liquids, and Gases

How is pressure measured?

Pressure is measured in a unit called **pascal**, the SI unit of pressure. The abbreviation for pascal is Pa. Because pressure is the amount of force divided by area, one pascal of pressure is one Newton per square meter, or $1N/m^2$. Because this is a very small pressure unit, most pressures are given in kilopascals, or 1,000 pascals. The abbreviation for kilopascals is kPa.

At sea level, atmospheric pressure is 101.3 kPa. This means that at Earth's surface at sea level, the atmosphere exerts a force of about 101,300 N on every square meter. This is about the weight of a large truck. The air in the atmosphere gets thinner as distance from sea level increases. This means that the air pressure decreases as you travel higher above sea level.

Boyle's Law

You have learned how gas creates pressure in a container. Suppose the container gets smaller. What happens to the gas pressure? The pressure of a gas depends on how often its particles hit the walls of the container. If you squeeze gas into a smaller space, its particles will strike the walls of their container more often. This increases the pressure. The opposite is true, too. If you give the gas particles more space, they will hit the walls less often. Gas pressure is reduced.

Robert Boyle was a British scientist who lived from 1627 to 1691. Boyle explained the property of gases described above. According to Boyle's law, if you decrease the volume of a container of gas and keep the temperature the same, the pressure of the gas will increase. If you increase the volume of the container and keep the temperature the same, the pressure of the gas will decrease.

Pressure Outside an Object Affects Volume

Boyle's law explains the behavior of weather balloons. Weather balloons are large balloons that carry instruments to high altitudes. The instruments detect weather information and send the data back to Earth. The balloons are inflated near Earth's surface with a low-density gas. As a balloon rises, the atmospheric pressure outside the balloon decreases. This allows the balloon to slowly get bigger. It can reach a volume of 30 to 200 times its original size. Eventually, the balloon expands so much it breaks. Boyle's law states that as pressure outside an object is decreased, the volume increases. The weather balloon demonstrates this law. The graph on the next page shows that the opposite is also true. As the pressure outside an object is increased, the volume will decrease.

FOLDABLES

E Organize Make the following Foldable to help you organize the important information about each law.

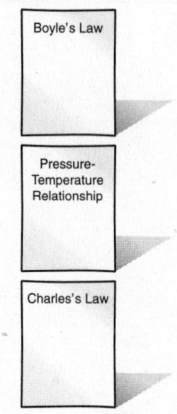

Reading Check

1. **Apply** If you increase the size of a container, will the pressure of the gas inside it increase or decrease?

Applying Math

2. **Interpret a Graph** What is the pressure of the gas at 300 L?

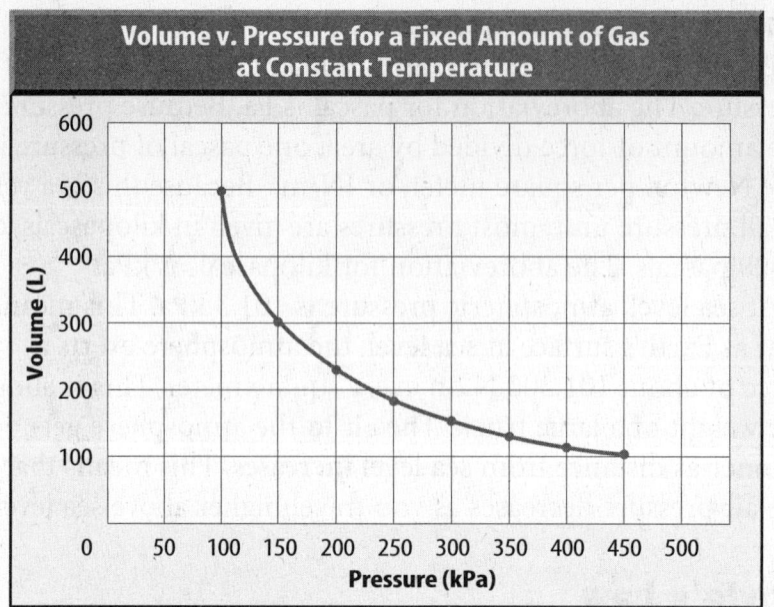

Volume v. Pressure for a Fixed Amount of Gas at Constant Temperature

Applying Math

3. **Solve an Equation** Use the equation $P_1V_1 = P_2V_2$. Assume that $P_1 = 101$ kPa and $V_1 = 10.0$ L. If $P_2 = 43.0$ kPa, what is V_2? Show your work.

Using Boyle's law

Using Boyle's law, you will find that the pressure inside an object multiplied by the volume of the object is always equal to the same number, or a constant. This is true only if the temperature stays the same. As the pressure and volume change, the constant number remains the same. You can use the equations $P_1V_1 = $ constant $= P_2V_2$ to express this mathematically. In this equation, P_1 and V_1 stand for the initial pressure and volume. P_2 and V_2 stand for the final pressure and final volume. The equation shows us that the product of the initial pressure and volume is equal to the product of the final pressure and volume. If you know three of these values, you can find the unknown fourth value.

The Pressure-Temperature Relationship

Have you ever seen the words "keep away from heat" on a spray can? If you heat a gas inside a closed container, the particles of gas will strike the walls of the container more often. Because the container is rigid, its volume cannot increase. Instead, its pressure increases. The pressure can become greater than the container can hold. If this happens, the container will explode. If the volume stays the same, an increase in temperature results in an increase in pressure. The opposite is also true. That is, a decrease in temperature results in a decrease in pressure, if the volume stays the same. Suppose the same spray can is placed in a very low temperature. The gas particles will slow down and decrease in pressure. The pressure decrease can buckle or crumple the container.

278 CHAPTER 16 Solids, Liquids, and Gases

Charles's Law

Have you ever seen a hot-air balloon being inflated? You know that gases expand when they are heated. Because particles in the hot air are farther apart than particles in the cool air, the hot air is less dense than the cool air. This difference in density allows the hot-air balloon to rise. Jacques Charles was a French scientist who studied gases. He lived from 1746 to 1823.

According to Charles's law, the volume of a gas increases with increasing temperature. This only happens if pressure does not change. The graph shows that the volume of a gas increases as temperature increases, but only at a constant pressure. As with Boyle's law, the reverse is true, also. The volume of a gas decreases as temperature decreases at a constant pressure.

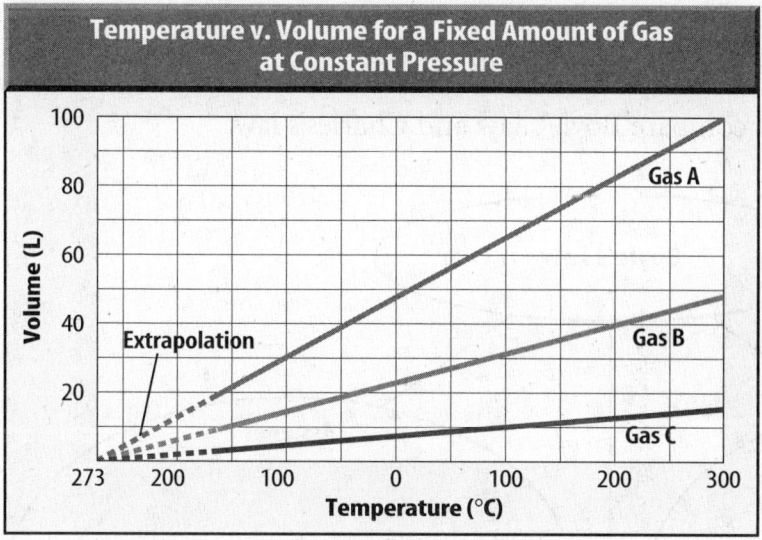

Applying Math

4. Interpret a Graph About how much did the volume of Gas B increase when the temperature rose from 100°C to 200°C?

Charles's law can be explained using the kinetic theory of matter. As a gas is heated, its particles move faster and faster and its temperature increases. Because the gas particles move faster, they begin to strike the walls of their container more often and with more force. In the hot-air balloon, the walls have room to expand. Instead of increased pressure, the volume of the gas increases.

How is Charles's law used?

In the same way that there is a formula for the relationship of pressure and volume, there is a formula for the relationship of temperature to volume. The temperature is given in kelvins. The formula for that relationship is $V_1/T_1 = V_2/T_2$. In this formula, V_1 and T_1 stand for initial volume and temperature. V_2 and T_2 stand for final volume and temperature. The pressure must be kept constant when you use Charles's law.

Reading Check

5. Infer When the temperature of the air in a hot-air balloon increases, what happens to the volume of the gas?

After You Read

Mini Glossary

pascal: the SI unit of pressure

1. Review the term *pascal* and its definition in the Mini Glossary. Use the term in a sentence that shows that you understand its definition.

   _____
   _____
   _____
   _____

2. Complete the graphic organizer to compare Boyle's law and Charles's law.

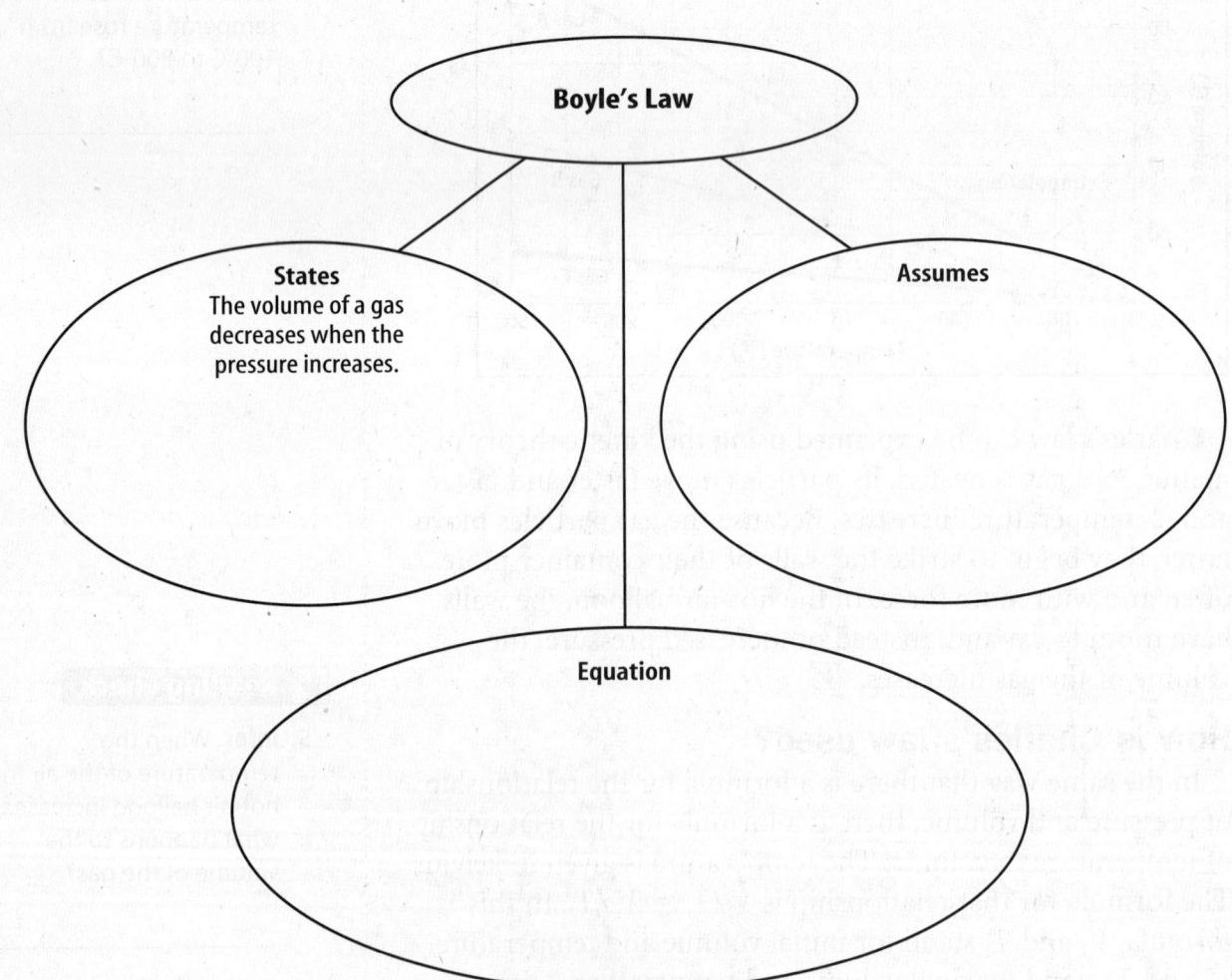

Boyle's Law

States
The volume of a gas decreases when the pressure increases.

Assumes

Equation

280　CHAPTER 16　Solids, Liquids, and Gases

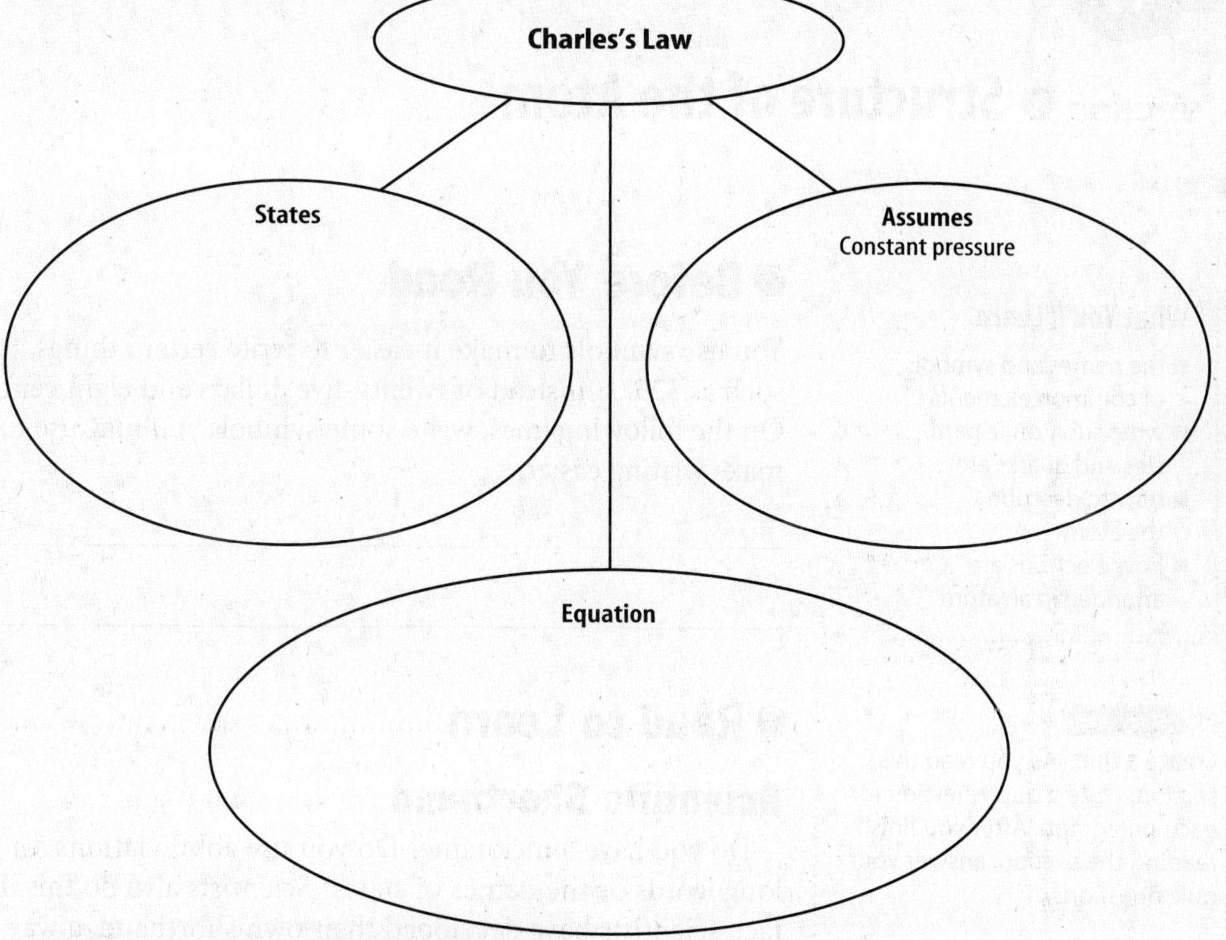

```
                    Charles's Law
                   /      |      \
              States      |      Assumes
                          |      Constant pressure
                          |
                       Equation
```

3. As you read this section, you wrote down the question headings and their answers. How could you use this to study for a quiz on this section?

_____

_____

_____

Science Online Visit **gpscience.com** to access your textbook, interactive games, and projects to help you learn more about the behavior of gases.

Reading Essentials **281**

Properties of Atoms and the Periodic Table

section ❶ Structure of the Atom

What You'll Learn
- the names and symbols of common elements
- what subatomic particles and quarks are
- how to describe the atom
- how electrons are arranged in an atom

Study Coach

Create a Quiz As you read this section, write a quiz question for each paragraph. After you finish reading the section, answer your quiz questions.

FOLDABLES

Ⓐ Organize Information Make the following Foldable to help organize information about scientific shorthand, atomic components, and quarks.

Scientific Shorthand	Atomic Components	Quarks

● Before You Read

You use symbols to make it easier to write certain things, such as $25.08 instead of twenty-five dollars and eight cents. On the following lines, write some symbols you may use to make writing easier.

_____

_____

● Read to Learn

Scientific Shorthand

Do you have a nickname? Do you use abbreviations for long words or the names of states? Scientists also do this. In fact, scientists have developed their own shorthand, a way to shorten long, complicated names.

C, Al, Ne, and Ag are all chemical symbols for different elements. A chemical symbol is shorthand for the name of an element. Chemical symbols make writing names of elements easier. Chemical symbols are either one capital letter or a capital letter plus one or two small letters. The table shows the chemical symbols for some elements. For some elements, the symbol is the first letter of the element's name. For example, C is for carbon. For other elements, the symbol is the first letter plus another letter from its name. For example, Ca is for calcium. Some symbols come from the Latin names of elements. *Argentum* is Latin for "silver." Silver's symbol is Ag.

Symbols of Some Elements

Element	Symbol	Element	Symbol
Aluminum	Al	Iron	Fe
Calcium	Ca	Mercury	Hg
Carbon	C	Nitrogen	N
Chlorine	Cl	Oxygen	O
Gold	Au	Potassium	K
Hydrogen	H	Sodium	Na

How have elements been named?

Elements have been named in many different ways. Elements have been named to honor scientists, for places, or for the elements' properties. Other elements have been named using rules made by an international committee. No matter what the origin of the name, scientists worldwide use the same system of element names and chemical symbols. People everywhere know that H means hydrogen, O means oxygen, and H_2O means dihydrogen oxide, or water. ☑

Atomic Components

An element is matter that is made up of one type of atom. An **atom** is the smallest piece of matter that still has the properties of the element. For example, the element silver is made up of only silver atoms. The element hydrogen is made up of only hydrogen atoms.

The figure below shows the structure of the atom. Atoms are made up of protons, neutrons, and electrons. **Protons** are particles with an electrical charge of 1+. **Neutrons** are particles with no electrical charge. **Electrons** are particles with an electrical charge of 1−. The **nucleus** is the small, positively charged center of the atom. It is made up of protons and neutrons. The nucleus is surrounded by a cloud containing electrons. The number of protons in an atom determines which element it is. For example, all atoms with 47 protons are silver atoms. All atoms with 1 proton are hydrogen atoms.

✓ Reading Check

1. **Identify** What is the symbol for dihydrogen oxide?

_____

Picture This

2. **Label** Write a plus sign on each proton shown in the nucleus.

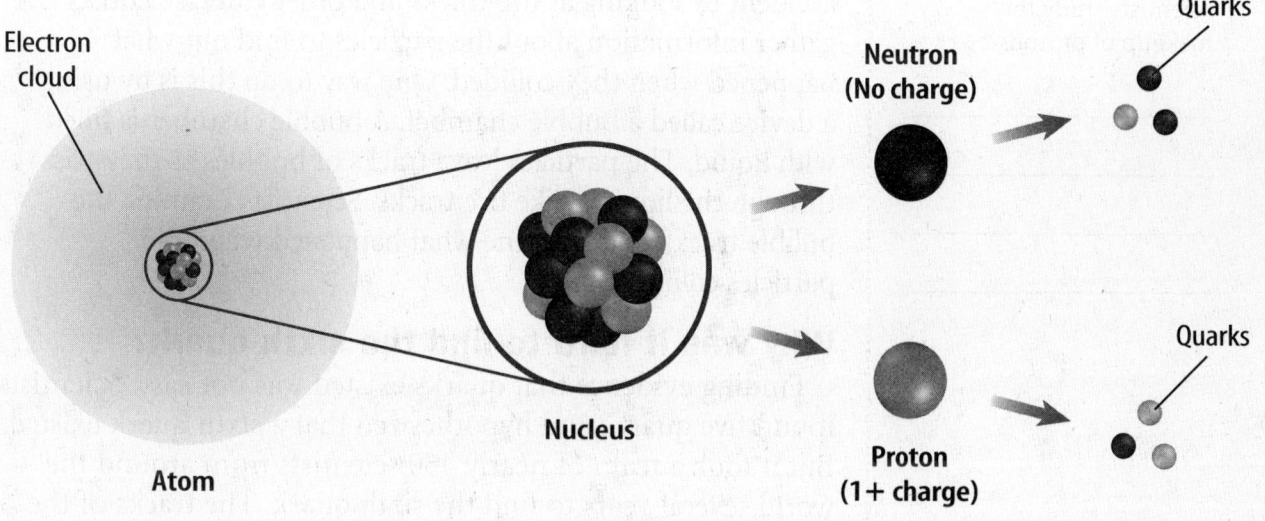

Quarks—Even Smaller Particles

Are protons, neutrons, and electrons the smallest particles that exist? Scientists hypothesize that electrons are not made up of smaller particles. If this is true, electrons are one of the most basic types of particles. But protons and neutrons are made up of smaller particles called **quarks**. So far, scientists have discovered six different quarks. Scientists theorize that protons are made up of three quarks. The quarks in a proton are held together with a force called the strong nuclear force. Neutrons are made up of another arrangement of three quarks. Scientists are still studying protons and neutrons to better understand them.

How do scientists find quarks?

To study quarks, scientists accelerate, or speed up, charged particles until they are moving extremely fast. Then they force the particles to collide with—or smash into—protons. The collision causes the protons to break apart. The Fermi National Accelerator Laboratory in Illinois has a machine that can accelerate particles fast enough to smash protons. This machine, called a Tevatron, is in a circular tunnel. The tunnel is 6.4 km in circumference. Scientists use electric and magnetic fields in the Tevatron to accelerate and smash particles.

How do scientists study quarks?

Scientists use different kinds of devices to detect the new particles that are made when particles are smashed together. Just as traffic investigators can tell what happened at an accident by looking at tire tracks and other clues, scientists gather information about the particles to find out what happened when they collided. One way to do this is by using a device called a bubble chamber. A bubble chamber is filled with liquid. The particles leave tracks of bubbles as they pass through the liquid—like tire tracks. Scientists examine the bubble tracks to determine what happened when the particles collided.

Why was it hard to find the sixth quark?

Finding evidence that quarks existed was not easy. Scientists found five quarks and hypothesized that a sixth quark existed. But it took a team of nearly 450 scientists from around the world several years to find the sixth quark. The tracks of the sixth quark were hard to detect. They were hard to detect because there was evidence of the sixth quark in only about one billionth of a percent of proton collisions. The sixth quark is called the *top* quark.

Reading Check

3. **Compare** Which is smaller, a proton or a quark?

Think it Over

4. **Describe** How do scientists study the makeup of protons?

Models—Tools for Scientists

Scientists use models to represent things that are difficult to visualize—or picture in your mind. Scaled-down models let you visualize something that is too large to see. Models of buildings, the solar system, and airplanes are scaled-down models. Scaled-up models are used to represent things that are too small to see. Scientists have developed scaled-up models to help them study the atom. To give you an idea of how small the atom is, it would take about 24,400 atoms stacked on top of each other to equal the thickness of a sheet of aluminum foil.

For a model of the atom to be useful, it must accurately represent everything we know about matter and how the atom behaves. As they learn more about atoms, scientists must change their models to include the new information.

How has the atomic model changed?

People have not always known that matter is made up of atoms. Around 400 B.C., a Greek philosopher named Democritus came up with the idea that atoms make up all substances. Another famous Greek philosopher, Aristotle, did not agree with Democritus' theory. Aristotle believed that each kind of matter was uniform, or the same all the way through, and not made of smaller particles. Aristotle's incorrect theory was accepted for about two thousand years. But in the 1800s, an English scientist named John Dalton was able to prove that atoms existed.

Dalton Model Dalton's model of the atom was a solid sphere, as shown in the first figure on the right. Dalton's model helped scientists explain why chemical reactions occur. Scientists then could use chemical symbols and equations to describe these reactions.

Dalton's Model

Thomson Model In 1904, English physicist Joseph John Thomson decided from his experiments that atoms contained small, negatively charged particles. He thought these "electrons" were spread out evenly throughout a positively charged sphere. His model, shown in the second figure, looks like a ball of chocolate chip cookie dough.

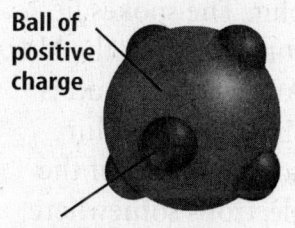

Ball of positive charge

Negatively charged electron

Thomson's Model

Think it Over

5. Explain Why have scientists developed scaled-up models to study the atom?

_____

_____

Picture This

6. Identify What part of Thomson's model are represented by the "chocolate chips" in the ball of cookie dough?

_____

Rutherford Model In 1911, another British physicist, Ernest Rutherford, thought that almost all the mass of an atom and all its positive charge were concentrated in the nucleus of an atom. He also thought the nucleus of an atom was surrounded by electrons, as shown in the first figure.

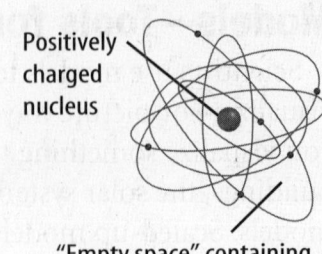

Rutherford's Model

Bohr Model In 1913, Danish physicist Neils Bohr hypothesized that electrons travel in fixed orbits around the nucleus of the atom, as the second figure shows. One of Bohr's students, James Chadwick, found that the nucleus contained positive protons and neutral neutrons. ☑

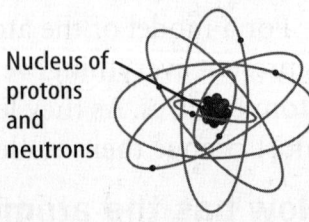

Bohr's Model

Reading Check

7. **Explain** What did Chadwick discover about the nucleus of an atom?

What is the electron cloud model?

The model of the atom has changed over time. By 1926, scientists had developed the electron cloud model of the atom in use today. An **electron cloud** is the area around the nucleus of an atom where its electrons are most likely found. The electron cloud is 100,000 times larger than the diameter of the nucleus. However, each electron in the cloud is much smaller than a single proton.

Scientists do not really know where in the electron cloud the electrons might be. Electrons move so fast and have such a small mass that it is impossible to describe exactly where they might be. It is best to describe their location as somewhere in the cloud. Think of the spokes on a spinning bicycle wheel. The spokes are moving so quickly that you can't tell exactly where any one spoke is. All you see is a blur. The spokes lie somewhere in the blur. An electron cloud is similar. It is a blur containing all of the electrons somewhere within it. The figure illustrates what an electron cloud might look like.

Picture This

8. **Compare and Contrast** How are Dalton's model (see figure on previous page) and electron cloud models of the atom similar? How are they different?

Similar:

Different:

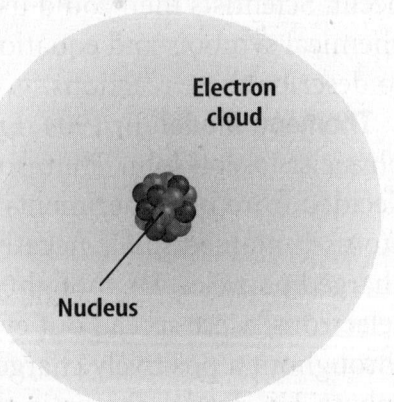

Electron Cloud Model

286 **CHAPTER 17** Properties of Atoms and the Periodic Table

After You Read

Mini Glossary

atom: the smallest piece of matter that still has the properties of the element
electron: particle with an electrical charge of 1−
electron cloud: the area around the nucleus of an atom where its electrons are most likely found
neutron: particle with no charge
nucleus: the small, positively charged center of the atom
proton: particle with an electrical charge of 1+
quark: smaller particle that makes up protons and neutrons

1. Review the terms and their definitions in the Mini Glossary. Write a sentence describing what parts make up an atom.

   _____
   _____

2. Below is a model of an atom. Label and describe each part of the atom. If any particles are made up of even smaller particles, list these also.

3. As you read this section, you created a quiz question for each paragraph. Did answering these quiz questions after you read the section help you learn the material? Why or why not?

   _____
   _____
   _____

 Visit **gpscience.com** to access your textbook, interactive games, and projects to help you learn more about the structure of atoms.

Chapter 17 Properties of Atoms and the Periodic Table

section 2 Masses of Atoms

What You'll Learn
- the difference between the atomic mass and the mass number of an atom
- how to identify components of isotopes
- how to interpret the average atomic mass of an element

Before You Read

Which metric unit do you use to measure the amount of gas that a car's gas tank holds? Which metric unit would you use to measure the distance to the next town? Explain why you would use these units and not smaller units.

___

___

___

Mark the Text

Highlight As you read the text under each heading, highlight the main ideas. After you finish reading the section, review the highlighted main ideas to help you learn the important topics of the section.

Read to Learn

Atomic Mass

Neutrons and protons are much more massive than electrons. Since the nucleus contains the neutrons and protons, it contains most of the mass of an atom. The mass of a proton is about the same as the mass of a neutron—about 1.6726×10^{-24} g, as shown in the table. The mass of a proton or a neutron is about 1,836 times greater than the mass of an electron. The mass of an electron is so small that it is not even considered when finding the mass of an atom.

Applying Math

1. **Comparing Decimals** Which has a larger mass, a proton or a neutron?

___

Subatomic Particle Masses

Particle	Mass (g)
Proton	1.6726×10^{-24}
Neutron	1.6749×10^{-24}
Electron	9.1093×10^{-28}

288 CHAPTER 17 Properties of Atoms and the Periodic Table

What is the atomic mass unit?

What unit would you use to estimate the height of your school building? Kilometers would be difficult to use. You probably would use a more appropriate unit, such as meters. Just as the kilometer is not the right unit for measuring the height of a building, scientists found that the gram was not the right unit for measuring the mass of an atom.

A useful unit gives numbers that are easy to work with. The unit used for measuring atomic particles is called the atomic mass unit (amu). The mass of a proton or neutron is almost equal to 1 amu. This is not a coincidence. The amu was defined that way. The amu is one-twelfth the mass of a carbon atom. A carbon atom contains six protons and six neutrons, or twelve particles. Since most of the mass of an atom is in the nucleus, each proton and neutron has a mass nearly equal to 1 amu.

How do protons identify elements?

Remember that atoms of different elements have different numbers of protons. In fact, the number of protons tells you what type of atom you have and vice versa. For example, every carbon atom has six protons. Also, every atom with six protons is carbon.

The **atomic number** of an element is the number of protons in an atom of the element. Since carbon has six protons, the atomic number of carbon is six. If you are given any one of the following for an element—its name, number of protons, or atomic number—you can find the other two.

What is the mass number?

The **mass number** of an atom is the sum of the number of protons and the number of neutrons in the nucleus of an atom. The table below shows this.

FOLDABLES

A Find the Main Idea Make the following Foldable to help take notes on the main ideas from this section.

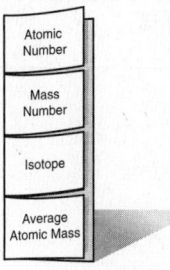

Picture This

2. **Complete** the table by finding the mass numbers for oxygen, sodium, and copper.

_____

Mass Numbers of Some Atoms							
Element	Symbol	Atomic Number	Protons	Neutrons	Mass Number	Average Atomic Mass*
Boron	B	5	5	6	11	10.81 amu
Carbon	C	6	6	6	12	12.01 amu
Oxygen	O	8	8	8		16.00 amu
Sodium	Na	11	11	12		22.99 amu
Copper	Cu	29	29	34		63.55 amu

*The atomic mass units are rounded to two decimal places.

Applying Math

3. **Apply** The element uranium has a mass number of 238, and an atomic number of 92. How many neutrons does an atom of uranium have?

Think it Over

4. **Determine** What is the same in two isotopes of an element? What is different?

Picture This

5. **Apply** How many years would it take half of the atoms in uranium-238 to change into lead-206?

How is the number of neutrons found?

If you know the mass number and atomic number of an atom, you can find the number of neutrons it contains.

number of neutrons = mass number − atomic number

Atoms of the same element with different numbers of neutrons can have different properties. For example, carbon with a mass number of 12 is called carbon-12. Carbon-14, with a mass number of 14 is radioactive. Carbon-12 is not radioactive.

Isotopes

Not all atoms of an element have the same number of neutrons. Atoms of the same element that have different numbers of neutrons are called **isotopes**. For example, boron atoms can have mass numbers of 10 or 11. To find the number of neutrons in an isotope, you can use the formula above. Look at the table on the previous page. Notice that boron has an atomic number of five. That means it has five protons. Substitute these numbers into the formula to get 11 − 5 = 6 and 10 − 5 = 5. So, boron isotopes have either five or six neutrons.

How can isotopes be used?

Atoms can be used to find the age of bones and rocks that are millions of years old. Radioactive isotopes release nuclear particles and energy as they decay into another element. The time it takes for half of the radioactive isotopes in a piece of rock or bone to change into another element is called its half life. Scientists use half lives of radioactive isotopes to measure time.

The table below lists the half-lives of some radioactive elements. It also lists the elements that the radioactive elements decay into. For example, it would take 5,715 years for half of the carbon-14 atoms in a rock to change into atoms of nitrogen-14. After another 5,715 years, half of the remaining carbon-14 atoms will change, and so on. These radioactive "clocks" can be used to measure different periods of time.

Half-Lives of Radioactive Isotope		
Radioactive Element	Changes to This Element	Half-Life
uranium-238	lead-206	4,460 million years
potassium-40	argon-40, calcium-40	1,260 million years
rubidium-87	strontium-87	48,800 million years
carbon-14	nitrogen-14	5,715 years

How do you identify isotopes?

The figure shows models of the two isotopes of boron. Because the numbers of neutrons in the isotopes is different, their mass numbers are different. To identify an isotope, use the name of the element followed by the mass number of the element. For example, the isotopes of boron are boron-10 and boron-11, because boron isotopes have mass numbers of either 10 or 11.

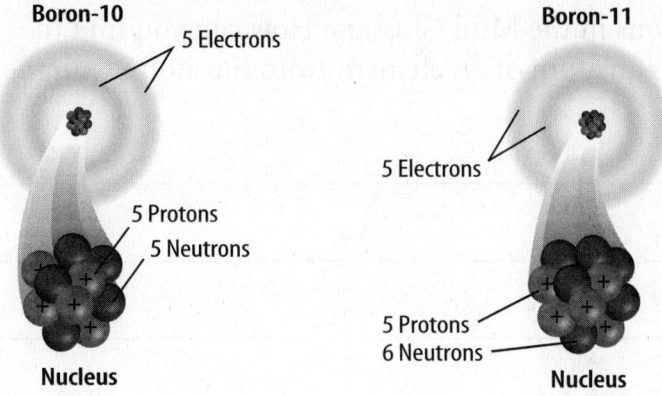

Most elements have more than one isotope. Because of this, each element has an average atomic mass. The **average atomic mass** of an element is the weighted-average mass of the mixture of its isotopes. For example, four out of five atoms of boron are boron-11. That means one out of five atoms is boron-10. To find the average atomic mass of boron, solve the following equation:

$$\frac{4}{5}(11 \text{ amu}) + \frac{1}{5}(10 \text{ amu}) = 10.8 \text{ amu}$$

The average atomic mass of boron is 10.8 amu. You round the average atomic mass to the nearest whole number to find the most abundant isotope of an atom. For example, the average atomic mass of boron, 10.8, rounds to 11. So, the most abundant isotope of boron is boron-11.

Picture This

6. Draw and Label Carbon-12 is an isotope with 6 protons and 6 neutrons. Draw a model of carbon-12. Label the protons and neutrons.

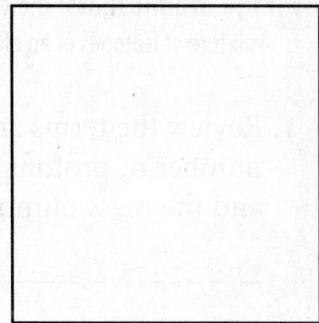

Applying Math

7. Apply The element magnesium has an average atomic mass of 24.305. What is the most abundant isotope of magnesium?

_____

After You Read

Mini Glossary

atomic number: a number equal to the number of protons in an atom

average atomic mass: the weighted-average mass of the mixture of isotopes of an element

isotopes: atoms of the same element that have different numbers of neutrons

mass number: the sum of the number of protons and the number of neutrons in an atom

1. Review the terms and their definitions in the Mini Glossary. How can you find the number of protons and neutrons in an atom of an element from the atomic number and the mass number?

____

____

____

2. Complete the Venn diagram by writing the given phrases in the correct area.

- determines which isotope
- equals the atomic number
- equals the mass number

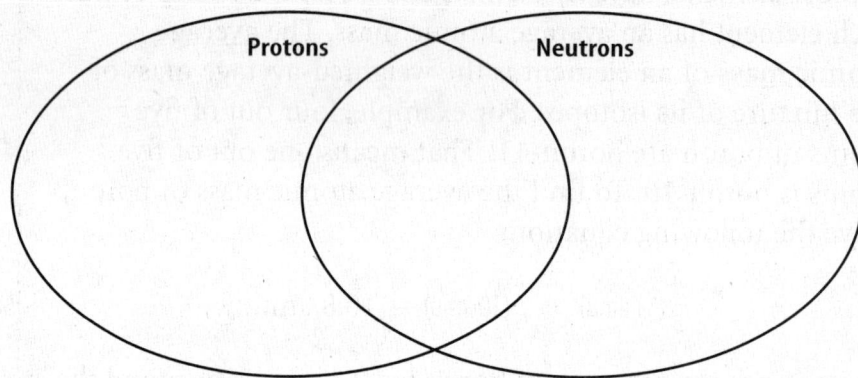

3. Tell how you could use a set of red and blue marbles to teach a friend about the atomic number and mass number of an element.

____

____

____

____

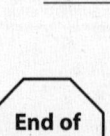

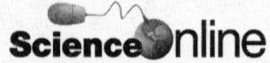

End of Section

Science Online Visit **gpscience.com** to access your textbook, interactive games, and projects to help you learn more about the masses of atoms.

Properties of Atoms and the Periodic Table

section ❸ The Periodic Table

● Before You Read

Many parts of our lives are affected by repeated patterns. For example, a calendar shows the patterns of weeks. Name some repeated patterns that you see happening all the time. How could you keep track of a pattern?

What You'll Learn
- the composition of the periodic table
- how to get information from the periodic table
- what metal, nonmetal, and metalloid mean

● Read to Learn

Organizing the Elements

When you look at the Moon, does it always appear the same? Each month, the Moon grows larger until it is full, then grows smaller until it seems to disappear. This type of change is called periodic. The word *periodic* means "repeated in a pattern." The days of the week are periodic because they repeat every seven days. Think of the calendar as a periodic table of days and months.

Who was Dmitri Mendeleev?

In the late 1800s, a Russian chemist named Dmitri Mendeleev wanted to find a way to organize the elements. He organized the elements known at the time into a table. He placed the elements in the table in order of increasing atomic mass.

Mendeleev discovered a pattern in his table. The properties of some lighter elements seemed to repeat in heavier elements. Because this pattern repeated, the pattern was considered to be periodic. Today, this arrangement is called a periodic table of elements. In the **periodic table**, the elements are arranged by increasing atomic number and by changes in physical and chemical properties. ☑

Mark the Text

Identifying the Main Point
Look for the main point of the paragraph or paragraphs under each heading in this section. When you have found the main point, write it down on a piece of paper. After you read the section, look over the main points again to help you learn the content of the section.

✓ Reading Check

1. **Determine** Who was the first person to organize the elements into a periodic table?

Reading Essentials **293**

How did Mendeleev's table predict properties?

Mendeleev left blank spaces in his table so that he could line up the elements. He looked at the elements surrounding the blank spaces. He predicted the properties and atomic masses of unknown elements to fit in the blank spaces.

The table shows Mendeleev's predictions for the element germanium. He called the element ekasilicon. His predictions proved to be accurate. Scientists eventually found all of the elements that were missing from Mendeleev's periodic table. The properties of these "missing" elements turned out to be extremely close to what Mendeleev predicted.

Applying Math

2. Use Decimals Find the difference between the predicted density of germanium and its actual density.

Mendeleev's Predictions	
Predicted Properties of Ekasilicon (Es)	**Actual Properties of Germanium (Ge)**
Existence Predicted—1871	Actual Discovery—1886
Atomic mass = 72	Atomic mass = 72.61
High melting point	Melting point = 938°C
Density = 5.5 g/cm^3	Density = 5.323 g/cm^3
Dark gray metal	Gray metal
Density of EsO$_2$ = 4.7 g/cm^3	Density of GeO$_2$ = 4.23 g/cm^3

How has the periodic table been improved?

Mendeleev's periodic table was very good for its time. However, scientists eventually found some problems with it. The elements on Mendeleev's table increased in atomic mass from left to right. Look at the modern periodic table at the back of this book. You will find examples, such as cobalt and nickel, that decrease in mass from left to right. However, notice that the atomic number always increases from left to right.

In 1913, the arrangement of the periodic table was changed. Instead of being arranged by increasing atomic mass, it was arranged by increasing atomic number. This change was due to the work of an English scientist named Henry G. J. Moseley. The new arrangement seemed to correct some of the problems of the old table. The current periodic table uses Moseley's arrangement and is shown in the back of this book. ✓

3. Observe How did Moseley arrange the periodic table?

The Atom and the Periodic Table

Objects are often sorted or grouped according to the properties they have in common. Elements on the periodic table are grouped according to their chemical properties. The vertical columns on the periodic table are called **groups**. The groups are numbered 1 through 18. Sometimes they are called families. Elements in each group have similar properties. For example, in Group 11, copper, silver, and gold have similar properties. Each is a shiny metal. Each is a good conductor of electricity and heat. Why are the elements in a group similar? Look at the structure of the atom to answer this question.

What is the structure of the electron cloud?

Where are the electrons located in an atom? How many are there? In a neutral atom, the number of electrons is equal to the number of protons. Carbon has an atomic number of six, which means it has six protons and six electrons. These electrons are located in the electron cloud that surrounds the nucleus.

Scientists have found that electrons in the electron cloud have different amounts of energy. Look at the figure. It shows a model of the energy differences that scientists use. Electrons fill the energy levels from the inner levels to the outer levels. The inner levels are closer to the nucleus and the outer levels are farther from the nucleus. Inner levels have less energy than the outer levels. Imagine that the nucleus is like a floor. Each energy level is a step up a flight of stairs above the floor. Each stair step represents an increase in energy. The figure shows the maximum number of electrons that will fit in each energy level of an atom. Not all atoms will have all levels filled. This depends on the number of electrons in an atom of that element.

Think it Over

4. Draw Conclusions Neon is a gas. Do you think neon is an element in Group 11? Explain.

_____
_____
_____

Picture This

5. Apply The element magnesium has 12 electrons. In how many energy levels are the electrons of magnesium?

_____

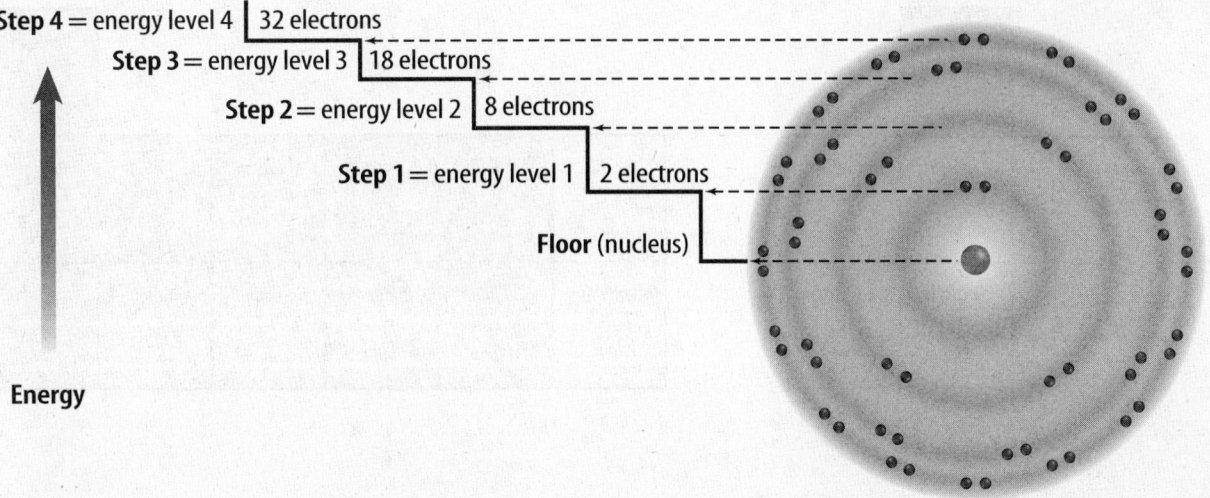

Energy Levels

How are electrons arranged in energy levels?

Elements that are in the same group have the same number of electrons in their outer energy level. The number of electrons in the outer energy level determines the chemical properties of the element. It is important to understand the link between the location on the periodic table, chemical properties, and the structure of the atom.

These energy levels are named using numbers one to seven. Electrons fill the energy levels starting with the inner level. For example, the element sulfur has 14 electrons. Two electrons will be in energy level 1 and eight electrons will be in energy level 2. The rest of the electrons will be in energy level 3.

Look again at the diagram on the previous page. Notice that energy levels 3 and 4 have increasingly large numbers of electrons. However, a stable outer energy level has eight electrons. How is this possible? In elements that have three or more energy levels, more electrons can be added to inner energy levels as long as the outer level contains eight electrons.

How are rows on the periodic table arranged?

Remember that the atomic number found on the periodic table is equal to the number of electrons in an atom. Look at the partial periodic table below.

Top Row The top row has hydrogen with one electron and helium with two electrons. Both of these electrons are in energy level 1. Energy level 1 is the outermost level in these elements. So, hydrogen has one outer electron and helium has two. Recall from the figure on the previous page that energy level 1 can hold only two electrons. Therefore, helium has a full outer energy level.

Picture This

6. Identify What is hydrogen's outermost energy level?

_____

Hydrogen 1 H							Helium 2 He
Lithium 3 Li	Beryllium 4 Be	Boron 5 B	Carbon 6 C	Nitrogen 7 N	Oxygen 8 O	Fluorine 9 F	Neon 10 Ne
Sodium 11 Na	Magnesium 12 Mg	Aluminum 13 Al	Silicon 14 Si	Phosphorus 15 P	Sulfur 16 S	Chlorine 17 Cl	Argon 18 Ar

Second Row The second row of the periodic table begins with lithium. Lithium has three electrons, two in energy level 1, and one in energy level 2. Next is beryllium with two outer electrons and boron with three outer electrons. The pattern continues until you reach neon. Neon has eight outer electrons. Look at the figure on the previous page. Energy level 2 can hold eight electrons. So, neon has a full outer energy level. Notice how a row in the table ends when an outer energy level is filled. The third row of elements, electrons begin filling energy level 3. The row ends with argon, which has a stable outer energy level.

What are electron dot diagrams?

Elements in the same group have the same number of electrons in their outer energy level. Outer electrons are used to determine the chemical properties of an element.

American chemist G. N. Lewis invented the electron dot diagram to show the outer electrons of an element. An **electron dot diagram** is the symbol of an element with dots representing the number of electrons in the outer energy level. The diagram for the elements sodium (Na) and chlorine (Cl) is shown below. These diagrams show how electrons in the outer energy level bond when elements combine to form compounds.

How are elements in the same group similar?

Elements in Group 17 are called halogens. They all have electron dot diagrams similar to chlorine, shown below. You can see that chlorine has seven electrons in its outer energy level. So do the other halogens. Since all elements in a group have the same number of electrons in their outer levels, those elements undergo chemical reactions in similar ways. ☑

How do halogens form compounds?

All halogens can form compounds with elements in Group 1. Group 1 elements, like sodium, all have one electron in their outer energy level. The figure shows an example of a compound formed by a reaction between sodium and chlorine. Sodium combines with chlorine to give each element a complete outer energy level. The result is the compound sodium chloride (NaCl), ordinary table salt.

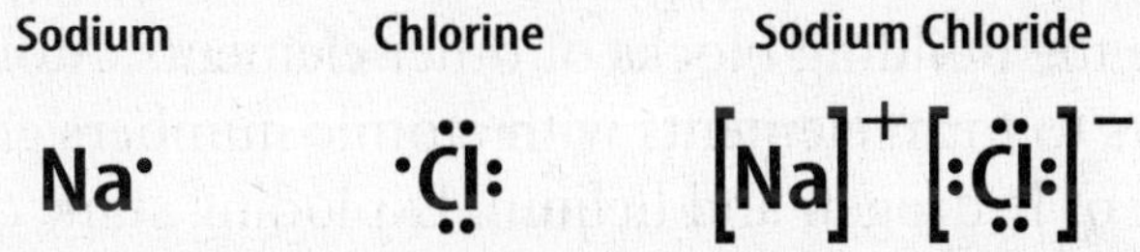

Sodium	Chlorine	Sodium Chloride
Na·	·C̈l:	[Na]$^+$ [:C̈l:]$^-$

Not all elements form compounds with other elements. Group 18 elements have completely filled outer energy levels. This makes group 18 elements unreactive.

Applying Math

7. **Use Numbers** How many electrons would fluorine need to gain in order to have a stable outer energy level?

_____

Reading Check

8. **Apply** What do all of the elements in a group have in common?

_____
_____
_____

Picture This

9. **Draw Conclusions** What seems to happen to the one electron in the outer level of sodium when it combines with chlorine to form sodium chloride?

_____
_____
_____

FOLDABLES

Compare and Contrast Use four quarter sheets of paper to compare and contrast metals, nonmetals, and metalloids.

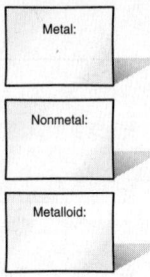

Think it Over

10. **Draw Conclusions** Silicon conducts electricity under some conditions but not under other conditions. In which region would you place silicon?

_____

Picture This

11. **Apply** On which side of the periodic table would you look for an element that definitely will not conduct electricity?

_____

Regions on the Periodic Table

The periodic table has several regions with specific names. The horizontal rows of elements are called **periods**. Recall that the elements increase by one proton and one electron as you go from left to right across a period. Also, each period represents a higher electron energy level.

All of the elements in the white squares in the diagram are metals. Iron, zinc, and copper are some examples of metals. Most metals are solids at room temperature. They usually are shiny. They can be drawn into wires and pounded into sheets. They are good conductors of heat and electricity.

The elements on the right side of the diagram in dark gray are classified as nonmetals. Oxygen, bromine, and carbon are nonmetals. Most nonmetals are gases. The elements in this region that are solids are brittle. Nonmetals are also poor conductors of heat and electricity at room temperature. The elements in light gray are metalloids, or semimetals. They have some properties of metals and some properties of nonmetals. Boron and silicon are examples of metalloids.

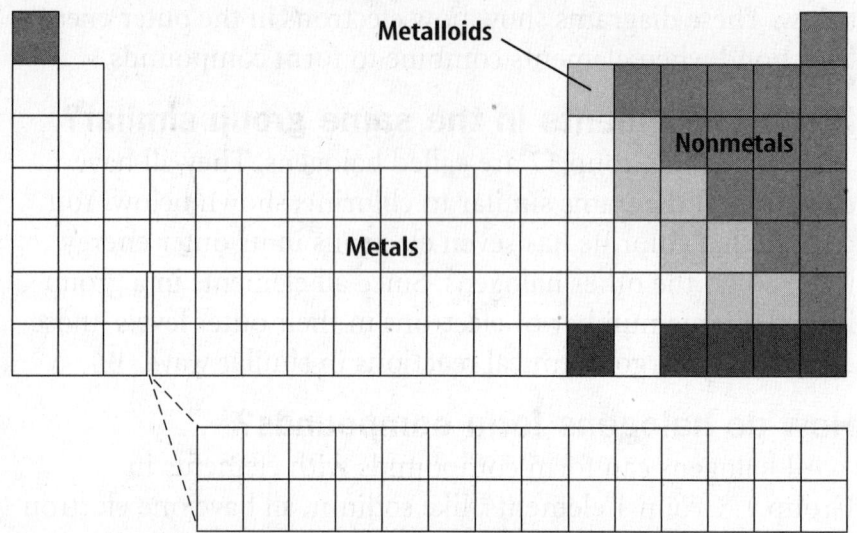

Regions on the Periodic Table

Elements in the Universe

Scientists have found the same elements throughout the universe. Many scientists hypothesize that hydrogen and helium are the building blocks of other elements. Atoms join within stars to form elements with atomic numbers greater than those of hydrogen and helium. Exploding stars, called supernovas, spread their mixture of elements throughout the universe. Scientists have made new elements in laboratories. These elements may have life spans less than a second.

After You Read

Mini Glossary

electron dot diagram: the symbol of an element with dots representing the number of electrons in the outer energy level

group: a vertical column of elements on the periodic table

period: a horizontal row of elements on the periodic table

periodic table: an arrangement of the elements by increasing atomic number and by changes in physical and chemical properties

1. Review the terms and their definitions in the Mini Glossary. Write a sentence using one of the terms that shows that you understand the term.

   _____

   _____

2. Below is a blank periodic table of elements. On this table, label the different sections as *metals, metalloids, nonmetals, period,* or *group.*

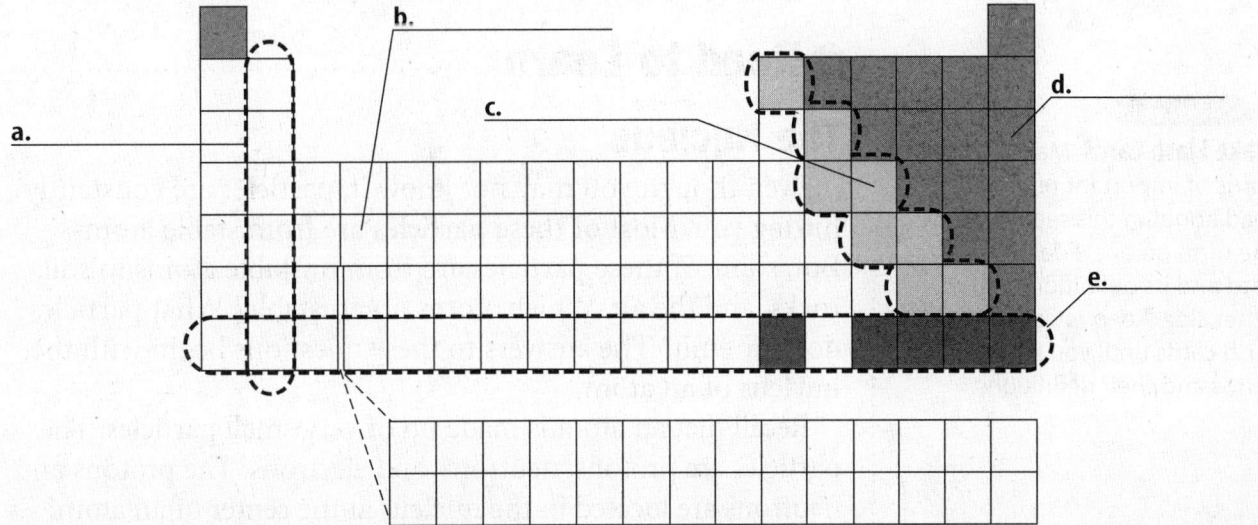

3. Explain what is important about the rows and columns on the periodic table.

   _____

   _____

   _____

 Visit **gpscience.com** to access your textbook, interactive games, and projects to help you learn more about the periodic table.

Reading Essentials **299**

chapter 18 Radioactivity and Nuclear Reactions

section ❶ Radioactivity

What You'll Learn
- what particles make up an atom and its nucleus
- how the nucleus is held together
- what radioactivity is
- the properties of radioactive and stable nuclei

● Before You Read

Have you ever heard of an object, such as chair, being described as unstable? In the space below, give examples of an object that is unstable and an object that is stable.

_____

_____

_____

● Read to Learn

The Nucleus

Even though you may not know it, particles are constantly hitting you. Most of these particles are from stable atoms. But, some of these particles are from unstable atoms in soil, rocks, and the air. Which atoms are unstable? What particles do they emit? The answers to these questions begin with the nucleus of an atom.

Recall that an atom is made up of very small particles. The particles are protons, neutrons, and electrons. The protons and neutrons are located in the nucleus at the center of an atom. Protons are positively charged particles. Neutrons have no electric charge. Neutrons are electrically neutral. Since the nucleus contains positively charged protons, it also has a positive charge. Each proton has one positive electrical charge, or +1. The total amount of positive charge in a nucleus is equal to the number of protons that the nucleus has. The number of protons in a nucleus is called its atomic number.

Atoms usually contain the same number of protons as electrons. Electrons have a negative charge. Electrons are attracted to the positively charged nucleus. This electric attraction pulls the electrons close to the nucleus. ☑

Study Coach

Make Flash Cards Make flash cards of important terms you read about in this section. Write the term on one side of a flash card and its definition on the other side. Keep reviewing the flash cards until you know all the terms and their definitions.

☑ Reading Check

1. **Identify** What type of charge does an electron have?

_____

_____

300 **CHAPTER 18** Radioactivity and Nuclear Reactions

Is the nucleus the largest part of an atom?

Protons and neutrons are packed together tightly in the nucleus of an atom. The particles in the nucleus are so close that the nucleus takes up only a tiny part of an atom. The remaining part is much larger than the nucleus. Think of an atom as a football stadium. It's nucleus would be the size of a marble.

Although the nucleus takes up very little space, it contains almost all the mass of an atom. Compared to an electron, protons and neutrons are quite heavy. A proton or a neutron has about 2,000 times the mass of an electron.

The Strong Force

Particles with the same charges repel, or push away from, each other. Why don't the positively charged protons in the nucleus repel each other? Another force holds the particles of the nucleus together. The **strong force** is the force that makes protons and neutrons attract each other and stay together.

The strong force is one of the four basic forces in nature. The strong force is 100 times stronger than the electric force, but it only works when particles are close together. When particles are close, like the protons and neutrons in a nucleus, the strong force is working. When particles are farther apart, the strong force weakens and the electric force takes over. So when protons are far apart, they are repelled by the electric force. The figure shows the strong force and the electric force.

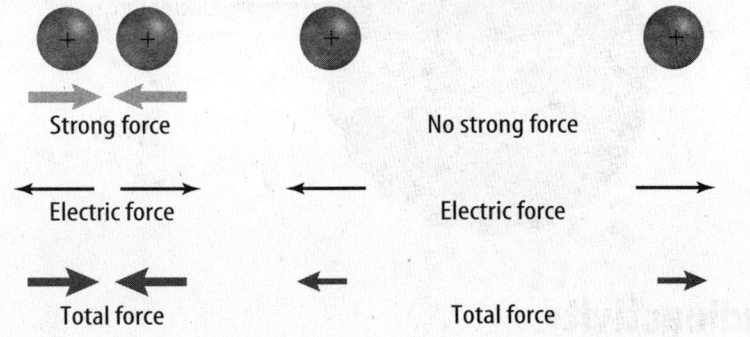

How do forces work in a small nucleus?

Not all nuclei (singular, *nucleus*) are the same size. Small nuclei have only a few protons and neutrons. Large nuclei have many protons and neutrons. Recall that the strong force is much stronger than the electric force when the particles are close together. In a small nucleus, the particles are close together. The strong force holding the particles together is greater than the electric force that is pushing the particles apart. The protons and neutrons are held tightly together.

FOLDABLES

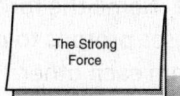

A **Find Main Ideas** Make a Foldable like the one shown below. As you read this section, write down main ideas about the strong force.

Picture This

2. **Interpret Scientific Illustrations** In the figure, circle the example where the strong force is greater than the electric force.

How do forces work in a large nucleus?

In a large nucleus, the strong force holds together only the particles that are closest to one another. Even though there are more particles in a large nucleus than in a small nucleus, the strong force in a large nucleus is about the same strength as the strong force in a small nucleus. In a nucleus with many protons, the electric force repels protons that are far apart. The electric force that pushes the protons apart is greater in a large nucleus. The increased repulsive force causes the particles in a large nucleus to be held together less tightly than those in a small nucleus. The figures compare the strong force in a small nucleus to the strong force in a large nucleus. ✔

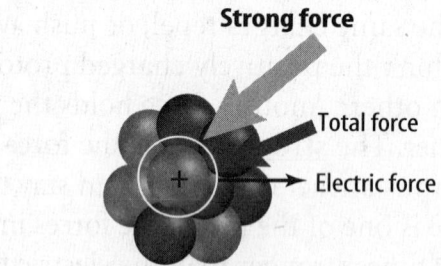

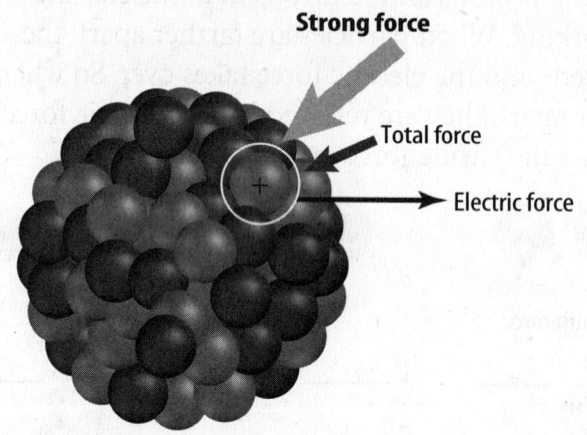

✔ Reading Check

3. Describe Name the force that causes protons to push away from each other.

_____

_____

Picture This

4. Identify In the large nucleus, circle two protons or neutrons on which the attractive strong force has little affect.

Radioactivity

When the strong force can hold a nucleus together forever, the nucleus is stable. If the strong force is not large enough to hold the nucleus together, the nucleus becomes unstable and can break apart or decay. When a nucleus decays, it emits, or gives off, particles and energy. **Radioactivity** is the process of a nucleus decaying and emitting particles and energy. Large nuclei are more unstable than small nuclei. All nuclei with more than 83 protons are radioactive. Some smaller nuclei are also radioactive. A nucleus with only one proton could be radioactive. ✔

What are isotopes?

All atoms of the same element have the same number of protons in the nucleus. But atoms of the same element do not always have the same number of neutrons in the nucleus. For example, all helium atoms have two protons. However, some atoms of helium can have two, three, or four neutrons. Atoms of the same element that have different numbers of neutrons are called isotopes. The atoms of all isotopes of the same element have the same number of protons and electrons and the same chemical properties. The figure shows the nuclei of two isotopes of helium. The number next to the name tells how many neutrons are in the nucleus.

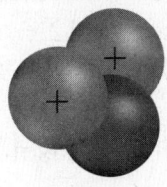
Helium-3

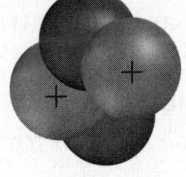
Helium-4

Picture This
5. Identify What is the mass number of the second helium isotope in the figure?

_____

What makes nuclei unstable?

The ratio of neutrons to protons determines whether a nucleus is stable or unstable. Isotopes with few particles in the nucleus are stable if the ratio of neutrons to protons is about 1 to 1. This means there are about the same number of neutrons as protons.

Isotopes with large numbers of particles in the nucleus can be stable if the ratio of neutrons to protons is about 3 to 2. In other words, for every three neutrons there are two protons in the nucleus. Generally, nuclei with too many or too few neutrons compared to the number of protons are unstable, or radioactive. ✓

How is a nucleus described?

You can describe a nucleus using its atomic number. The atomic number of an atom is the number of protons in the nucleus. The atomic number of carbon is 6, which means that a carbon atom has six protons in its nucleus. The nucleus contains almost all the mass of an atom. The total number of protons and neutrons in an atom is the mass number of that atom. The mass number of the first helium isotope in the figure on the previous page is 5. The sum of the two protons and three neutrons is five.

✓ Reading Check
6. Explain What is the ratio of neutrons to protons in a stable isotope?

_____

Think it Over

7. Apply What is the name of the isotope of oxygen that has a mass number of 18?

_____

Applying Math

8. Calculate How many neutrons are there in the isotope of hydrogen shown below?

$^{3}_{1}H$

_____

Reading Check

9. Recall Who discovered the elements polonium and radium?

_____

How is an atom's information shown?

Scientists use symbols to write information about atoms. Look at the symbol below for the stable isotope of carbon.

mass number → $^{12}_{6}C$ ← element symbol
atomic number →

The symbol for carbon is C. The name for this isotope of carbon is carbon-12. It is named for its mass number. Recall that the mass number tells you the total number of protons and neutrons in the nucleus. The atomic number tells you only the number of protons in the nucleus.

If you subtract the atomic number from the mass number, you get the number of neutrons in the nucleus. In an atom of carbon-12, there are six neutrons because $12 - 6 = 6$. Because the ratio of neutrons to protons is 6 to 6, or 1 to 1, this isotope is stable.

Now look at the symbol for a radioactive isotope of carbon.

mass number → $^{14}_{6}C$ ← element symbol
atomic number →

The name of this isotope is carbon-14. The mass number tells you it has a total of 14 neutrons and protons in its nucleus. The atomic number tells you that carbon-14 has six protons. If you subtract the atomic number (6) from the mass number (14) you find that carbon-14 has eight neutrons in its nucleus. The ratio of neutrons to protons is 8 to 6, which is not a 1-to-1 ratio, so this isotope is unstable, or radioactive.

Who discovered radioactivity?

In 1896, Henri Becquerel made an interesting discovery. He left pieces of uranium salt in a drawer on a photographic plate. When he developed the plate, he saw an outline of the uranium salt on it. Becquerel realized that the uranium must have given off rays that darkened the film.

Two years after Henri Becquerel's discovery of radioactivity, there was another important discovery. Marie and Pierre Curie discovered two new elements. The elements are polonium and radium. Both new elements were radioactive. Marie and Pierre Curie wanted to get a large sample of radium so they could study it. It took them more than three years to obtain about 0.1 g of radium from several tons of the mineral pitchblende. ✓

After You Read
Mini Glossary

radioactivity: the process of a nucleus decaying and emitting particles and energy

strong force: the force that makes protons and neutrons in the nucleus attract each other and stay together

1. Review the terms and their definitions in the Mini Glossary. Write a sentence on the lines below that shows your understanding of the term radioactivity.

   _____
   _____

2. Complete the table below to organize information about stable and unstable nuclei.

Type of Nucleus	Comparison of Strong Force v. Electric Force	Radioactive?	Example
Stable			
Unstable			

3. Think about what you have learned. How did making flash cards of important terms help you learn the content?

   _____
   _____

 Visit **gpscience.com** to access your textbook, interactive games, and projects to help you learn more about radioactivity.

End of Section

Radioactivity and Nuclear Reactions

section ❷ Nuclear Decay

What You'll Learn
- how alpha, beta, and gamma radiation are similar and different
- what the half-life of a radioactive material is
- how radioactive dating is used

Mark the Text

Identify the Main Point As you read the section, highlight the main point in each paragraph.

● Before You Read

When you read the word *radiation*, what do you think of? Brainstorm some words or phrases that come to mind and write them on the lines below.

_____

_____

● Read to Learn

Nuclear Radiation

When an unstable nucleus decays, it breaks apart. As it decays the nucleus emits particles and energy, called nuclear radiation. The three types of nuclear radiation are alpha, beta (BAY tuh), and gamma radiation. Alpha and beta radiation are particles. Gamma radiation is an electromagnetic wave.

Alpha Particles

An **alpha particle** is made of two protons and two neutrons. During alpha radiation, the decaying nucleus emits an alpha particle. The symbol for an alpha particle is ^4_2He. An alpha particle has a mass number of 4 and an atomic number of 2. This means it has two protons and two neutrons. An alpha particle is the same as the nucleus of a helium (He) atom. ☑

Alpha particles have much more mass than beta or gamma radiation. An alpha particle also has an electric charge of +2. Alpha particles can penetrate, or pass through, matter. When alpha particles penetrate matter, they attract negatively charged electrons from the atoms that they pass. This electric force pulls electrons away from the atoms. The alpha particles lose energy quickly and slow down. Alpha particles are heavier and move more slowly than the other two types of radiation. Alpha radiation cannot penetrate material as deeply as beta and gamma radiation. A sheet of paper can stop alpha particles.

✓ Reading Check

1. **Determine** An alpha particle is made up of _____ and _____.

How can alpha particles harm you?

Alpha particles are heavy and move slowly. You can think of them as bowling balls moving in slow motion. They may not be able to penetrate material deeply, but they can do a lot of damage to whatever they hit. Alpha particles can be harmful if they are released inside the human body. These particles can damage cells. Damaged cells do not work properly and can cause illness and disease.

How can alpha particles help you?

Smoke detectors work by emitting alpha particles. The alpha particles collide with molecules in the air. The molecules break apart and form atoms with positive and negative charges. These charged particles flow within the smoke detector, creating an electric circuit. If smoke particles enter the smoke detector, they break this circuit. Once the circuit is broken, the smoke detector sounds an alarm.

What is transmutation?

Recall that an alpha particle is made up of two protons and two neutrons. A decaying nucleus emitted these protons and neutrons. The nucleus now has two fewer protons and two fewer neutrons than it had originally. When an atom loses protons, it becomes a different element. **Transmutation** is the process of changing one element to a different element by the decaying process.

The nucleus that emitted the alpha particle has changed. It now has two fewer protons. Its atomic number is two less than the atomic number of the original element. The new element also has two fewer neutrons. Its mass number is four less than the original element.

The figure shows a polonium nucleus. Polonium has 84 protons and a total of 210 protons and neutrons. During transmutation, the nucleus gives off an alpha particle. The nucleus now has 82 protons and a total of 206 protons and neutrons. The element with an atomic number of 82 and a mass number of 206 is lead. The polonium atom has become a lead atom.

$$^{210}_{84}Po - {}^{4}_{2}He = {}^{206}_{82}Pb$$

Transmutation

FOLDABLES

B Compare and Contrast Make a Foldable like the one shown below. As you read this section, take notes on how alpha, beta, and gamma radiation are similar and how they are different.

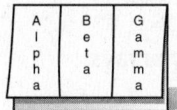

Picture This

2. Identify What is the symbol for lead?

_____

Reading Check

3. Classify Which of the following is emitted when a neutron decays into a proton? Circle your answer.

a. alpha particle

b. beta particle

c. gamma rays

Picture This

4. Identify In the figure, circle the mass numbers and underline the atomic numbers.

Beta Particles

A second type of radioactive decay is beta radiation. Sometimes in an unstable nucleus, a neutron decays into a proton and emits an electron. The electron leaves the nucleus. This electron is a beta particle. A beta particle is an electron that a neutron emits when it decays into a proton. The symbol for a beta particle is $_{-1}^{0}e$. Beta decay is caused by another basic force called the weak force.

An atom that loses a beta particle undergoes transmutation. It changes into a different element. The figure below shows an iodine nucleus giving off a beta particle. Before the nucleus decays, it has 53 protons and a total of 131 protons and neutrons. One of the neutrons becomes a proton by giving off a beta particle. Now, the number of protons increases by one and the number of neutrons decreases by one. The new element, xenon, has 54 protons. Notice that the mass number has not changed. There is still a total of 131 protons and neutrons in the nucleus.

$$_{53}^{131}I \rightarrow \, _{-1}^{0}e \; + \; _{54}^{131}Xe$$

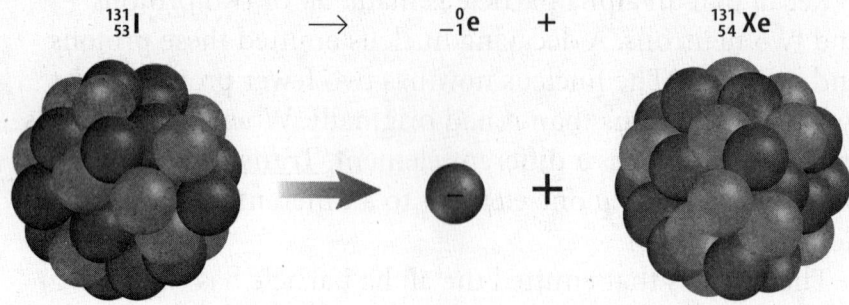

How can beta particles harm you?

Beta particles move much faster than alpha particles because they are smaller and lighter. Beta particles can also penetrate deeper into the material they hit. A beta particle can pass through a sheet of paper. A sheet of aluminum foil will stop a beta particle. Beta particles, like alpha particles, can damage cells if they are released inside the human body.

Gamma Rays

The third type of radiation is gamma radiation. Gamma radiation is not emitted as particles, like alpha and beta radiation. Gamma radiation is emitted as electromagnetic waves called gamma rays. **Gamma rays** are electromagnetic waves with the highest frequencies and the shortest wavelengths in the electromagnetic spectrum. The symbol for a gamma ray is γ, which is the Greek letter gamma.

Gamma rays have no mass and no charge. They travel at the speed of light. A nucleus usually emits gamma rays when an alpha or beta particle is created. Gamma rays can pass easily through a sheet of paper or aluminum foil because they travel very fast and have no mass. It takes thick blocks of a material such as concrete or lead to stop gamma rays. Gamma rays cause less damage to cells inside the human body than alpha and beta particles.

Radioactive Half-Life

Not all radioactive isotopes decay in the same amount of time. Some decay in less than a second. Others continue to decay for millions of years. The measure for the time it takes a radioactive nucleus to decay is called a half-life. The **half-life** of a radioactive isotope is the amount of time it takes for half of the nuclei in a sample of the isotope to decay. The nucleus left after the isotope decays is called the daughter nucleus.

Look at the figure. It shows radioactive hydrogen-3 (H-3) decaying into helium-3 (He-3). Notice that there are eight H-3 atoms in the beginning. The half-life of H-3 is 12.3 years. After the first 12.3 years have passed, the amount of H-3 left has been reduced by half. Now there are four H-3 atoms and 4 He-3 atoms. After the second 12.3 years have passed, there are two H-3 atoms and six He-3 atoms. The amount of H-3 in the sample has been reduced by half again.

Picture This

5. **Illustrate** In the space below, make a drawing that shows the number of H-3 and He-3 atoms there will be after a third half-life of 12.3 years passes.

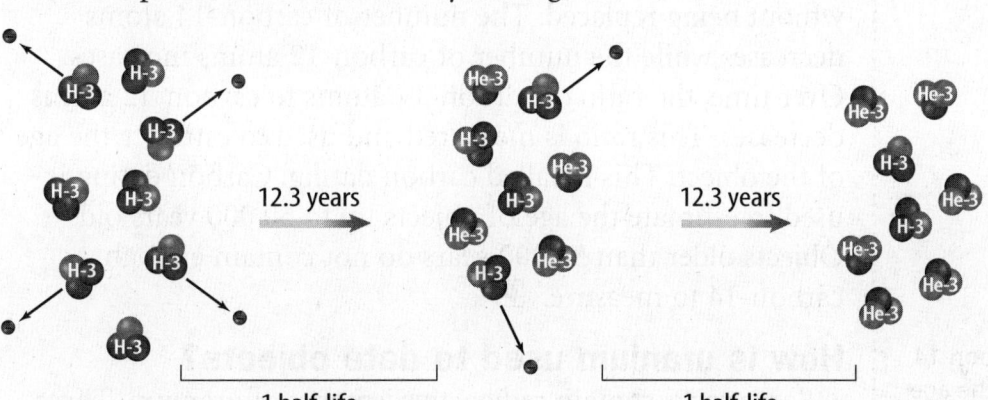

Radioactive Dating

Scientists often want to know the ages of the rocks and fossils. The ages of these objects can be found using radioactive isotopes and their half-lives. First, the amount of the radioactive isotope in the object is measured. Then, the amount of the isotope's daughter nuclei is measured. With these measurements, scientists know how much of the radioactive material has decayed and how much is still radioactive.

Think it Over

6. **Infer** Name two objects scientists might want to date.

The number of half-lives it took for the radioactive isotope to decay can now be calculated. The number of half-lives is the amount of time that has passed since the isotope began to decay. This is usually also the age of the object. Different isotopes are used to date different materials.

How is carbon used to date objects?

Scientists often use the radioactive isotope carbon-14 to help them estimate the age of plant and animal remains. Carbon-14 has a half-life of 5,730 years. Carbon dioxide contains carbon-14. When plants use carbon dioxide to make food, some carbon-14 stays in the plant. When an animal eats the plant, carbon-14 is added to its body. Therefore, most plants and animals and the objects made from them contain carbon-14. For this reason, carbon-14 is useful in dating many objects.

What happens when carbon-14 decays?

As carbon-14 decays, it becomes carbon-12. Carbon-12 is another isotope of carbon. Carbon-12 is stable and nonradioactive. As the carbon-14 in a plant or animal decays, more carbon-14 replaces it when a plant makes food or an animal eats a plant. This means the ratio of carbon-14 atoms to carbon-12 atoms in a plant or animal stays the same as long as the plant or animal is alive.

Once a plant or animal dies, the carbon-14 atoms decay wthout being replaced. The number of carbon-14 atoms decreases while the number of carbon-12 atoms increases. Over time, the ratio of carbon-14 atoms to carbon-12 atoms decreases. This ratio is measured and used to estimate the age of the object. This is called carbon dating. Carbon dating is used to estimate the age of objects up to 50,000 years old. Objects older than 50,000 years do not contain enough carbon-14 to measure.

How is uranium used to date objects?

Some rocks contain radioactive isotopes of uranium. These radioactive isotopes can be used to estimate the age of rocks. Uranium has two radioactive isotopes that have long half-lives. Each isotope decays into a different isotope of lead. The amounts of these uranium isotopes and their daughter nuclei are measured. The ratios of these amounts are used to calculate the number of half-lives that have passed since the rock was formed. This gives an estimate of the age of the rock.

Think it Over

7. Explain Why is carbon-14 useful in dating objects?

_____
_____
_____

Reading Check

8. Apply Why can't carbon-14 be used to estimate the age of objects older than 50,000 years?

_____
_____
_____

After You Read

Mini Glossary

alpha particle: particle made of two protons and two neutrons that are emitted from a decaying atomic nucleus

beta particle: an electron that a neutron emits when it decays into a proton.

gamma rays: electromagnetic waves with the highest frequencies and the shortest wavelengths in the electromagnetic spectrum

half-life: the amount of time it takes for half of the nuclei in a radioactive sample to decay

transmutation: the process of changing one element to a different element by the decaying process

1. Review the terms and definitions in the Mini Glossary. Choose two of the terms that are related and write a sentence using both terms.
   _____
   _____

2. Complete the outline to help you organize what you learned about nuclear radiation.

   Nuclear Radiation
   I. Alpha radiation
      A. Given off by the decaying nucleus
      B. _____
      C. _____
         _____

   II. Beta radiation
      A. _____
      B. Made of an electron
      C. _____

   III. Gamma radiation
      A. _____
      B. _____
      C. Moves at the speed of light and has no mass

Science Online Visit gpscience.com to access your textbook, interactive games, and projects to help you learn more about nuclear decay.

End of Section

Chapter 18: Radioactivity and Nuclear Reactions

section ③ Detecting Radioactivity

What You'll Learn
- how cloud and bubble chambers are used to detect radioactivity
- how an electroscope is used to detect radiation
- how a Geiger counter measures radiation

Before You Read

Many devices warn you about something that could cause a problem or be dangerous. For example, a smoke detector makes a loud noise to let you know there may be a fire. On the lines below, list three devices and their warning signals.

_____

_____

_____

Mark the Text

Locate Information Underline every heading in the section that asks a question. Then, highlight the answers to the questions as you find them.

Read to Learn

Radiation Detectors

You can't see or feel alpha particles, beta particles, or gamma rays. You must use special instruments to tell if they are present. These instruments detect radioactivity. Some radioactive particles have an electric charge. The charged particles form ions in the matter they pass through. Radiation detectors are instruments that detect newly formed ions.

How does a cloud chamber detect radiation?

A <u>cloud chamber</u> can be used to detect alpha or beta particle radiation. A cloud chamber is a rectangular box with transparent sides. It contains water vapor or ethanol vapor.

A radioactive sample placed in the cloud chamber gives off charged alpha or beta particles. These charged particles move through the chamber and pull electrons off atoms in the air. When an atom looses electrons, it becomes an ion. The radioactive particles leave a trail of ions as they travel through the cloud chamber. The water vapor or ethanol vapor condenses around these ions, forming small drops. A trail of small drops can be seen along the path of the radioactive particle. Beta particles leave long, thin trails. Alpha particles leave shorter, thicker trails.

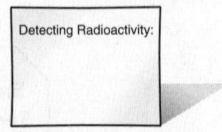

Make a List Make a Foldable like the one shown below. As you read this section, list the instruments used to detect or measure radiation. Beside each instrument, briefly describe how it works.

Detecting Radioactivity:

312 CHAPTER 18 Radioactivity and Nuclear Reactions

What is a bubble chamber?

A bubble chamber can be used to detect radioactive particles. A **bubble chamber** contains superheated liquid. The liquid does not boil because the pressure in the chamber is very high. When a radioactive particle passes through a bubble chamber, it leaves behind a trail of ions, just like in a cloud chamber. In a bubble chamber, the superheated liquid boils along the ion trail. The path of the radioactive particle shows up as a path of bubbles.

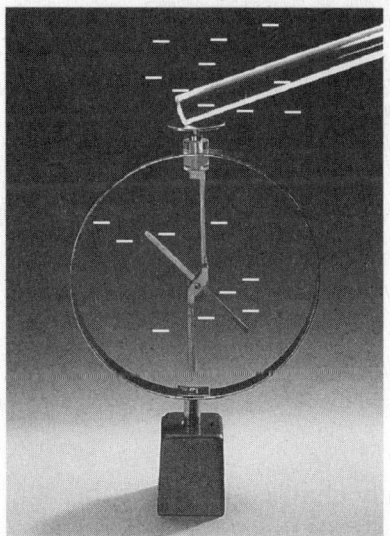
Negatively charged leaves

Alpha particles create positive ions.

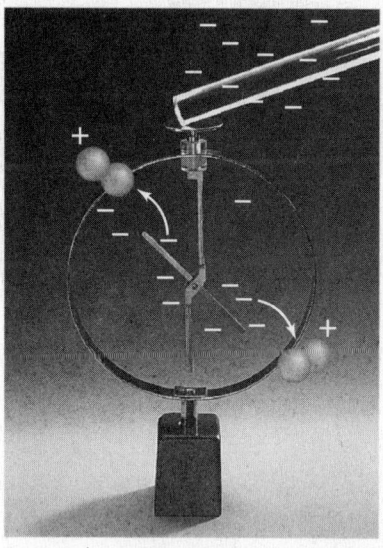
Negative charges move to positively charged ions.

How does an electroscope detect radiation?

Nuclear radiation can cause an electroscope to lose its charge. In the first figure above, the leaves of the electroscope are negatively charged. The leaves repel each other. The leaves stay apart until their extra negative charges can combine with positive charges. The negatively charged electrons can move to particles in the air that have a positive charge. When radioactive particles move through the air, they remove electrons from some molecules in air. Molecules that lose electrons become positive ions. Radioactive particles can make other molecules in air gain electrons and become negative ions. The second figure shows radioactive particles creating positive ions in air.

Positive ions that form near the negatively charged leaves attract electrons from the leaves. The third figure shows negatively charged electrons moving from the leaves of the electroscope to positive ions in the air. When the leaves of the electroscope lose their negative charge, they come together.

Picture This

1. **Make a Drawing** In the space below, draw the leaves of the electroscope above without an electric charge.

Reading Essentials **313**

Picture This

2. **Identify** In the figure, circle what attracts the knocked off electrons.

Measuring Radiation

It is important to keep track of the amount of radiation a person receives. Large amounts of radiation can harm the human body. A **Geiger counter** is a device that measures radiation by making an electric current when it detects a charged particle. A Geiger counter is shown below.

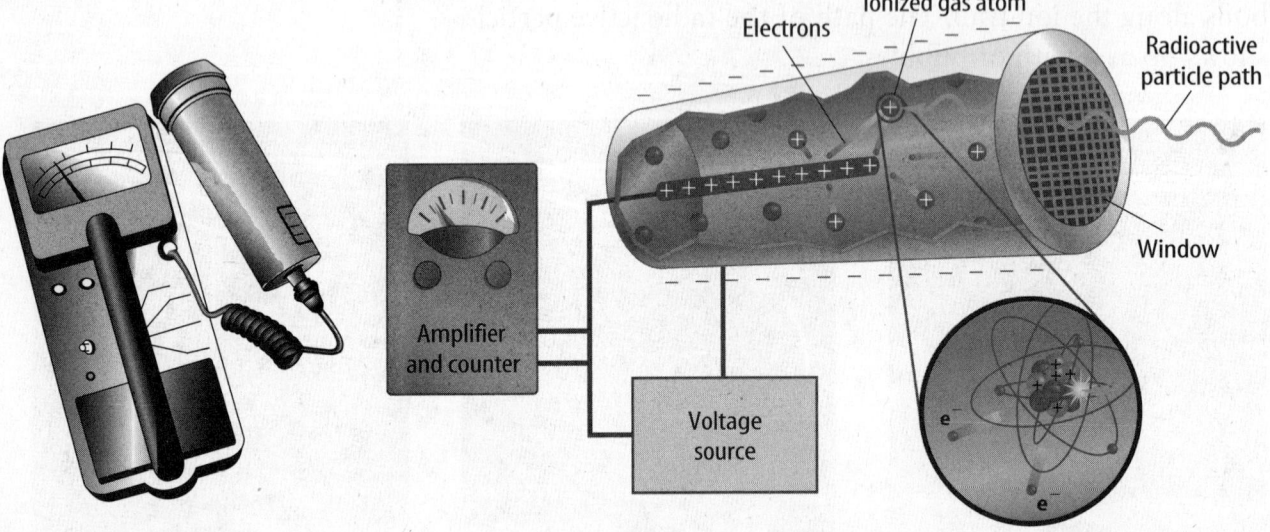

A Geiger counter has a negatively charged copper tube. A positively charged wire runs through the tube. The tube is filled with gas at a low pressure. When radiation enters the tube at one end, it knocks electrons off the gas atoms. These electrons then knock more electrons off other gas atoms. The process continues and produces an "electron avalanche." The positively charged wire in the tube attracts the free electrons. When a large number of electrons touches the wire, a short, strong current is produced in the wire. An amplifier strengthens the current, creating a clicking sound or a flashing light. The number of clicks or flashes of light that occur each second tells how strong the radiation is.

Background Radiation

The air, the ground, and even the walls of your house give off radiation. This type of radiation is background radiation. Background radiation is found in small amounts. Radioactive isotopes that occur in nature emit background radiation. Rocks, soil, and air contain these isotopes. Bricks, wood, and stone also contain small amounts of radioactive isotopes. Even the food, water, and air used by animals and plants have them. As a result, all animals and plants have small amounts of these isotopes.

3. **Explain** What emits background radiation?

_____

_____

Where does background radiation come from?

Background radiation comes from several sources. The circle graph shows the sources of background radiation received on average by a person living in the United States. The largest source comes from the decay of radon gas. Decay of uranium-238 in soil and rocks produces radon gas. Radon gas can move from the soil and rocks into houses and basements.

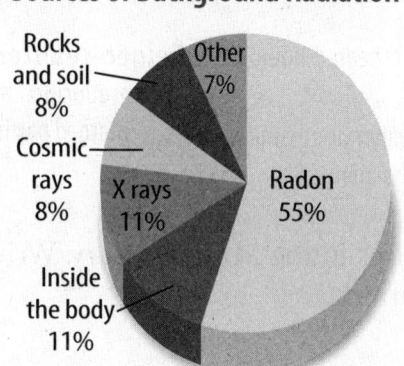

Sources of Background Radiation
- Rocks and soil 8%
- Other 7%
- Cosmic rays 8%
- X rays 11%
- Inside the body 11%
- Radon 55%

Some background radiation comes from high-speed nuclei that hit Earth's atmosphere. These nuclei are called cosmic rays. When cosmic rays hit the atmosphere, they give off alpha, beta, and gamma radiation. Because the atmosphere absorbs most of this radiation, it does not reach Earth's surface. At very high altitudes, the atmosphere is thinner than it is at Earth's surface. At high altitudes, there is less atmosphere to absorb this radiation. As a result, background radiation from cosmic rays is greater at higher altitudes.

Is there radiation in your body?

Some of the elements in your body contain radioactive isotopes that occur naturally. Remember, carbon-14 is a radioactive isotope that emits a beta particle when it decays. About one out of every trillion carbon atoms is a carbon-14 atom. Every time you breathe, you inhale about 3 million carbon-14 atoms.

One person can receive a different amount of background radiation than another. The total amount of background radiation your body receives depends on many things. It depends on the type of rocks that are in the ground where you live, the materials used to build your house, the elevation where you live, and many other things. Background radiation comes from processes that happen naturally. You never can remove all background radiation from your surroundings.

Applying Math

4. **Use a Graph** List the three largest sources of background radiation in the United States.

_____

_____

_____

Reading Check

5. **Describe** Name two things that affect the amount of background radiation you receive.

_____

_____

After You Read

Mini Glossary

bubble chamber: a device containing superheated liquid that detects radioactive particles

cloud chamber: a device containing water vapor or ethanol vapor that detects alpha or beta particle radiation

Geiger counter: a device that measures radiation by producing an electric current when it detects a charged particle

1. Review the terms and definitions in the Mini Glossary. Write a sentence about one of the devices used to measure radiation.

   _____

2. Complete the graphic organizer below to organize what you have learned about detecting radiation with a cloud chamber. Fill in the boxes with the following sentences. Place them in the order that they happen.
   - Water or ethanol vapor condenses around the ions.
   - Condensed water leaves a path of small drops.
   - Radioactive sample creates ions in the air.

Detecting Radiation with a Cloud Chamber

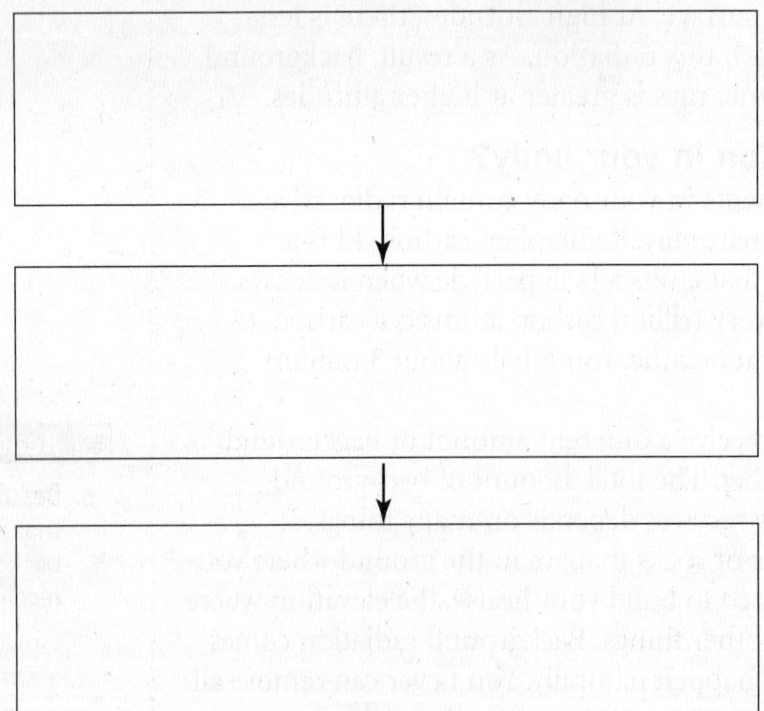

End of Section

 Visit gpscience.com to access your textbook, interactive games, and projects to help you learn more about radioactivity.

316 CHAPTER 18 Radioactivity and Nuclear Reactions

# Radioactivity and Nuclear Reactions

section ❹ Nuclear Reactions

● Before You Read

Nuclear reactions produce power in many parts of the world. In the space below, list two things that you would like to learn about nuclear reactions.

___

___

What You'll Learn
- what nuclear fission is
- what nuclear fusion is
- how radioactive tracers can be used in medicine
- how nuclear reactions can help treat cancer

● Read to Learn

Nuclear Fission

In 1934, the physicist Enrico Fermi tried bombarding uranium (U) nuclei with rapidly moving neutrons. He incorrectly thought the neutrons would combine with the nuclei and form larger, heavier nuclei. In 1938, Otto Hahn and Fritz Strassmann found that when a neutron hits a uranium-235 (U-235) nucleus, the nucleus splits apart into smaller nuclei.

In 1939, Lise Meitner theorized that when a neutron hits a uranium-235 nucleus, the nucleus becomes so unstable that it splits into two smaller nuclei. The process of splitting a nucleus into smaller nuclei is called **nuclear fission**. The word *fission* means "to divide."

What nuclei can split during nuclear fission?

Only large nuclei, such as the nuclei of uranium and plutonium, can split apart during nuclear fission.

Nuclear fission begins with a neutron hitting a U-235 nucleus. This produces a nucleus of U-236. U-236 is so unstable that it immediately splits into a barium nucleus and a krypton nucleus. The process also produces several neutrons. The total mass of the products of a fission reaction is a little less than the total mass of the original nucleus and the neutron. This small amount of missing mass becomes another product of nuclear fission. The missing mass is changed into a large amount of energy. This process is shown in the figure at the top of the next page.

Study Coach

State the Main Ideas As you read the section, stop after each paragraph and put what you have just read into your own words.

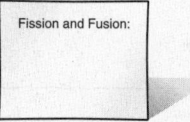

❿ Find Main Ideas Make a Foldable like the one below to take notes on the main ideas of each reaction.

Fission and Fusion:

Reading Essentials **317**

Picture This

1. Calculate What is the sum of the mass numbers for Kr and Ba? Why does this not equal the mass number of U-236?

_____

_____

_____

Reading Check

2. Interpret Data Fill in the blanks to make a true statement: Einstein's special theory of relativity tells us that a _____ amount of mass can be changed into a _____ amount of energy.

Think it Over

3. Illustrate In the space below, draw a diagram of a chain reaction.

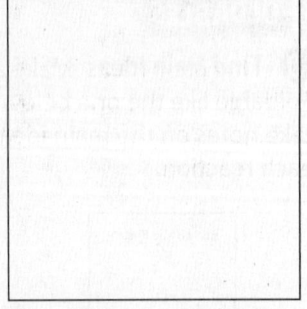

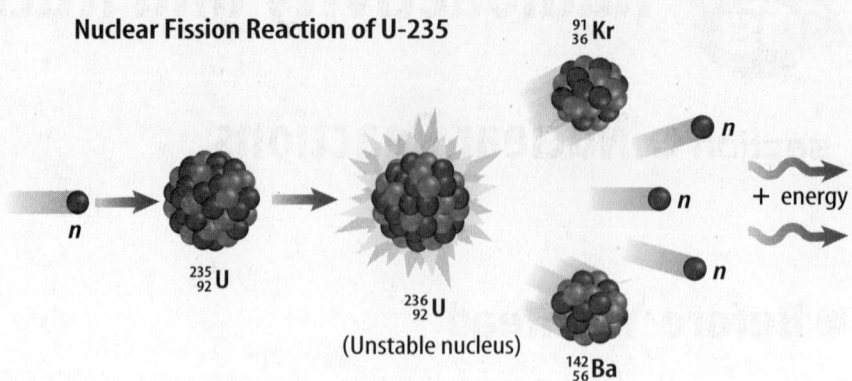

Nuclear Fission Reaction of U-235

How are mass and energy related?

Albert Einstein proposed that mass and energy are related. He said that mass can be changed into energy and that energy can be changed into mass. This is Einstein's special theory of relativity. The theory says that energy in joules is equal to mass in kilograms multiplied by the speed of light squared. Einstein expressed his theory in the mass-energy equation.

Energy (joules) = **mass** (kg) × [**speed of light** (m/s)]^2
$$E = mc^2$$

Einstein's theory of relativity tells us that a small amount of mass can be changed into a huge amount of energy. For example, if one gram of mass is changed into energy, it releases about 100 trillion joules of energy.

What is a chain reaction?

A nuclear fission reaction produces neutrons. These free neutrons can hit other nuclei. Once hit, these nuclei split and emit more neutrons. These neutrons hit other nuclei, and the fission reaction can repeat. A reaction that repeats over and over is a chain reaction. A **chain reaction** is a series of fission reactions caused by neutrons released in each reaction.

If a chain reaction continues and is uncontrolled, a huge amount of energy is released in a very short amount of time. Chain reactions can be controlled. Adding materials to the reaction that absorb the free neutrons is one way to control them. This keeps the neutrons from hitting other nuclei and continuing the reaction. If the right number of neutrons is absorbed, the reaction is controlled.

Not all fission reactions repeat themselves in a chain reaction. A chain reaction cannot occur if there is less than critical mass to provide enough free neutrons. The **critical mass** is the amount of starting material that makes each fission reaction produce at least one more fission reaction.

318 CHAPTER 18 Radioactivity and Nuclear Reactions

Nuclear Fusion

You have learned that nuclear fission releases great amounts of energy. When one nucleus of uranium-235 splits, it releases about 30 million times more energy than when one molecule of dynamite explodes. Another type of nuclear reaction called nuclear fusion releases even more energy than nuclear fission. In a **nuclear fusion** reaction, two small, light nuclei combine to form one larger, heavier nucleus. Fusion combines atomic nuclei and fission splits nuclei apart.

How are temperature and fusion related?

In a fusion reaction, two nuclei combine into one. How can two nuclei get close enough to combine? The nuclei have to be moving very fast. All nuclei are positively charged. Positively charged objects repel each other. If nuclei move fast enough, their kinetic energy overcomes the electric force pushing them apart. Then the two nuclei get close enough to combine.

A particle's kinetic energy increases as temperature increases. The temperature must be millions of degrees Celsius for nuclei to move fast enough to combine in a fusion reaction. These high temperatures exist on the Sun and other stars.

How does the Sun produce energy?

The Sun is made mostly of hydrogen. It produces its energy by the fusion of hydrogen nuclei. The figure shows one stage of the fusion process: A proton (H-1) and a hydrogen isotope (H-2) combine to form an isotope of helium (He-3). To complete the process, four hydrogen nuclei combine into one helium nucleus during which a small amount of mass changes into a huge amount of energy. The heat and light Earth receives comes from this process.

About one percent of the Sun's hydrogen has been changed into energy. Scientists estimate the Sun has enough hydrogen to continue fusion reactions for another 5 billion years.

Think it Over

4. **Explain** how two nuclei can overcome the electric force that tries to keep them apart.

   _____
   _____
   _____
   _____

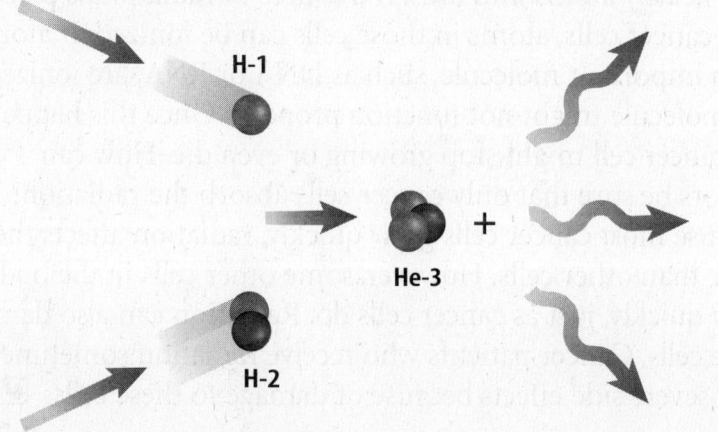

Picture This

5. **Label** the proton, hydrogen isotope, and helium isotope.

Using Nuclear Reactions in Medicine

Suppose you planned to meet a friend in a crowded place. How could you find her? It would be easier to find her if you knew she would be wearing a red hat. In a similar way, scientists can find one molecule in a large group of molecules if they know something special about the molecule. A molecule can't wear a red hat, but scientists can add something to a molecule to help them find it in a large group of molecules. Scientists can put a radioactive isotope inside a molecule. Then they can find the molecule by detecting the radiation that the isotope emits.

A <u>tracer</u> is a radioactive isotope used to find or keep track of a molecule. Doctors can use tracers to follow a molecule as it moves through your body. Tracers can be used to see how plants use nutrients and fertilizers.

How are iodine tracers used?

The thyroid gland is located in your neck. If the thyroid gland is not working properly, you could get sick. When iodine enters the body, it goes to the thyroid and stays there. The radioactive isotope iodine-131 is used to see if a person's thyroid is working properly. In the thyroid, iodine-131 atoms decay and give off gamma rays. The gamma rays can be detected. Doctors can use the information to tell whether a thyroid is healthy.

How can cancer be treated with radioactivity?

When a person has cancer, a group of cells in that person's body grows uncontrollably. These cells can form a tumor. Radiation can be used to stop some cells from growing into tumors. Sometimes, a radioactive isotope can be placed inside or near a tumor. Other times, tumors can be treated from outside the body.

Remember that radiation emitted when particles decay can turn nearby atoms into ions. If a source of radiation is placed near cancer cells, atoms in those cells can be ionized. If atoms in an important molecule, such as DNA or RNA, are ionized, the molecule might not function properly. Once this happens, the cancer cell might stop growing or even die. How can doctors be sure that only cancer cells absorb the radiation? Because most cancer cells grow quickly, radiation affects them more than other cells. However, some other cells in the body grow quickly, just as cancer cells do. Radiation can also damage these cells. Cancer patients who receive radiation sometimes have severe side effects because of damage to these cells.

Think it Over

6. Apply Give an example of something people do that is similar to a doctor using a tracer.

Reading Check

7. Determine How can radiation kill cancer cells?

After You Read

Mini Glossary

chain reaction: a series of fission reactions caused by neutrons released in each reaction

critical mass: the amount of starting material that makes each fission reaction produce at least one more fission reaction

nuclear fission: the process of splitting a nucleus into smaller nuclei

nuclear fusion: a reaction where two small, light nuclei combine to form one larger, heavier nucleus

tracer: a radioactive isotope used to find or keep track of a molecule

1. Review the terms and definitions in the Mini Glossary above. Write a sentence using the terms chain reaction and critical mass.

2. Complete the Venn diagram by listing one thing that fission and fusion have in common, one thing that applies only to fission, and one thing that applies only to fusion.

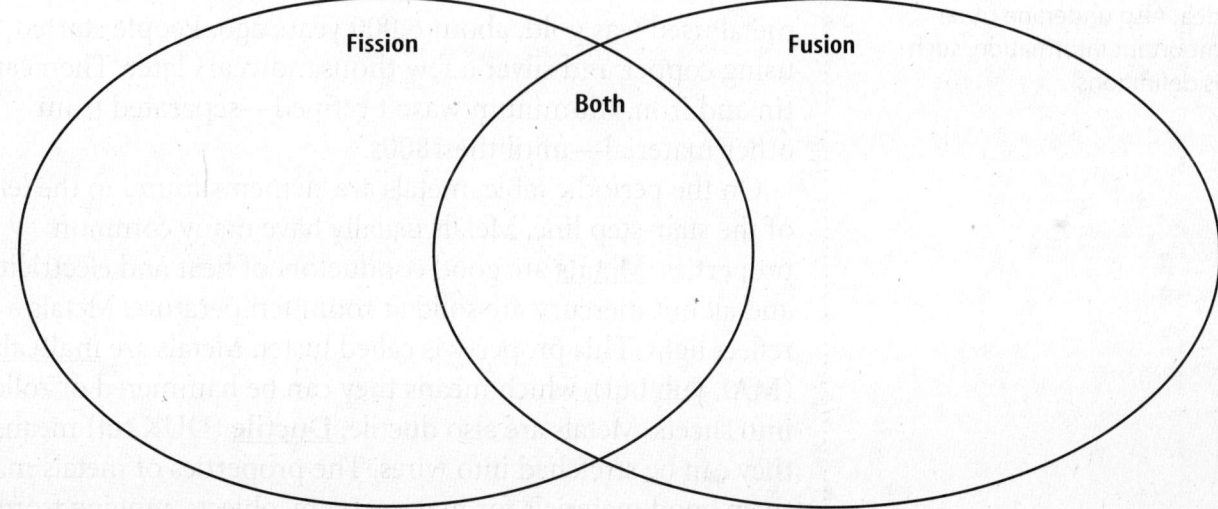

3. Think about what you have learned. You stopped after each paragraph and put what you have just read into your own words. Would you recommend this strategy to a friend? Why or why not?

 Visit gpscience.com to access your textbook, interactive games, and projects to help you learn more about nuclear reactions.

chapter 19 Elements and Their Properties

section ❶ Metals

What You'll Learn
- the properties of metals
- how to identify alkali metals and alkaline earth metals
- the difference among transition elements

Mark the Text

Underline As you read each paragraph, underline the main idea. Also underline other important information, such as definitions.

Picture This
1. **Circle** the atom that gave away one or more electrons.

● Before You Read

Think about what you are wearing. On the lines below, list all of the metal objects that are on your body.

_____

_____

● Read to Learn

Properties of Metals

People have used metals for thousands of years. The first metal used was gold, about 6,000 years ago. People started using copper and silver a few thousand years later. Then came tin and iron. Aluminum wasn't refined—separated from other material—until the 1800s.

On the periodic table, metals are elements found to the left of the stair-step line. Metals usually have many common properties. <u>Metals</u> are good conductors of heat and electricity, and all but mercury are solid at room temperature. Metals reflect light. This property is called luster. Metals are <u>malleable</u> (MAL yuh bul), which means they can be hammered or rolled into sheets. Metals are also ductile. <u>Ductile</u> (DUK tul) means they can be stretched into wires. The properties of metals make them good materials for making many objects, ranging from eyeglass frames to computers to building materials.

How do metals form ionic bonds?

When a metal atom gives one or more electrons in its outer energy level to a nonmetal atom in a chemical reaction, the atoms bond. This electron loss and gain results in both atoms becoming ions. The ions form ionic bonds, such as shown in the figure. Both atoms become more chemically stable because each ion has a full outer energy level.

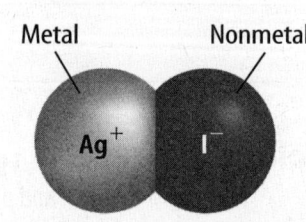

What is metallic bonding?

The atoms in metals have a special property called metallic bonding. In **metallic bonding**, a cloud of electrons surrounds positively charged metallic ions. Metallic bonding happens because electrons in the outer energy level of a metal atom are not held tightly to the nucleus. This property lets the electrons move freely among many positively charged metal ions. The electrons form a cloud around the metal ions. ☑

Metallic bonding explains many of the properties of metals. For example, metals can be hammered, rolled, and stretched without breaking because of metallic bonding. The ions are in layers that slide past each other without losing their attraction to the electron cloud. Metallic bonding also makes metals good conductors of electricity. Electrons in an electric current flow easily through the electron cloud.

Where are metals on the periodic table?

How many elements are classified as metals? On the periodic table, all the elements in Groups 1 through 12, except hydrogen, are metals. Also, the elements under the stair-step line in Groups 13 through 15 are metals.

The Alkali Metals

The elements in Group 1 on the periodic table shown in the figure are called alkali (AL kuh li) metals. Like other metals, Group 1 metals are shiny, malleable, and ductile. They are also good conductors of electricity. However, alkali metals are softer than most other metals. They also are the most reactive of all the metals. They react rapidly with oxygen and water. Because they are so reactive, alkali metals are stored in substances that are not reactive, such as oil.

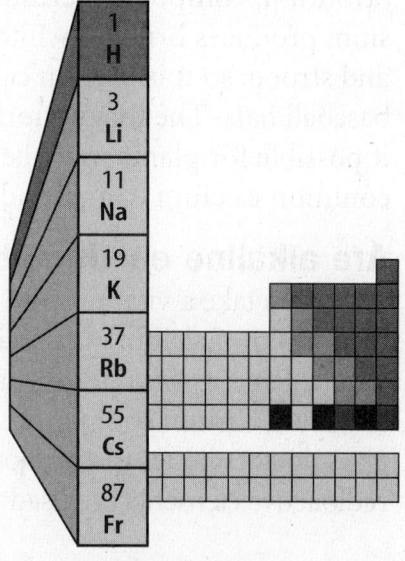

Alkali in Compounds Alkali metals are always found in chemical compounds in nature. Each alkali metal atom has one electron in its outer energy level. Alkali metals give up the electron and form ionic bonds with other elements. Ionic compounds with alkali metals include sodium chloride, NaCl, and potassium bromide, KBr.

Reading Check

2. **Infer** Why does metallic bonding happen?

_____

_____

_____

Picture This

3. **Determine** Circle the element in Group 1 that is not an alkali metal. What is the name of that element?

_____

FOLDABLES

Ⓐ **Organize Information** Make the following Foldable to help you organize information about the differences among the alkali metals, alkaline earth metals, and transition elements.

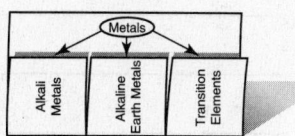

Alkali metals and their compounds have many uses. All living things need potassium and sodium compounds to stay healthy. Doctors use lithium compounds to regulate bipolar disorder. One alkali metal, francium, is radioactive. A <u>radioactive element</u> is an element in which the nucleus of the atom breaks down and gives off particles and energy.

The Alkaline Earth Metals

The alkaline earth metals are the elements in Group 2 on the periodic table shown in the figure. Alkaline earth metals have properties similar to most other metals. Like alkali metals, they form ionic bonds easily with other elements and are not found as free elements in nature. Each alkaline earth metal atom has two electrons in its outer energy level.

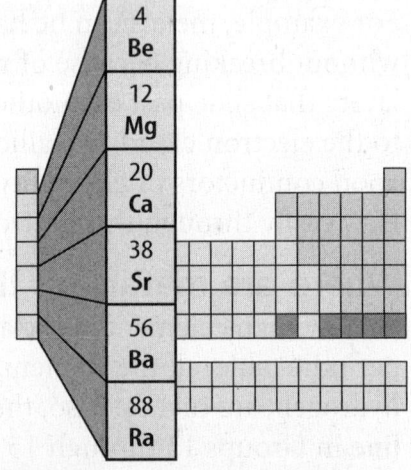

Picture This
4. Interpret a Table How many elements are classified as alkaline earth metals?

_____

_____

What are some uses of alkaline earth metals?

Some alkaline earth metals give fireworks their colors. Strontium compounds create bright red flashes and magnesium produces brilliant white sparks. Magnesium is very light and strong, so it is used in cars, airplanes, spacecraft, and even baseball bats. The magnesium compound chlorophyll makes it possible for plants to make food. Calcium carbonate is a common calcium compound found in chalk and limestone.

Are alkaline earth metals used in your body?

Do you take a vitamin with calcium? Calcium is important to your health. Calcium phosphate in your bones helps make them strong. Patients who need X rays of their digestive systems swallow a barium compound. It absorbs X-ray radiation and gives doctors better pictures of the digestive system. Some radioactive elements are used in treating cancer.

Transition Elements

<u>Transition elements</u> are the elements in Groups 3 through 12 on the periodic table. They are called transition elements because they are in the middle, or transition, between Groups 1 and 2 and Groups 13 through 18. Titanium bike frames and tungsten lightbulb filaments are some of the things that are made from transition elements.

✓ Reading Check
5. Identify Where are the transition elements located on the periodic table?

_____

_____

_____

Transition elements are often found in nature in their pure, or elemental, form. The elements we think of as being typical metals are transition elements. Iron, copper, and gold are some examples. Transition elements often form colored compounds. Cadmium yellow and cobalt blue paints are made from compounds of transition elements.

What are the uses of iron, cobalt, and nickel?

Iron, cobalt, and nickel are the first elements in Groups 8, 9, and 10 on the periodic table. They are sometimes called the iron triad. These elements are used in the process to create steel and other metal mixtures. Iron is the main ingredient in steel and is the most often used metal. Iron is the second most common metallic element found in Earth's crust, after aluminum. Some steels contain nickel and cobalt. Nickel is added to some metals to give them strength or a shiny protective coating.

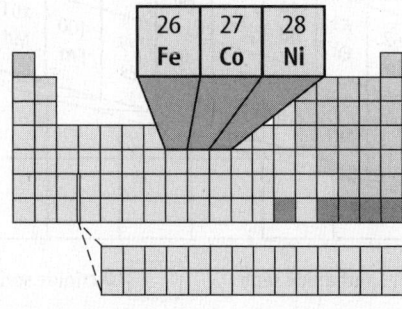

Picture This
6. **Explain** Use the periodic table to explain why iron, cobalt, and nickel are not in the same group of elements.

_____
_____
_____

How are copper, silver, and gold used?

The three elements in Group 11 on the periodic table are copper, silver, and gold. These elements are stable, malleable, and found as free elements in nature. They are known as the coinage metals because they were once used to make coins. Gold and silver are too expensive to be used in most coins today, but copper is still used. Copper also is used for electrical wiring. It is inexpensive and conducts electricity very well. Silver compounds are used to make photographic film. Silver and gold have an attractive color, do not corrode or wear away, and are rare, so they often are used to make jewelry.

Think it Over
7. **Apply** Pennies used to be called "coppers." How do you suppose they got this nickname?

_____
_____
_____

What are uses for zinc, cadmium, and mercury?

Zinc, cadmium, and mercury are the elements in Group 12 on the periodic table. Zinc and cadmium are often used to coat other metals. Cadmium also is used in rechargeable batteries. Since mercury is a liquid at room temperature, it is used in thermometers, thermostats, switches, and batteries. Mercury is poisonous and can accumulate in the body.

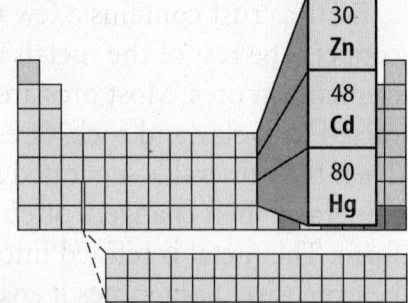

Picture This
8. **Infer** On the periodic table to the left, highlight the group that contains the coinage metals. What are the atomic numbers of copper, silver, and gold?

_____

Reading Essentials **325**

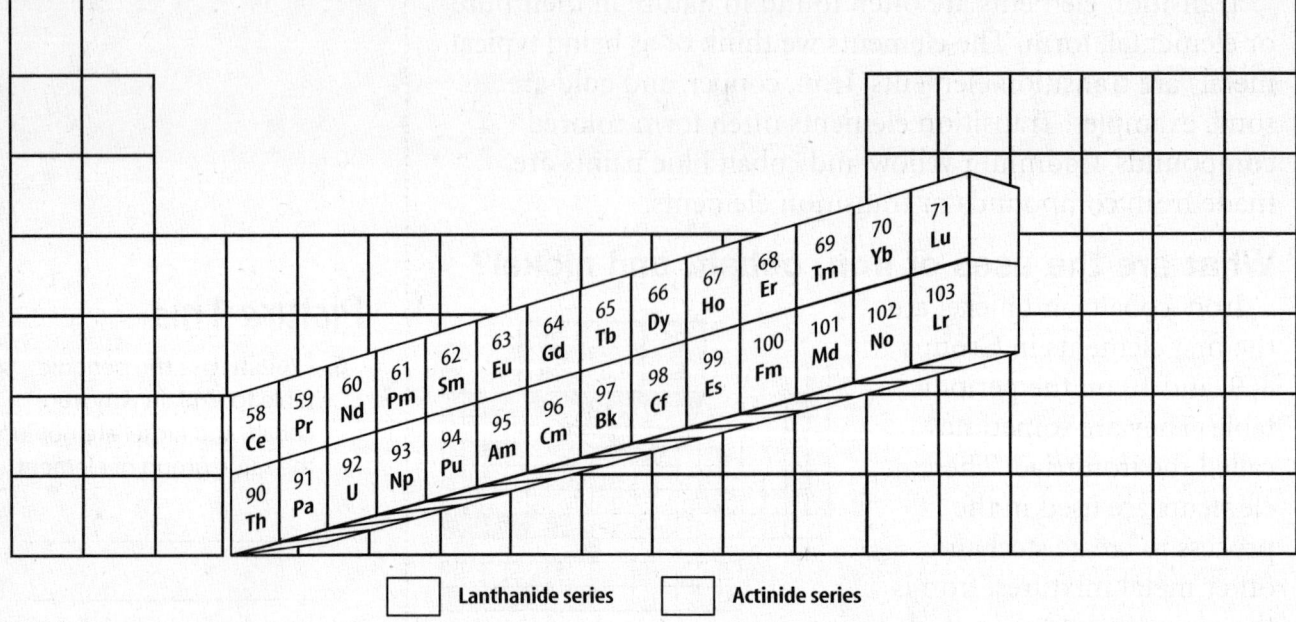

Lanthanide series   Actinide series

Picture This

9. Locate On the periodic table, highlight the lanthanides in one color. Highlight the actinides in a second color. Make the key match your colors.

The Inner Transition Metals

Two rows of elements seem to be apart from the rest on the periodic table shown on the figure. These are the inner transition elements, listed below the main periodic table to save room.

Lanthanides The first row of inner transition elements are called lanthanides (LAN thuh nidez) because they follow lanthanum on the table. Europium, gadolinium, and terbium are used to produce the colors on a TV screen.

Actinides In the second row are the actinides (AK tuh nidez) because they follow actinium. All actinides are radioactive and unstable. Thorium and uranium are the only actinides found in Earth's crust in usable quantities. Thorium is used to make high-quality glass for camera lenses. Uranium is best known for its use in nuclear reactors and weapons.

Metals in the Crust

Earth's crust contains a few pure metals, such as gold and copper. The rest of the metals are combined with other elements in ores. Most ores are made up of a metal, or mineral, mixed with clay and rock. Ores are mined from Earth's crust. Then the mineral is separated from the clay and rock. The mineral is then changed, often with heat, to another physical form. The metal is refined into a pure form. This process can be expensive. Sometimes it costs more to separate a metal from the waste rock than the metal is worth. When this is true, the mineral is no longer called an ore.

Reading Check

10. Determine How are metals obtained from ores?

_____
_____
_____
_____

326 CHAPTER 19 Elements and Their Properties

After You Read

Mini Glossary

ductile: the ability to be stretched into a wire
malleable: the ability to be hammered or rolled into a sheet
metal: an element that is a good conductor of heat and electricity and is usually a solid at room temperature
metallic bonding: occurs because a cloud of electrons moves freely among positively charged metallic ions

radioactive element: an element in which the nucleus of the atom breaks down and gives off particles and energy
transition element: one of the elements in Groups 3 through 12 on the periodic table

1. Review the terms and their definitions in the Mini Glossary. Write a sentence that shows your understanding of one of the properties of metals.

   _____

   _____

2. In the graphic organizer below, list characteristics that are common to all metals. Then list characteristics of the three types of metals. Give examples of each type of metal.

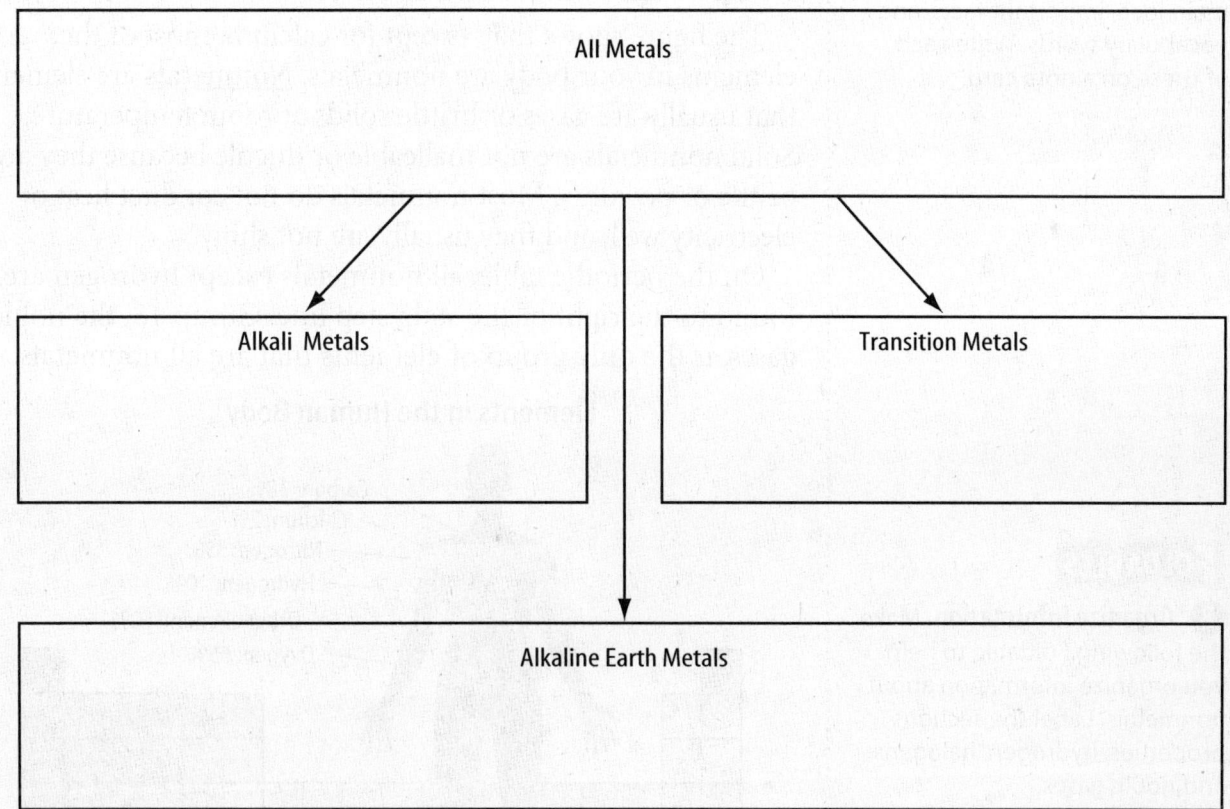

 Visit **gpscience.com** to access your textbook, interactive games, and projects to help you learn more about metals.

chapter 19 Elements and Their Properties

section ❷ Nonmetals

What You'll Learn
- recognize hydrogen as a nonmetal
- the similarities and differences between the halogens
- the properties and uses of noble gases

Study Coach
Make Flash Cards As you read each paragraph, identify the main idea, important facts, and vocabulary words. Write each of these on a note card.

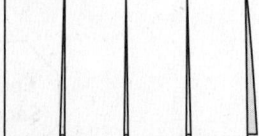

❼ Organize Information Make the following Foldable to help you organize information about nonmetals. Label the sections properties, hydrogen, halogens, and noble gases.

● Before You Read

Think of an object that is metal. Now think of an object that is not metal. What are some differences between these objects? How do they look different? How do they feel different?

_____

_____

● Read to Learn

Properties of Nonmetals

The figure shows that, except for calcium, most of the elements in your body are nonmetals. **Nonmetals** are elements that usually are gases or brittle solids at room temperature. Solid nonmetals are not malleable or ductile because they are brittle or powdery. Most nonmetals do not conduct heat or electricity well, and they usually are not shiny.

On the periodic table, all nonmetals except hydrogen are found to the right of the stair-step line. Group 18, the noble gases, is the only group of elements that are all nonmetals.

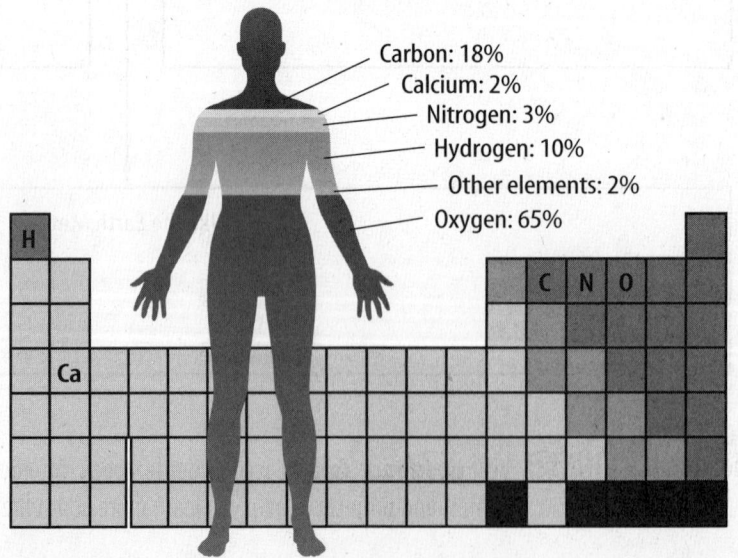

Elements in the Human Body

Carbon: 18%
Calcium: 2%
Nitrogen: 3%
Hydrogen: 10%
Other elements: 2%
Oxygen: 65%

328 CHAPTER 19 Elements and Their Properties

How do nonmetals bond?

Most nonmetals can form ionic and covalent compounds. When nonmetals gain electrons from metal atoms, the nonmetals become the negative ions in ionic compounds. One example is sodium chloride (NaCl), or table salt, combined from the metal sodium and the nonmetal chlorine.

Many covalent compounds are made up of nonmetals. When nonmetals bond with other nonmetals, the atoms usually share electrons to form covalent compounds. Ammonia, NH_3, is a covalent compound. Ammonia produces a strong odor that you may have noticed in some household cleaning products.

Hydrogen

About 90 percent of all the atoms in the universe are hydrogen atoms. Most hydrogen on Earth is found in the compound water. As an element, hydrogen is a gas that is made up of diatomic molecules. A **diatomic molecule** consists of two atoms of the same element in a covalent bond.

Hydrogen is very reactive. A hydrogen atom has a single electron. It shares this electron when it combines with other nonmetals, such as oxygen to form water, H_2O. Hydrogen also can gain an electron when it combines with alkali and alkaline earth metals. These compounds are hydrides. An example is sodium hydride, NaH.

The Halogens

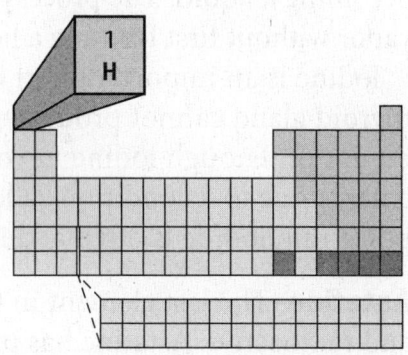

The Group 17 elements are called halogens. They are very reactive and their compounds have many uses. Halogen atoms have seven electrons in their outer energy level. Only one electron is needed to fill the outer level. If a halogen gains an electron from a metal, an ionic compound, called a **salt**, is formed. An example is NaCl. Halogens also can share electrons to form covalent compounds that can be identified by their colors. Chlorine is greenish yellow, bromine is reddish orange, and iodine is violet, or purple.

✓ **Reading Check**

1. **Identify** About what percent of the atoms in the universe are hydrogen?

___

Picture This

2. **Explain** In which group is hydrogen located on the periodic table?

___

💡 **Think it Over**

3. **Conclude** In a salt, what charge does the halogen element have when it gains an electron?

___

What are some uses of halogens?

Fluorine The halogen fluorine is the most chemically active of all the elements. Hydrofluoric acid, a mixture of hydrogen fluoride and water, is used to etch glass and frost the inner surface of lightbulbs. It is also used when making semiconductors. Fluorine compounds are added to toothpaste and drinking water to prevent tooth decay.

Chlorine The most common halogen is chlorine. It is obtained from seawater. Chlorine compounds are added to swimming pools and drinking water to disinfect the water. Many bleaches also contain chlorine compounds. Chlorine bleaches are used to whiten clothing, flour, and paper.

Bromine The only nonmetal compound that is liquid at room temperature is bromine. Bromine also is produced from compounds found in seawater. Bromine compounds are used as dyes and in cosmetics.

Iodine The halogen iodine is a shiny solid at room temperature. When heated, iodine changes to a vapor without first becoming a liquid. The process of a solid changing directly to a vapor without first forming a liquid is called **sublimation**.

Iodine is an important part of your diet. Without it, your thyroid gland cannot produce the hormone thyroxin. If you do not get enough iodine in your diet, your thyroid gland will enlarge due to a condition called goiter. To prevent goiter, potassium iodide is often added to table salt.

Astatine The last element in Group 17 is astatine. It is rare and radioactive. Astatine has properties similar to those of the other halogens. Because it is so rare, it has no known uses.

The Noble Gases

The noble gases are elements that do not form compounds in nature. They are stable because atoms of noble gases have full outer energy levels. Because they are stable, noble gases are useful. Helium is used in blimps and balloons because it is lighter than air. Neon and argon are used in "neon" lights. Argon and krypton are used in lightbulbs and in some lasers.

Reading Check

4. **Define** What is sublimation?

_____

_____

_____

Picture This

5. **Infer** What can you tell about the noble gases from their position on the periodic table?

_____

_____

_____

After You Read

Mini Glossary

diatomic molecule: a molecule consisting of two atoms of the same element in a covalent bond

nonmetals: elements that usually are gases or brittle solids at room temperature

salt: an ionic compound formed when a halogen gains an electron from a metal

sublimation: the process by which a solid changes directly to a vapor without first forming a liquid

1. Review the terms and their definitions in the Mini Glossary. In the space below, draw a diagram that will help you remember the definition of a diatomic molecule.

2. Complete the graphic organizer below to organize properties of nonmetals.

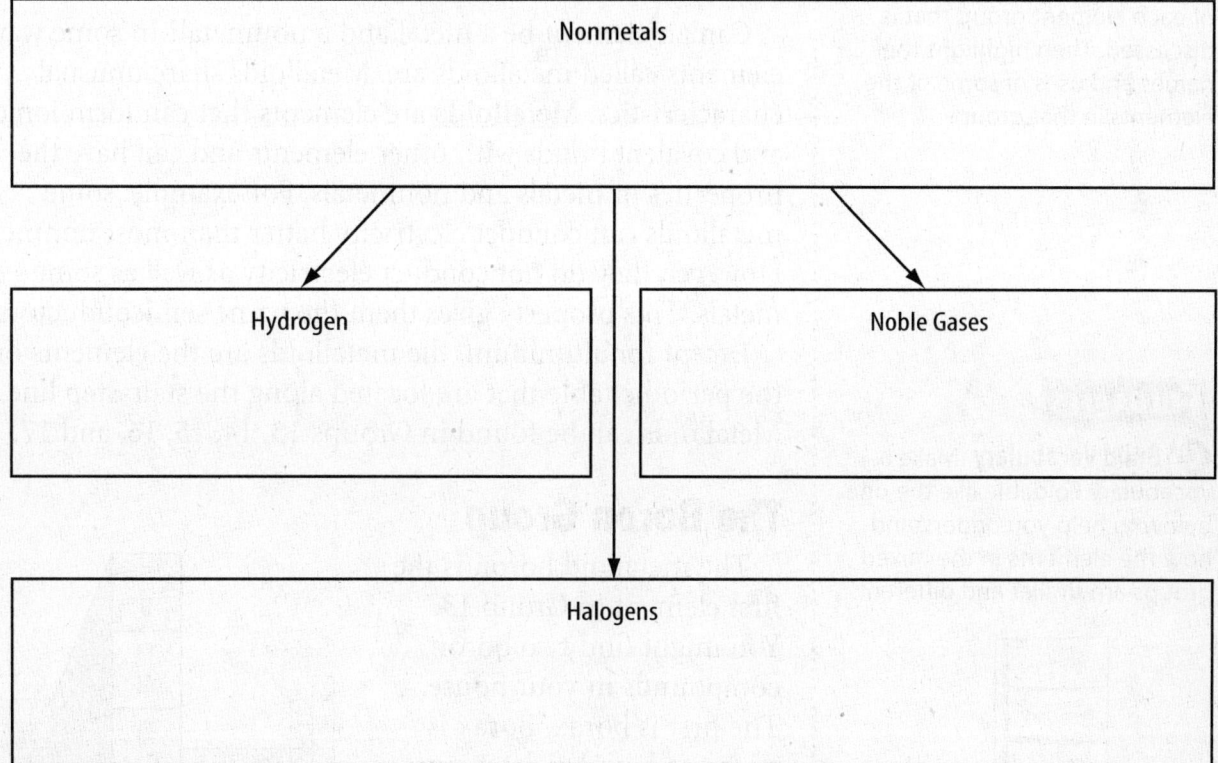

Science Online Visit gpscience.com to access your textbook, interactive games, and projects to help you learn more about nonmetals.

Chapter 19 Elements and Their Properties

section ❸ Mixed Groups

What You'll Learn
- the differences between metals, nonmetals, and metalloids
- what allotropes are
- the different crystal structures of carbon
- why synthetic elements are important

Mark the Text

Identify Elements As you read the section, highlight the name of each element group that is discussed. Then highlight the names and uses of some of the elements in that group.

FOLDABLES

Build Vocabulary Make a vocabulary Foldable like the one below to help you understand how the elements in the mixed groups are similar and different.

- Metalloids
- Boron Group
- Carbon Group
- Nitrogen Group
- Oxygen Group
- Synthetic Elements

● Before You Read

Have you ever seen a mixed-breed dog? What characteristics of different breeds did the dog have?

_____

_____

_____

● Read to Learn

Properties of Metalloids

Can an element be a metal and a nonmetal? In some ways, elements called metalloids are. Metalloids share unusual characteristics. **Metalloids** are elements that can form ionic and covalent bonds with other elements and can have the properties of metals and nonmetals. For example, some metalloids can conduct electricity better than most nonmetals. However, they do not conduct electricity as well as some metals. This property gives them the name semiconductors.

Except for aluminum, the metalloids are the elements on the periodic table that are located along the stair-step line. Metalloids can be found in Groups 13, 14, 15, 16, and 17.

The Boron Group

The metalloid boron is the first element in Group 13. You might find two boron compounds in your house. The first is borax. Borax is added to laundry detergents to soften water. The other is boric acid, an antiseptic. Compounds called boranes are used for jet and rocket fuel.

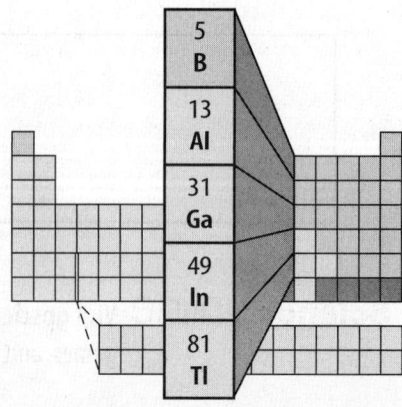

332 CHAPTER 19 Elements and Their Properties

Aluminum is a metal in Group 13. It is the most common metal in Earth's crust. You have seen aluminum in soft drink cans, foil wrap, and cooking pans. Aluminum is strong and light and is used in making airplanes.

The Carbon Group

Atoms of the elements in Group 14 have four electrons in their outer energy levels. Other than this, the elements in the carbon group are very different. Carbon is a nonmetal, silicon and germanium are metalloids, and tin and lead are metals.

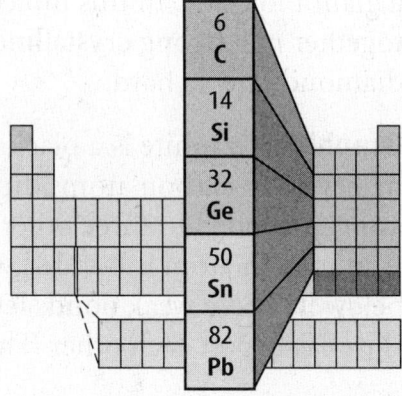

Picture This

1. **Identify** Circle the metalloids in the figure.

Carbon Carbon occurs as an element in coal. Compounds of carbon make up oil and natural gas. Carbon can combine with oxygen to produce carbon dioxide, CO_2. Plants use CO_2 and sunlight to make food. All organic compounds contain carbon, but not all carbon compounds are organic.

Silicon and Germanium The metalloid silicon is the second most common element in Earth's crust. Most silicon is found in sand (SiO_2). It also is found in almost all rocks and soil. <u>Allotropes</u> are different forms of the same element that have different molecular structures. Molecular structure is how the atoms in a molecule are arranged.

Silicon has two allotropes. One allotrope of silicon is a hard, gray substance. The other is a brown powder. Because it is a semiconductor, silicon is used in many electronic devices, such as transistors and computer chips. A <u>semiconductor</u> is an element that conducts electric current under certain conditions. Germanium is the other metalloid in the carbon group. It is used with silicon to make semiconductors. ✓

Reading Check

2. **Explain** What are semiconductors?

Tin and Lead Tin is used to coat other metals to prevent corrosion. It also is mixed with other metals to make bronze and pewter. Tin cans are made of steel coated with tin. Lead is a soft metal that was once used to make paints. Lead is no longer used in paint because it is toxic.

What are the allotropes of carbon?

What do a diamond ring and a pencil have in common? The diamond in a ring and the graphite in a pencil are carbon. There are three known allotropes of carbon. They are diamond, graphite, and buckminsterfullerene (BUK mihn stur ful ur een).

Think it Over

3. Apply Why do you think some saw blades are coated in diamond powder?

_____

_____

Diamonds Diamonds are clear and extremely hard. In fact, they are the hardest things in the world. In a diamond, each carbon atom is bonded to four other carbon atoms. The bonded carbon atoms form a geometric shape called a tetrahedron. Many of these tetrahedrons join together to form a giant molecule. In this molecule, the atoms are held tightly together in a strong crystalline structure. This strucure is why diamonds are so hard.

Graphite Graphite is a black powder that is made of hexagonal layers of carbon atoms. In the hexagons, each carbon atom is bonded to three other carbon atoms. The fourth electron in each atom is bonded weakly to the layer above or below it. These weak bonds let the layers of carbon atoms slide easily past each other. This makes graphite very slippery.

Buckminsterfullerene Buckminsterfullerene was discovered in the 1980s. It forms molecules that are shaped like soccer balls. It is named after R. Buckminster Fuller, who designed structures with similar shapes. Scientists have used buckministerfullerene to make tiny tubes called nanotubes. They are one-billionth of a meter in diameter. You can stack tens of thousands of nanotubes to get the thickness of one sheet of paper. Nanotubes might be used some day to make smaller, faster computers.

Applying Math

4. Use Percentages Assume you inhale in about 6 L of air per minute. How much nitrogen do you breathe in one minute?

_____

The Nitrogen Group

Group 15 is known as the nitrogen family. Atoms of each element in the group have five electrons in their outer energy levels. Group 15 elements usually share their electrons in covalent bonds with other elements.

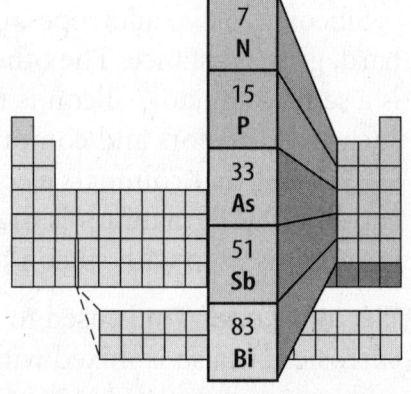

About 80 percent of the air you breathe is nitrogen. In the air, nitrogen exists as diatomic molecules, N_2. All organisms need nitrogen compounds to live. Even though you breathe it, your body cannot use nitrogen in its diatomic form. It must be in the form of a nitrogen compound. Nitrogen often is used to make compounds called nitrates and ammonia, NH_4. Nitrates are compounds that contain the nitrate ion, NO_3. Nitrates and ammonia are important ingredients of fertilizers.

How are elements of the nitrogen group used?

Phosphorus is a nonmetal that has three allotropes. Its compounds are used to make many things, including fertilizers, water softeners, match heads, and fine china. Antimony is a metalloid, and bismuth is a metal. These elements are added to metals to lower their melting points. Automatic fire sprinkler heads sometimes contain bismuth. The metal melts from the heat of a fire and turns on the sprinkler.

The Oxygen Group

Oxygen Group 16 on the periodic table is the oxygen group. Oxygen, a nonmetal, makes up about 21 percent of air. Oxygen exists in the air as a diatomic molecule, O_2. During electrical storms, some oxygen molecules, O_2, change into ozone, O_3. Nearly all living things on Earth need O_2 to live. Living things also depend on a layer of O_3 in the atmosphere, called the ozone layer. The ozone layer protects living things from the Sun's radiation.

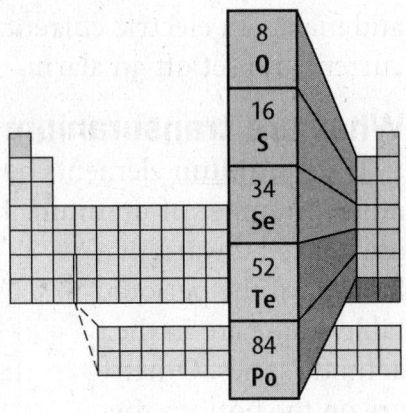

Sulfur The second element in the oxygen group is sulfur. It is a nonmetal and has several allotropes. Sulfur combines with metals to form compounds called sulfides. Some sulfides are colorful and are used as pigments in paints.

Other Elements Selenium, a nonmetal, and tellurium and polonium, metalloids, are the other elements in Group 16. You need a tiny amount of selenium in your body, so it is included in multivitamins. But too much selenium is toxic. Selenium also is used in photocopiers.

Synthetic Elements

Scientists have created elements that usually do not exist on Earth. These are called synthetic elements. To create them, scientists smash existing elements with particles from a heavy ion accelerator. Except for some isotopes of natural elements, all synthetic elements have more than 92 protons. Many synthetic elements disintegrate, or fall apart, soon after they are made. It may seem strange to make elements that fall apart, or disintegrate, but scientists are learning about atoms this way.

Picture This

5. Use a Diagram What is the atomic number of oxygen?

_____

Reading Check

6. Apply What happens to many of the synthetic elements after they are made?

_____

How are synthetic elements used?

Scientists smash protons into uranium to make neptunium, element 93. In about two days, half of neptunium atoms disintegrate. When neptunium atoms disintegrate, they form plutonium, element 94. Plutonium has been produced in nuclear reactors and is used in bombs. Plutonium also can be changed to americium, element 95. You probably have some americium in your home. There is a small amount of the element in smoke detectors. An electric plate in smoke detectors attracts some of the charged americium particles and makes an electric current. A lot of smoke will break the current and set off an alarm.

What are transuranium elements?

<u>Transuranium</u> elements have more than 92 protons, the atomic number of uranium. The transuranium elements are located toward the bottom of the periodic table. Some are in the actinide series. Others are on the bottom row of the main periodic table. All of the transuranium elements are synthetic. They also are unstable and many disintegrate quickly.

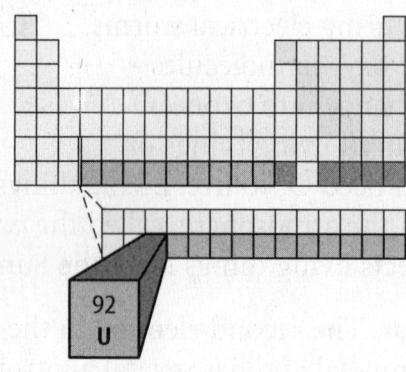

Why make elements?

For centuries, scientists discovered new natural elements. Now, all new elements are created in laboratories. When these atoms disintegrate, they are said to be radioactive. Radioactive elements can be useful. For example, technetium has medical uses.

If most transuranium elements break down quickly, why make them? Scientists make new elements to study the forces that hold the nucleus together. In the 1960s, scientists theorized that stable synthetic elements exist. Maybe scientists will one day find a way to make transuranium elements that do not break down. Perhaps a transuranium element will be found that has everyday uses.

Picture This

7. Infer The atomic number of einsteinium is 99. Is einsteinium a transuranium element? Why or why not?

_____

_____

Think it Over

8. Draw Conclusions Why do you think scientists have not discovered new natural elements in many years?

_____

_____

After You Read

Mini Glossary

allotropes: different forms of the same element that have different molecular structures

metalloids: elements that can form ionic and covalent bonds with other elements and can have the properties of metals and nonmetals

semiconductor: an element that can conduct electricity under certain conditions

transuranium element: an element with more than 92 protons

1. Review the terms and their definitions in the Mini Glossary. Write a sentence that tells what semiconductors are used for.

   _____

   _____

2. Use the graphic organizer below to list the properties of metalloids. Then list examples of some important metalloids from each group.

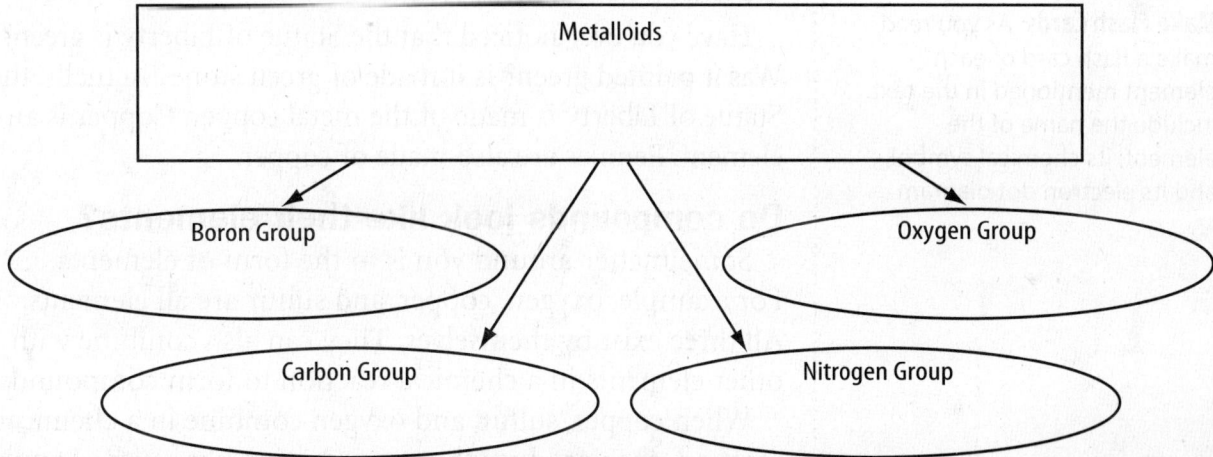

3. As you read the section, you highlighted the element groups that were discussed and the names and uses of some of the elements in each group. Was this a good strategy for learning the information? Why or why not?

   _____

   _____

   _____

 Visit **gpscience.com** to access your textbook, interactive games, and projects to help you learn more about the mixed groups of elements.

chapter 20 Chemical Bonds

section ❶ Stability in Bonding

What You'll Learn
- about elements in a compound
- chemical formulas
- how electric forces help form compounds
- why a chemical bond forms

Study Coach

Make Flash Cards As you read, make a flash card of each element mentioned in the text. Include the name of the element, its chemical symbol, and its electron dot diagram.

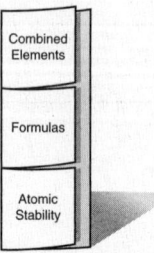

A Find Main Ideas Make a Foldable like the one below. As you read this section, write down the main ideas about combined elements, formulas, and atomic stability.

● Before You Read

Have you ever seen an object with rust on it? What was the object made of? Why do you think rust formed on it?

_____

_____

● Read to Learn

Combined Elements

Have you ever noticed that the Statue of Liberty is green? Was it painted green? Is it made of green stone? Actually, the Statue of Liberty is made of the metal copper. Copper is an element. Pennies are also made of copper.

Do compounds look like their elements?

Some matter around you is in the form of elements. For example, oxygen, copper, and sulfur are all elements. All three exist by themselves. They can also combine with other elements in a chemical reaction to form compounds.

When copper, sulfur, and oxygen combine in a chemical reaction, they produce the green coating like on the Statue of Liberty. This compound is copper sulfate, which is a green solid. It is nothing like the elements that combined to make it. Copper is a shiny, copper-colored solid. Sulfur is a yellow solid. Oxygen is a colorless, odorless gas. When elements combine, they produce compounds with their own special properties. To form rust, iron combines with water and oxygen.

Do compounds have new properties?

Compounds may have different properties from the elements that form them. Sodium chloride, or table salt, is a compound made from the elements sodium and chlorine. Sodium is a shiny, soft, silvery metal that reacts violently with water. Chlorine is a greenish-yellow, poisonous gas. These two elements combine to form the salt that people use in their food.

338 CHAPTER 20 Chemical Bonds

Formulas

The formula for sodium chloride is NaCl. Na is the chemical symbol for the element sodium. Cl is the chemical symbol for the element chlorine. Together they make up the formula for sodium chloride. A **chemical formula** tells what elements are in a compound and how many atoms of each element are in one unit of the compound. ✓

Let's look at the formula for a compound you use every day. H_2O is the chemical formula for water. H is the symbol for the element hydrogen. O is the symbol for the element oxygen. The number 2 in the formula is called a subscript. Subscript means "written below." A subscript written after a symbol tells how many atoms of that element are in one unit of the compound. In H_2O, the 2 tells you there are two atoms of hydrogen in one unit of water. If there is no subscript after a symbol in a formula, there is only one atom of that element in the compound. So, in one unit of H_2O there are two hydrogen atoms and one oxygen atom. The table shows some familiar compounds and their formulas.

Reading Check

1. **Explain** What does a chemical formula tell you?

_____
_____
_____
_____

Some Familiar Compounds

Familiar Name	Chemical Name	Formula
Sand	Silicon dioxide	SiO_2
Milk of magnesia	Magnesium hydroxide	$Mg(OH)_2$
Cane sugar	Sucrose	$C_{12}H_{22}O_{11}$
Vinegar	Acetic acid	CH_3COOH

Picture This

2. **Describe** The formula for cane sugar, or sucrose, is $C_{12}H_{22}O_{11}$. Describe what one unit of cane sugar is made of.

_____
_____
_____

Atomic Stability

Recall that protons have a positive charge and electrons have a negative charge. These opposite electric forces attract each other. They are the forces that pull atoms together to form compounds. ✓

Reading Check

3. **Determine** What are the forces that pull atoms together to form compounds?

_____
_____

Reading Essentials   **339**

Reading Check

4. Explain How are noble gases different from other elements?

_____

_____

_____

Why do atoms form compounds?

Look at the periodic table on the inside back cover of this book. It lists all the known elements. Most of these elements can combine with others to form compounds that are more stable. Notice the six elements in Group 18. These elements are gases called the noble gases. Atoms of the noble gases are very stable. They are different from the other elements because they almost never combine to form compounds. Compounds that are formed with a noble gas are less stable than the original atom. ✔

Why are the noble gases different?

A helpful way to picture the stability of the noble gases is to look at the electron dot diagrams of those elements. An electron dot diagram shows the symbol of the element. It also shows the electrons in the outer energy level of an atom. The number of electrons in the outer energy level of an atom determines if that atom will combine to form a compound.

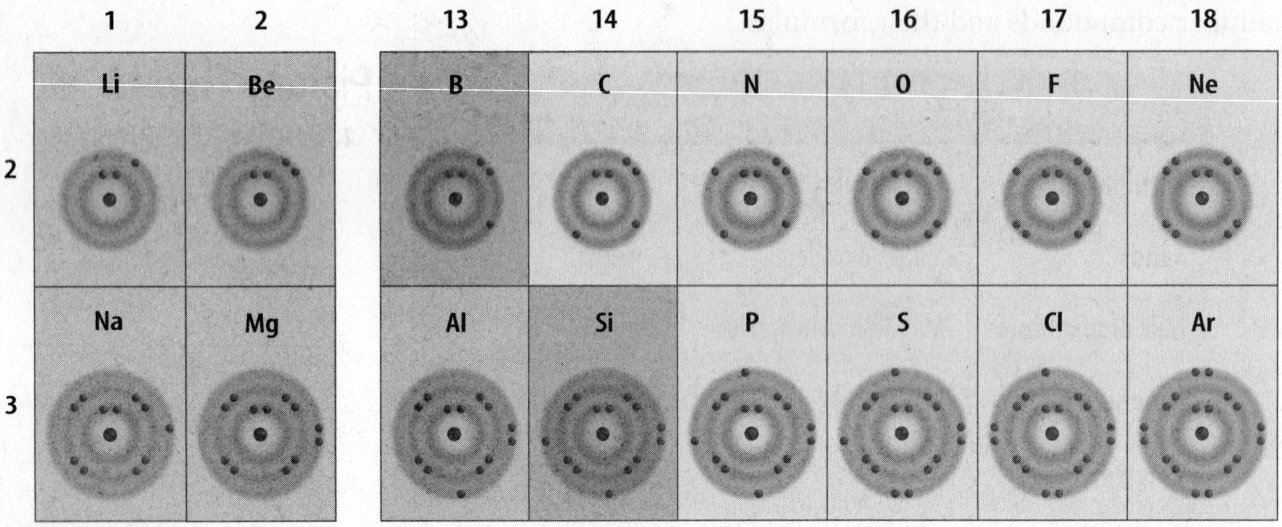

	1	2	13	14	15	16	17	18
2	Li	Be	B	C	N	O	F	Ne
3	Na	Mg	Al	Si	P	S	Cl	Ar

Picture This

5. Interpret Scientific Illustrations How many electrons are in the outer energy level of an atom of carbon?

_____

_____

How do you know how many dots to put in an electron dot diagram? For elements in Groups 1 and 2 and 13 through 18, you can use the periodic table. The figure above shows two rows of the periodic table. Look at the outer ring of each element in Group 1. All elements in Group 1 have one outer electron. The elements in Group 2 have two outer electrons. Group 13 elements have three outer electrons, Group 14 has four, and so on. The noble gases in Group 18 have eight outer electrons.

What makes an atom stable?

An atom is chemically stable when it has a complete outer energy level. If an atom is chemically stable, it does not easily form compounds with other atoms.

Stable Noble Gases The figure to the right shows electron dot diagrams of the noble gases. The first electron dot diagram is for helium. Notice that there are two electrons around the symbol for helium, He. Remember that hydrogen and helium need only two electrons in their outer level to be stable. The outer energy levels of all the other elements are stable when they contain eight electrons.

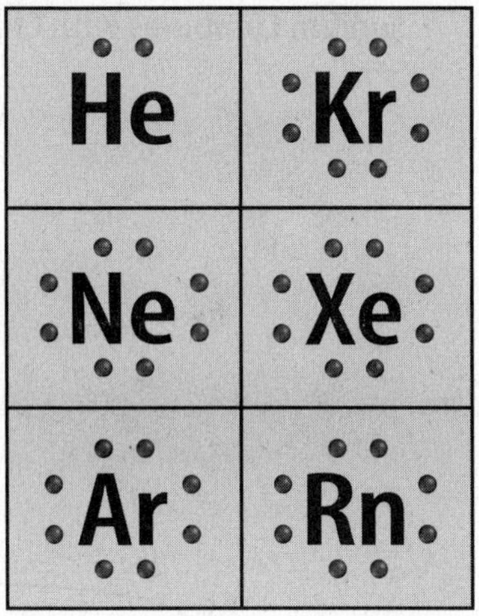

Now look at the electron dot diagram for neon, Ne. There are eight electrons in neon's outer energy level. Neon is stable as an atom. Neon does not become more stable if it forms a compound. All noble gases have a complete outer energy level which makes them stable.

What elements have incomplete outer energy levels?

Hydrogen and helium are the only elements in the first row, or period, of the periodic table. Both hydrogen and helium need only two outer electrons to be stable. Helium has two electrons in its outer energy level and is stable. But hydrogen has only one electron. Its outer level is not full. Therefore, hydrogen is more stable when it is part of a compound.

Look at the periodic table again. You can see that none of the elements in Groups 1 through Group 17 have full outer energy levels. These elements are more stable when they form compounds.

How do atoms become more stable?

Atoms that do not have a stable outer energy level can do one of three things to complete their outer level. They can gain electrons, lose electrons, or share electrons to make a full outer level. Atoms combine with other atoms that also do not have complete outer levels. This way, each atom becomes more stable.

The figure on the next page shows electron dot diagrams for sodium and chlorine. Sodium has one outer electron and chlorine has seven. When these two elements combine, sodium gives its outer electron to chlorine. Chlorine now has eight electrons in its outer level. This is a full, stable energy level. But what about sodium?

Picture This
6. Describe How many electrons are in the outer energy level of xenon?

_____

_____

Reading Check
7. Infer What makes noble gases stable?

_____

_____

Picture This

8. Use a Model Draw an electron dot diagram of sodium showing its new outer energy level after it has given its outer electron to chlorine.

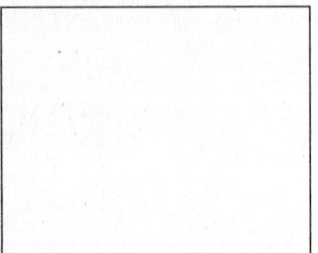

Sodium Combines with Chlorine to Form Sodium Chloride

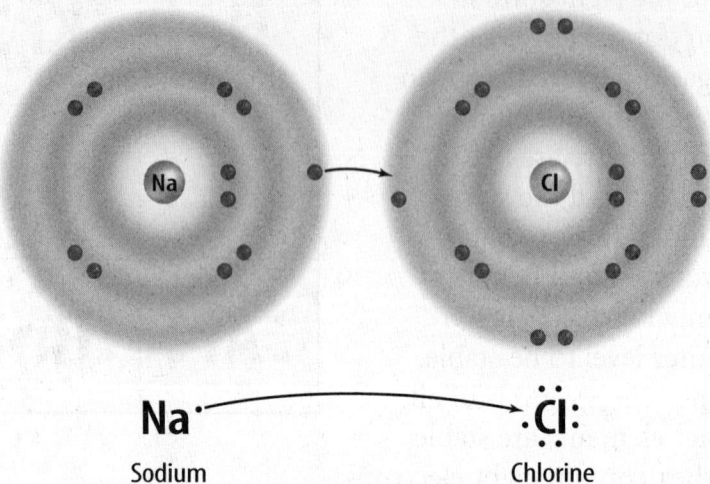

How can giving up an electron make sodium more stable?

Sodium gave up the only electron in its outer level when it combined with chlorine to form sodium chloride. If sodium loses the electron in the outer energy level, what is left? The next energy level of the sodium atom has eight electrons. When the one outer electron is removed, the next level becomes the new outer level. Sodium now has a complete, stable outer energy level. Sodium and chlorine exchanged an electron and are now both stable in the compound they formed.

What is another way atoms can become stable?

Remember the formula for the compound water, H_2O. Hydrogen atoms need one electron in their outer level to be stable. Oxygen atoms need two electrons to be stable. In this case, neither atom can give up an electron. Instead, they share electrons. Each hydrogen atom shares one electron from an oxygen atom to complete its outer level. In turn, the oxygen atom shares each of the hydrogen atoms' electrons to complete its outer level. The compound water is more stable than a hydrogen atom or an oxygen atom.

An attractive force forms when atoms gain, lose, or share electrons to make a compound. This force is a chemical bond. A **chemical bond** is the force that holds atoms together in a compound.

Reading Check

9. Explain What forms when atoms gain, lose, or share electrons to make a compound?

After You Read

Mini Glossary

chemical bond: the force that holds atoms together in a compound

chemical formula: a formula that tells what elements and how many atoms of each element are in a compound

1. Review the terms and their definitions in the Mini Glossary. Use the term *chemical bond* in a sentence that explains how atoms become more stable.

   _____

   _____

   _____

2. Complete the concept web to organize the information from this section.

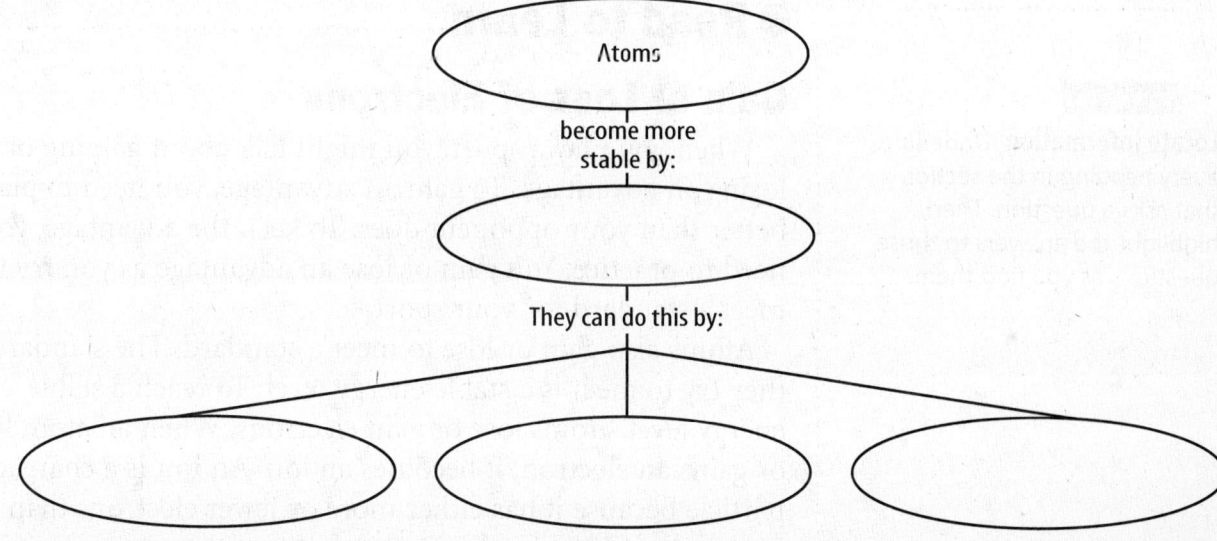

3. As you read this section, you made flash cards of the elements mentioned in the text. Do you think that making flashcards would be a good way to learn the symbols of all the elements in the periodic table? Why or why not?

   _____

   _____

   _____

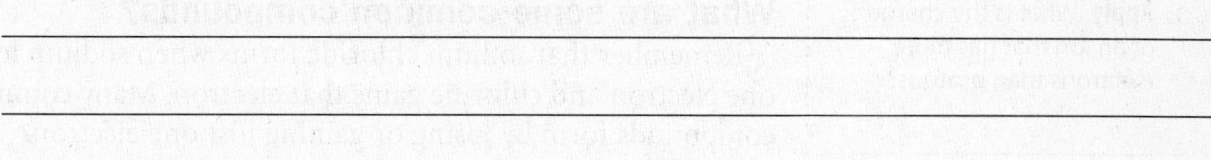

Visit **gpscience.com** to access your textbook, interactive games, and projects to help you learn more about stability in bonding.

End of Section

Reading Essentials **343**

chapter 20 Chemical Bonds

section ❷ Types of Bonds

What You'll Learn
- what ionic bonds are
- what covalent bonds are
- what particles are produced by ionic and covalent bonding
- about polar and nonpolar covalent bonds

Mark the Text

Locate Information Underline every heading in the section that asks a question. Then, highlight the answers to those questions as you find them.

Reading Check

1. **Apply** What is the charge of an ion that has more electrons than protons?

_____

● Before You Read

Some atoms share electrons and become more stable. Describe a situation in which people share something and everyone benefits.

_____

_____

● Read to Learn

Gain or Loss of Electrons

When you play a sport, you might talk about gaining or losing an advantage. To gain an advantage, you need to play better than your opponent does. To keep the advantage, you need to practice. You gain or lose an advantage as you try to meet a standard for your sport.

Atoms also gain or lose to meet a standard. The standard they try to meet is a stable energy level. To reach a stable energy level, atoms lose or gain electrons. When an atom loses or gains an electron, it becomes an ion. An **ion** is a charged particle because it has either more or fewer electrons than it has protons. If an ion has more electrons than protons, it is a negative ion. If it has more protons than electrons, it is a positive ion. The electric forces between positive and negative particles hold compounds together. ☑

What are some common compounds?

Remember that sodium chloride forms when sodium loses one electron and chlorine gains that electron. Many common compounds form by losing or gaining just one electron. These compounds are made from a Group 1 element, such as sodium, and a Group 17 element, such as chlorine. Other examples of these compounds are sodium fluoride, an ingredient added to some toothpaste to fight cavities, and potassium iodide, an ingredient in iodized table salt.

344 CHAPTER 20 Chemical Bonds

How is potassium iodide formed?

What happens when potassium and iodide bond to form potassium iodide? The figure below shows the dot diagrams and electron distribution for this process.

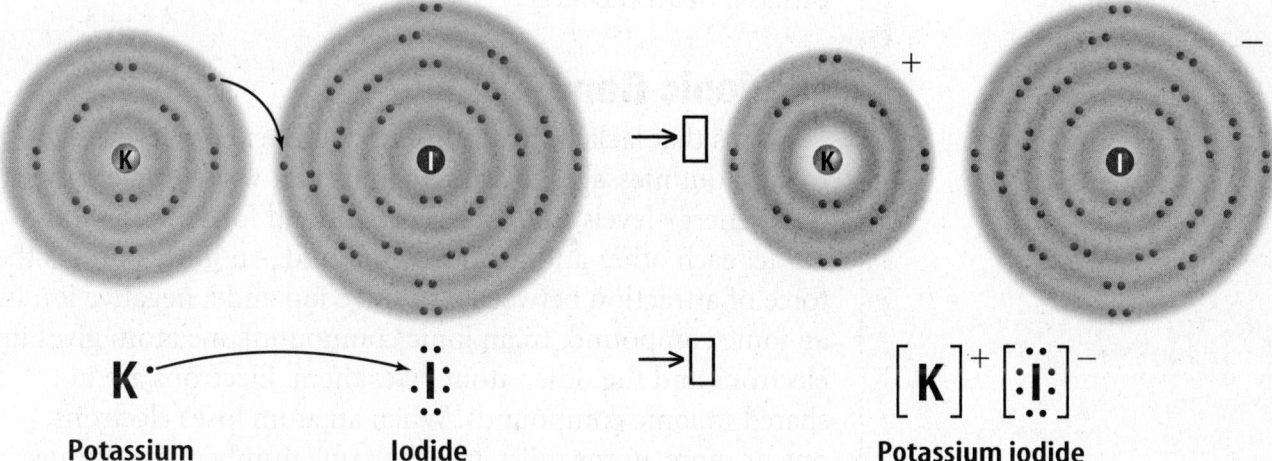

An atom of potassium has one electron in its outer level. This is not a stable outer energy level. When a potassium atom combines with an iodine atom, the potassium atom loses the one electron in its outer level. The third energy level now becomes a complete outer level. The potassium atom now has one less electron than it has protons. The positive and negative charges are no longer equal. The potassium atom is now a positive ion. It has more protons with positive charges than electrons with negative charges.

The potassium ion has a 1+ charge. The symbol for a positive potassium ion is K^+. The plus sign shows its positive charge. In the symbol K^+, the plus sign is a superscript. Superscript means "written above."

How does the iodine atom change?

The iodine atom also changes when it combines to form potassium iodide. An iodine atom has seven electrons in its outer energy level. If it had eight electrons, it would have a stable outer energy level. When the iodine atom reacts with the potassium atom, the iodine atom gains one electron from potassium. There are now eight electrons in iodine's outer level.

The iodine atom now is stable, but it has one more electron than it has protons. It has more negative particles than positive particles. The iodine atom has become a negative ion with a charge of 1−. It is called an iodide ion and its symbol is I^-.

Picture This

2. Describe How many electrons are there in potassium's outer energy level *after* the bond has formed?

_____

FOLDABLES

B Build Vocabulary Make a Foldable like the one below. As you read this section, add the definitions of ionic bonds, covalent bonds, polar bonds, and nonpolar bonds to your Foldable. Include a compound that contains each bond.

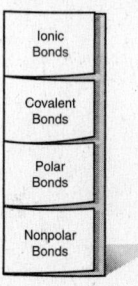

An ionic compound forms when two or more ions combine. Notice that the ionic compound potassium iodide does not have a charge. The 1+ charge and the 1− charge cancel each other out. This means the compound is neutral or has a neutral charge.

The Ionic Bond

Atoms that lack electrons in their outer energy level seem to send out messages to attract atoms that will complete their outer energy levels, and vice versa. A bond forms when ions attract each other and form a compound. An **ionic bond** is the force of attraction between a positive ion and a negative ion in an ionic compound. In an ionic compound, one atom gives up electrons and the other atom takes them. Electrons are not shared in ionic compounds. When an atom loses electrons, one or more atoms must gain the same number of electrons that were lost. That way, the compound stays neutral.

Potassium iodide forms an ionic bond when one electron is transferred. What happens when more than one electron is transferred? The figure shows the formation of another ionic compound, magnesium chloride, $MgCl_2$.

$$Mg: \quad \begin{matrix} \cdot \ddot{\underset{\cdot \cdot}{Cl}}: \\ \cdot \ddot{\underset{\cdot \cdot}{Cl}}: \end{matrix} \quad \rightarrow \quad [:\ddot{\underset{\cdot \cdot}{Cl}}:]^- \; Mg^{2+} \; [:\ddot{\underset{\cdot \cdot}{Cl}}:]^-$$

Magnesium   2 chlorine atoms   Magnesium chloride

When magnesium reacts with chlorine, a magnesium atom loses the two electrons in its outer energy level. The atom becomes a positively charged magnesium ion. The symbol for this ion is Mg^{2+} because it has lost two electrons and now has a charge of 2+.

Two chlorine atoms each take one of the electrons and complete their outer levels. Each chlorine atom becomes an ion with a 1− charge. Cl^- is the symbol for the chloride ion.

In this case, the magnesium atom has two electrons to give. However, a chlorine atom can only use one electron. So, it takes two chlorine atoms to take the two electrons from magnesium and combine to form the ionic compound magnesium chloride.

Picture This

3. Think Critically Explain why magnesium chloride cannot be made from one atom of magnesium and one atom of chlorine.

_____

_____

Does an ionic compound have a charge?

The ionic compound magnesium chloride is neutral. It does not have a charge. The compound is neutral because the sum of the charges on the ions is zero. The 2+ charge on the magnesium ion is exactly equal to two 1− charges on the chloride ions. When atoms form an ionic compound, they transfer their electrons. The total number of electrons and protons stays the same. The compound is neutral.

Ionic bonds usually form when a metal bonds with a nonmetal. Elements on the far left side of the periodic table tend to form ionic bonds with elements on the far right. Ionic compounds are often crystalline solids with high melting points.

Sharing Electrons

Some atoms of nonmetal elements become more stable when they share electrons. Look at the elements in Group 14 of the periodic table. These elements have four electrons in their outer levels. They would have to gain or lose four electrons to have a stable outer energy level. It takes a lot of energy for an atom to lose or gain that many electrons. Each time an ion loses an electron, the nucleus holds the remaining electrons even more tightly. Once one electron is removed, it takes more energy to remove a second electron. It takes even more energy to remove a third, and so on. For elements in Group 14, it is much easier for them to become stable by sharing electrons.

Atoms that share electrons form a covalent bond. A **covalent bond** is the force of attraction between two atoms that share electrons. A **molecule** is the neutral particle that forms when atoms share electrons.

How are single covalent bonds formed?

A single covalent bond forms when two atoms share two electrons. Usually one electron comes from each atom in the covalent bond. Look at the water molecule in the figure. There are two single covalent bonds in a water molecule. In each single bond, a hydrogen atom and an oxygen atom each give one electron, which the atoms share.

Covalent bonds also help atoms fill their outer energy levels and become more stable. In a water molecule, each hydrogen atom now has two electrons in its outer level. The oxygen atom is also stable, because it has eight outer electrons.

Reading Check

4. **Explain** Why is the ionic compound magnesium chloride neutral?

_____
_____
_____

Picture This

5. **Locate** Circle the electrons that could have belonged only to the hydrogen atoms before they bonded with the oxygen atom.

What are multiple bonds?

A covalent bond can have more than two electrons. Look at the dot diagram of the two nitrogen atoms. Each nitrogen atom has five electrons in its outer energy level. A nitrogen atom needs to gain three electrons to be stable.

When two nitrogen atoms combine, they share three electrons with each other. The bond between the two atoms has six electrons, or three pairs of electrons. Each electron pair is a covalent bond. Three pairs of electrons form a triple bond. By sharing the electrons, each nitrogen atom now has eight electrons in its outer energy level. The symbol for the nitrogen molecule is N_2.

Molecules also can have double bonds. A double bond is two pairs of electrons shared between two atoms. In the carbon dioxide molecule, CO_2, the carbon atom shares two electrons with one oxygen atom and two electrons with the other oxygen atom. In return, each oxygen atom shares two electrons with the carbon atom. In this way, all three atoms have eight electrons in their outer energy levels.

Covalent bonds form between nonmetals. Nonmetals are found in the upper right-hand corner of the periodic table. Many covalent compounds are liquids or gases at room temperature.

Are electrons always shared equally?

Atoms in a covalent bond do not always share electrons equally. The positive charge of an atom's nucleus attracts the electrons in a bond. Some nuclei attract electrons more strongly than others. If a shared electron is closer to one nucleus, that nucleus could attract the electron more strongly.

You can see this with a magnet and a piece of metal. When the magnet is closer to the metal, it attracts the metal more strongly. Some nuclei have a greater positive charge than others. Nuclei with a greater positive charge attract electrons more strongly. In the same way, a strong magnet holds the metal more firmly than a weak magnet.

Picture This

6. **Use Models** Draw an electron dot diagram showing two oxygen atoms combining with one carbon atom to form a molecule of carbon dioxide, CO_2.

What is an example of unequal electron sharing?

The covalent bond in a molecule of hydrogen chloride, HCL, is an example of unequal electron sharing. When HCl mixes with water, it becomes hydrochloric acid. Hydrochloric acid is used to clean metal in factories. It is also found in your stomach where it helps digest food.

The chlorine atom attracts the electrons in the bond more strongly than the hydrogen atom. That means the electrons in the bond spend more time closer to the chlorine atom than to the hydrogen atom. The chlorine atom has a partial negative charge when the bonding electrons are closer to it. The symbol for a partial negative charge is the lower case Greek letter delta followed by a negative superscript, δ^-. The hydrogen atom has a partial positive charge when the bonding electrons are farther away from it. The symbol for a partial positive charge is δ^+.

Reading Check

7. Apply Why does the chlorine atom have a partial negative charge in a molecule of hydrogen chloride?

What are polar and nonpolar molecules?

The figure shows the energy levels of a water molecule. The oxygen atom forms a covalent bond with each hydrogen atom. The oxygen atom has a stronger attraction for these bonding electrons. The two electron pairs spend more time closer to the oxygen atom than to the hydrogen atom. This gives the oxygen atom a partial negative charge. The hydrogen atoms have a partial positive charge.

$\delta-$

$\delta+$ $\delta+$

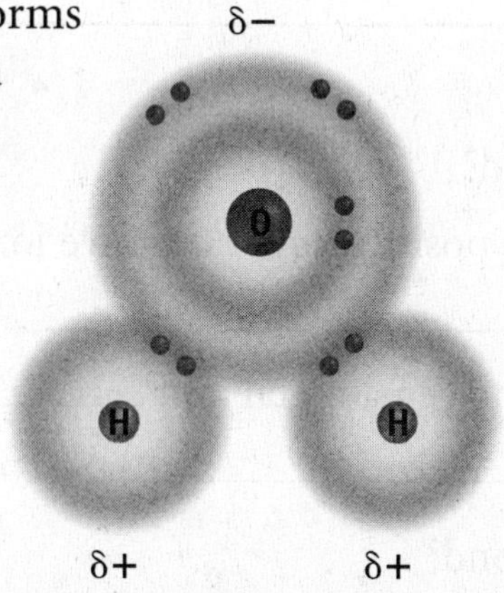

H_2O
Water

Because it has an end that is partially positive and an end that is partially negative, water is a polar molecule. *Polar* means "having opposite ends." A **polar molecule** is a molecule that has a slightly positive end and a slightly negative end, but the molecule itself is neutral.

When two atoms that are exactly alike form a covalent bond, they share the bonding electrons equally. In a **nonpolar molecule**, the electrons are shared equally in the bond. A nonpolar molecule does not have partial charges. It does not have oppositely charged ends. Atoms of the same element can form nonpolar molecules.

Picture This

8. Determine Circle the electrons that are not used to form covalent bonds in the water molecule.

After You Read
Mini Glossary

covalent bond: the force of attraction between two atoms that share electrons

ion: a charged particle that has either more or fewer electrons than it has protons

ionic bond: the force of attraction between a positive ion and a negative ion in an ionic compound

molecule: the neutral particle that forms when atoms share electrons

nonpolar molecule: a molecule where the electrons are shared equally in the bond

polar molecule: a molecule that has a slightly positive end and a slightly negative end, but the molecule itself is neutral

1. Review the terms and definitions in the Mini Glossary. Choose two terms that are related and write a sentence that uses those two terms.

   _____

   _____

2. Complete the outline to help you organize what you learned about chemical bonds.

   Types of Bonds

   I. What is an ionic bond?
      A. Forms between a positive and a negative ion
      B. Electrons are _____
      C. Usually forms between a metal and _____
      D. Forms a(n) _____ compound

   II. What is a covalent bond?
      A. _____
      B. Electrons are _____
      C. _____
      D. Forms a _____

3. You underlined every heading in the section that asks a question and then highlighted the answers. How did this help you learn the content of this section?

   _____

 Visit **gpscience.com** to access your textbook, interactive games, and projects to help you learn more about types of bonds.

chapter 20 Chemical Bonds

section ❸ Writing Formulas and Naming Compounds

● Before You Read

Shakespeare asked, "What's in a name?" In this section, you are going to learn what the names of chemical compounds can tell you. On the lines below, explain what your name tells about you.

_____

_____

What You'll Learn
- how to determine oxidation numbers
- how to write formulas and names for ionic compounds
- how to write formulas and names for ionic covalent compounds

● Read to Learn

Binary Ionic Compounds

Would you like to turn lead into gold? A long time ago, some scientists, called alchemists, spent much of their time trying to do just that. They never succeeded, but they did develop some laboratory methods that scientists still use today. Alchemists also developed symbols to represent formulas of chemical compounds. These symbols don't look anything like the symbols scientists use today.

The first formulas you will learn are for binary ionic compounds. A **binary compound** is a compound made of two elements. An example of a binary compound is potassium iodide, a compound added to table salt.

Are electrons gained or lost?

Before you can write a correct formula of a compound, you need to know which elements combine to make that compound. You also need to know how many electrons the elements lose, gain, or share when they combine. The oxidation number of an element gives you that information. The **oxidation number** of an element tells you how many electrons an atom gains, loses, or shares to become stable.

The charge on the ion in an ionic compound is the same as its oxidation number. For example, a sodium ion has a charge of 1+ and the oxidation number of sodium is 1+. A chloride ion has a charge of 1− and its oxidation number is 1−.

Study Coach

Make a Quiz As you read the text under each heading, write a question that your teacher might ask on a quiz. Exchange your questions with a partner and take each other's quizzes.

FOLDABLES

C Build Vocabulary Make two quarter sheet Foldables as shown below. Record information about the oxidation number of the elements and rules for writing formulas as you read.

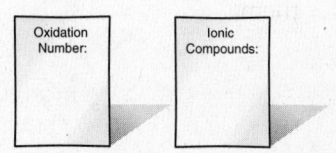

Reading Essentials 351

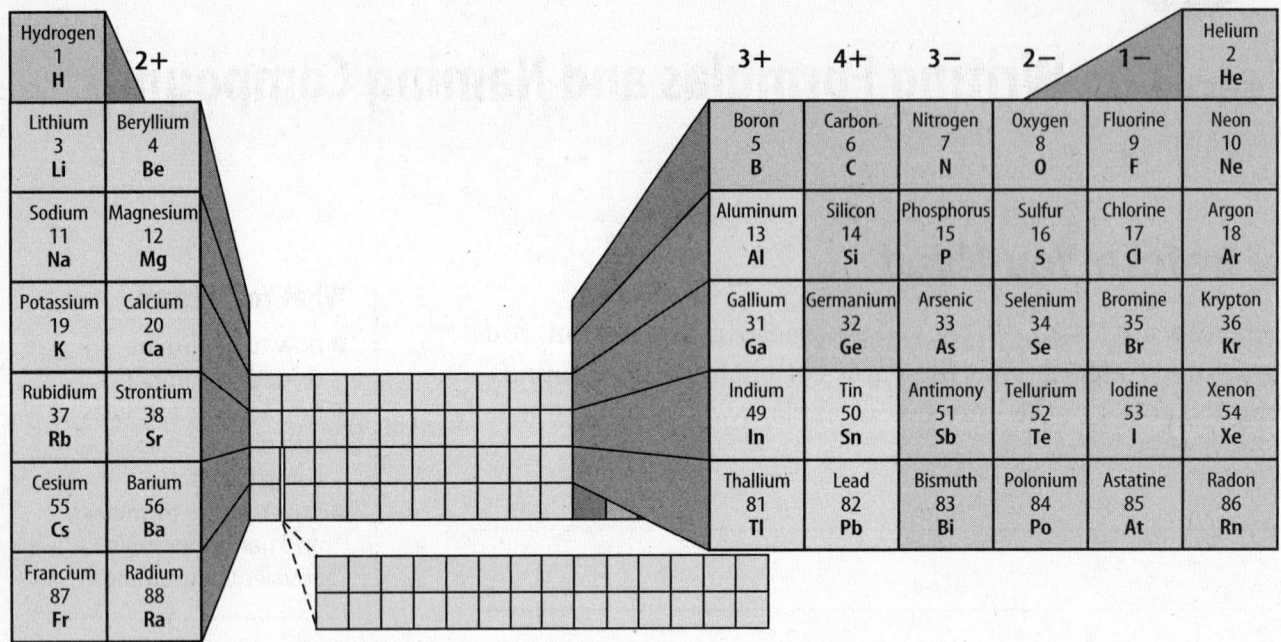

Picture This

1. **Identify** Highlight the oxidation numbers on the periodic table.

Picture This

2. **Compare** Circle the Roman numerals and the oxidation numbers for each element in the table and compare them.

How are oxidation numbers related to the periodic table?

Look at the periodic table above. Notice the numbers above each column. These are the oxidation numbers for the elements in the column. Notice how the oxidation numbers fit with the columns of elements shown.

A part of the periodic table is not included in the figure. Some elements in this section can have more than one oxidation number. The table shows some of these elements and their oxidation numbers. Because these elements can have more than one oxidation number, you must include more information when you name them. Include the oxidation number as a roman numeral in the names of these elements. For example, when iron with an oxidation number of 3+ combines with oxygen, the compound they form is iron(III) oxide.

Special Ions	
Name	Oxidation Number
Copper (I)	1+
Copper (II)	2+
Iron (II)	2+
Iron (III)	3+

352 CHAPTER 20 Chemical Bonds

How many positive and negative ions must a formula have?

Remember that ionic compounds are neutral. The ions in an ionic compound have charges, but the compound itself does not. The formula for an ionic compound must have the right number of positive and negative ions to make the charges balance. For example, sodium chloride has one sodium ion with a charge of 1+ and one chloride ion with a charge of 1−. The 1+ balances the 1−. The correct formula for sodium chloride is NaCl. The formula tells you that one positively charged ion combines with one negatively charged ion and makes a neutral compound.

What about the ionic compound calcium fluoride? A calcium ion has a charge of 2+. A fluoride ion has a charge of 1−. The charge 1− does not balance 2+. You need to have two fluoride ions for every calcium ion to balance the charges. The formula for the neutral compound calcium fluoride, CaF_2, shows exactly that. There is one calcium ion and two fluoride ions in the compound. ☑

How do you find the correct subscripts?

Sometimes you need to use math skills to write a formula correctly. In the compound aluminum oxide, there are aluminum ions and oxygen ions. But, how many of each one? By its place on the periodic table, you can tell that aluminum has an oxidation number of 3+. That means the aluminum ion has a charge of 3+. Oxygen has an oxidation number of 2−, so its ion has a charge of 2−. You must find the least common multiple of 3 and 2 to balance the charges. The least common multiple of 3 and 2 is 6.

Multiply 3+ by 2 to equal 6+. You need two aluminum ions to have a charge of 6+. Multiply 2− by 3 to equal 6−. You need three oxygen ions to have a charge of 6−. The charges balance. The correct formula for the neutral compound aluminum oxide is Al_2O_3.

How do you write the formula of an ionic compound?

You learned how to use the oxidation number to find the charge on an ion. You also learned that the charges on the ions must balance to form a neutral compound. Now you can write formulas for ionic compounds. Follow rules 1, 2, and 3 on the next page.

Reading Check

3. **Infer** What does the formula CaF_2 tell you about the compound?

_____

_____

_____

Applying Math

4. **Calculate** What is the least common multiple of 5 and 2?

_____

Reading Essentials 353

✓ **Reading Check**

5. Identify Which is an example of a polyatomic ion? Circle your answer.

a. NH_4^+

b. Al^{3+}

c. O^{2-}

d. $2O^{2-}$

💡 **Think it Over**

6. Apply Write the formula for calcium chloride. Use the periodic table to identify the positive and negative ions.

_____

Rules for writing formulas of ionic compounds

1. Write the symbol of the ion with the positive charge first. Sometimes an ion contains more that one atom. This is a polyatomic ion. One polyatomic ion with a positive charge is the ammonium ion. Its symbol is NH_4^+. Hydrogen, the ammonium ion (NH_4^+), and all metals form ions that have a positive charge. ✓

2. Write the symbol of the ion with the negative charge. Nonmetals other than hydrogen form ions that have a negative charge. All polyatomic ions except NH_4^+ have a negative charge.

3. The number of the charge (without the sign) of one ion becomes the subscript of the other ion.

Now try these rules to write the formula for lithium nitride. Lithium and nitrogen are the two atoms that make this compound. Look on the periodic table to see which element forms a positive ion. Lithium is in Group 1, so it forms ions with a 1+ charge. Write the symbol for lithium, Li, first.

Find the oxidation number of nitrogen. Nitrogen is in Group 15. It forms ions with a charge of 3−. You now can write LiN. Can you stop now? Look at the charges of the two ions. Do 1+ and 3− balance? No, you cannot stop yet.

Use the number of the charge of a nitrogen ion as the subscript for Li. Use the number of the charge of a lithium ion as the subscript for N. That gives Li_3N. When an element has no subscript, it means only one ion is in the compound. Do 3(1+) and 3− balance? Yes, this is the correct formula.

How do you write the name of a binary ionic compound?

When you know the formula, you can write the name of a binary ionic compound by following these rules.

1. Write the name of the positive ion.

2. Look to see if the positive ion is listed in the Special Ions table. If it is not in the table, go right to Step 3. If it is, the ion can have more than one oxidation number. To find the correct oxidation number, look at the formula of the compound. The charge of the compound is always zero. The negative ion can only have one possible charge. From the table, pick the charge of the positive ion that balances the negative charge. Write that positive ion's symbol along with the correct roman numeral in parentheses.

3. Write the root name of the negative ion. The root is the first part of the element's name. For example, the root name of chlorine is *chlor-*. The root name of oxygen is *ox-*.

4. Add the ending *-ide* to the root name. For example, write *oxide*.

Do not use subscripts in the name of an ionic compound. You may need to use the subscripts in the formula to help you figure out the charge on a metal ion that has more than one charge.

How do you use these rules?

Use these rules to write the name of the compound CuCl. Find the name of the positive ion on the periodic table. Cu is the symbol for copper. Is copper in the Special Ions table? Yes, the copper ion can have a 1+ or a 2+ charge. To find the charge on the copper in CuCl, look at the negative ion. Cl is the symbol for chlorine. Chlorine is in Group 17 of the periodic table. That means it has an oxidation number of 1−. From the formula, you see there is only one chloride ion in the compound. To balance a 1− charge on the chloride ion, the copper ion must have a 1+ charge. Now you can write copper(I) as the first part of the name.

Write the root name of the negative ion. The root name for chlorine is *chlor-*. Add *-ide* to the root. That gives you chloride. The correct name of CuCl is copper(I) chloride.

Compounds with Complex Ions

Not all compounds are binary compounds. The formula for baking soda used in cooking is $NaHCO_3$. Baking soda is an ionic compound that is not binary. Compounds like baking soda are made of more than two elements. These compounds have polyatomic ions. A **polyatomic ion** is a charged group of atoms that are bonded together by a covalent bond. The prefix *poly-* means "many" and *polyatomic* means "many atoms." The polyatomic ion in baking soda is the bicarbonate or hydrogen carbonate ion. Its symbol is HCO_3^-.

 **Reading Check**

7. **Determine** What is the root name of oxygen?

_____

 **Think it Over**

8. **Apply** Write the name of the compound FeI_2.

_____

How do you name a complex compound?

To write the name of a compound with a polyatomic ion, first write the name of the positive ion. If the positive ion is polyatomic, use the table to the right to find its name. Next, write the name of the negative ion. Again, if it is polyatomic, look up its name in the table. What is the name of the compound K_2SO_4? K is the symbol of the positive ion. It is not polyatomic. K is the symbol for potassium. The negative ion is SO_4^{2-}. It is polyatomic. Use the table again. K_2SO_4 is potassium sulfate.

Polyatomic Ions		
Charge	Name	Formula
1+	ammonium	NH_4^+
1−	acetate	$C_2H_3O_2^-$
	chlorate	ClO_3^-
	hydroxide	OH^-
	nitrate	NO_3^-
2−	carbonate	CO_3^{2-}
	sulfate	SO_4^{2-}
3−	phosphate	PO_4^{3-}

Now try naming the compound $Sr(OH)_2$. You can see that the positive ion is not polyatomic. Find the name for the symbol Sr on the periodic table. It is strontium. The negative ion is polyatomic. From the table of polyatomic ions, you see that OH^- is the hydroxide ion. The name of $Sr(OH)_2$ is strontium hydroxide.

How do you write the formula of a complex compound?

To write the formula for a complex compound, use the rules for writing the formula of a binary compound, but add one more thing. If you need to show more than one polyatomic ion, put parentheses around the formula for the ion before you write the subscript.

Write the formula for barium chlorate. First, write the symbol of the positive ion. The symbol for barium is Ba. Barium is in Group 2, so it forms a 2+ ion.

Now write the formula for the negative ion. The table shows you the formula for the chlorate ion is ClO_3^-. Are the charges on these two ions balanced? No, 2+ does not balance 1−. It takes two chlorate ions to balance the 2+ charge on the barium ion. The formula for barium chlorate is $Ba(ClO_3)_2$.

Applying Math

9. **Determine** How many 2+ ions does it take to balance the charge on one 4− ion?

Compounds with Added Water

Some ionic compounds have water molecules as part of their structure. These compounds are hydrates. A **hydrate** is a compound that has water chemically attached to its ions and written into its formula. The word *hydrate* means "water."

What are common hydrates?

If you evaporate a solution of cobalt chloride, pink crystals form. The crystals have six water molecules for each unit of cobalt chloride. The formula for this compound is $CoCl_2 \cdot 6H_2O$. The compounds name is cobalt chloride hexahydrate. The prefix *hexa-* means "six," so hexahydrate means "six waters."

Naming Binary Covalent Compounds

Remember that covalent compounds form between elements that are nonmetals. Some nonmetals can form more than one compound. For example, nitrogen and oxygen can form N_2O, NO, NO_2, and N_2O_5. If you used the rules you learned earlier, all these compounds would be called nitrogen oxide. Now you will learn how to give each of these compounds a different name.

How are prefixes used to name covalent compounds?

The table on the next page lists some Greek prefixes used to name covalent compounds made with the same elements. These prefixes tell how many atoms of each element are in a compound. For example, the compound NO_2 is nitrogen *di*oxide. The prefix *di-* tells you that there are two oxygen atoms in the compound. N_2O is dinitrogen oxide. The compound has two nitrogen atoms. The name of the compound N_2O_5 is dinitrogen pentoxide. This name uses two prefixes. There are two nitrogen atoms, so use dinitrogen. There are also five oxygen atoms, which is called pentoxide. ✓

Think it Over

10. Analyze How many water molecules does the hydrate $LiNO_2 \cdot H_2O$ have?

_____

Reading Check

11. Explain What do the prefixes in the names of covalent compounds tell you?

_____
_____
_____

Reading Essentials **357**

Picture This

12. Use Numbers What does the prefix *tetra-* mean?

_____

What rules apply in naming covalent compounds?

Prefixes for Covalent Compounds	
Number of Atoms	Prefix
1	mono-
2	di-
3	tri-
4	tetra-
5	penta-
6	hexa-
7	hepta-
8	octa-

Drop the last vowel of the prefix when the second element of the compound begins with a vowel. In pentoxide, the *a* is dropped from *penta-*.

There is a prefix to use when a compound has only one atom of an element. The prefix is *mono-*. Many times *mono-* is not used. Instead, it is understood that if no prefix is used, there is only one atom of that element in a compound. In some cases, *mono-* is used for emphasis. Carbon monoxide is one example.

After You Read

Mini Glossary

binary compound: a compound made of two elements
hydrate: a compound that has water chemically attached to its ions and written into its formula
oxidation number: the number that tells how many electrons an atom gains, loses, or shares to become stable
polyatomic ion: a charged group of atoms that are bonded together by a covalent bond

1. Review the terms and definitions in the Mini Glossary. Write a sentence that explains in your own words what a polyatomic ion is.

2. Complete the flow chart with the steps used in writing the name of a binary ionic compound.

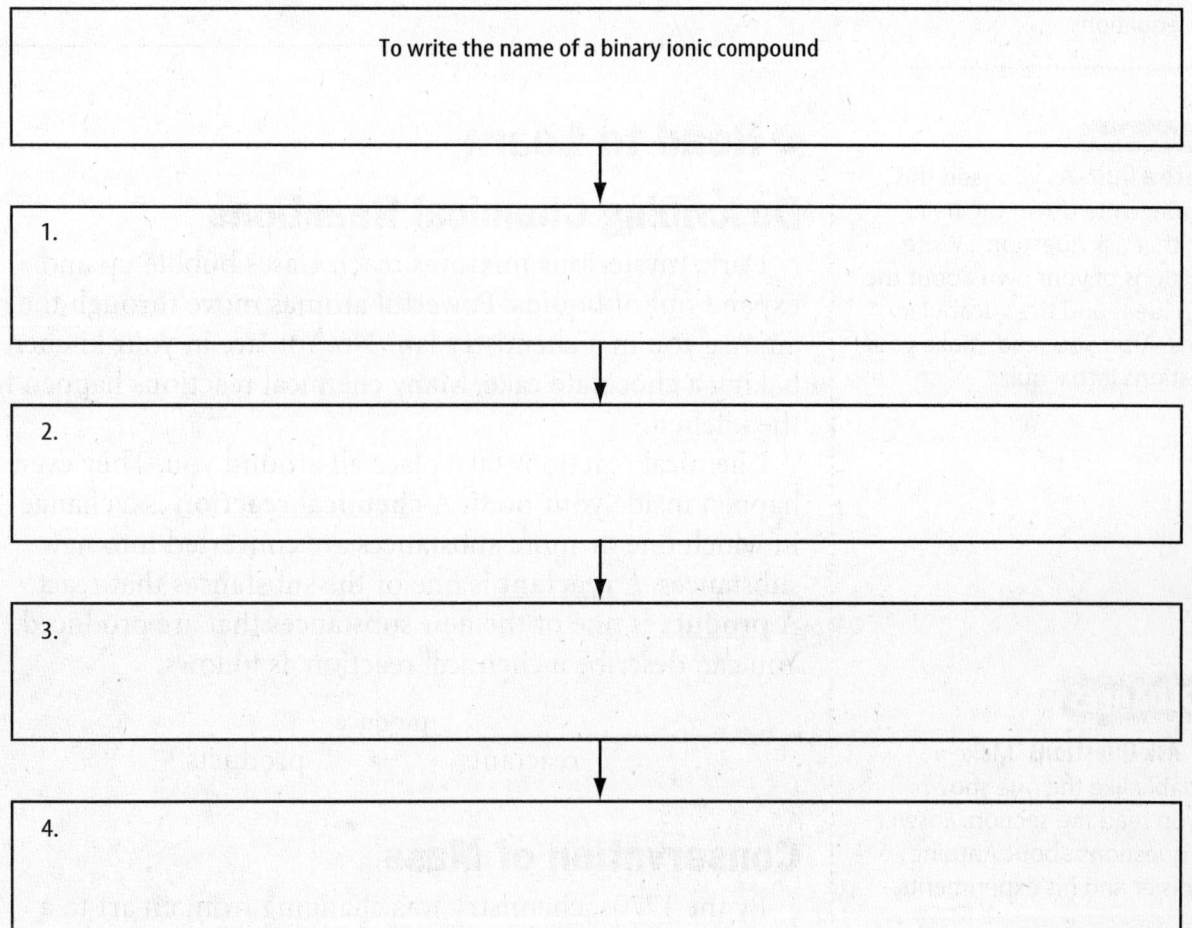

Visit **gpscience.com** to access your textbook, interactive games, and projects to help you learn more about writing formulas and naming compounds.

Reading Essentials 359

chapter 21 Chemical Reactions

section ● Chemical Changes

What You'll Learn
- how to identify the reactants and products in a chemical reaction
- how a chemical reaction follows the law of conservation of mass
- how chemists describe chemical changes with equations

Study Coach

Create a Quiz As you read this section, write down the headings that ask questions. Write questions of your own about the main ideas and the vocabulary terms. After you read, make your questions into a quiz.

FOLDABLES

A **Ask Questions** Make a Foldable like the one shown. As you read the section, answer the questions about Antoine Lavoisier and his experiments.

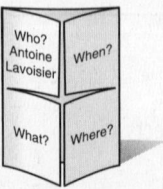

● Before You Read

Think about what happens when you bake a cake. On the lines below, describe how the ingredients change form throughout the process of baking a cake.

_____

_____

_____

● Read to Learn

Describing Chemical Reactions

Dark, mysterious mixtures react. Gases bubble up and expand out of liquids. Powerful aromas move through the air. Are you in a chemistry lab? No. You are in your kitchen baking a chocolate cake. Many chemical reactions happen in the kitchen.

Chemical reactions take place all around you. They even happen inside your body. A **chemical reaction** is a change in which one or more substances are converted into new substances. A **reactant** is one of the substances that react. A **product** is one of the new substances that are produced. You can describe a chemical reaction as follows:

$$\text{reactants} \xrightarrow{\text{produce}} \text{products}$$

Conservation of Mass

By the 1770s, chemistry was changing from an art to a science. Scientists began to study chemical reactions more carefully. The French chemist Antoine Lavoisier discovered an important rule. He found that the total mass of the products of a chemical reaction always equals the mass of the reactants. This is called the conservation of mass.

360 CHAPTER 21 Chemical Reactions

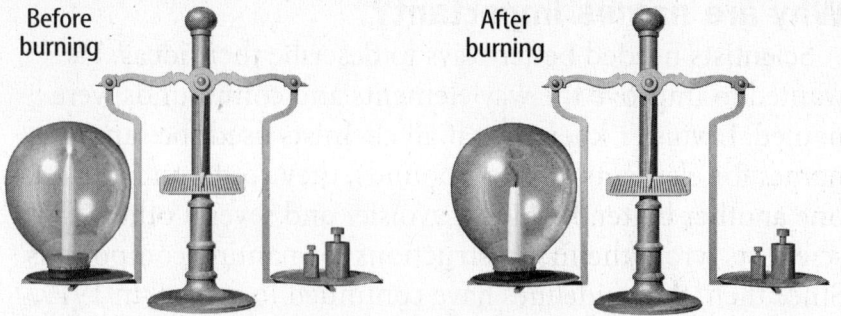

Before burning / After burning

Picture This

1. **Compare** How does the height of the right side of the scale in the first figure compare to the height of the right side of the scale in the second figure?

_____

_____

The figures above show an experiment he performed. The mass of the candle and the air in the jar (the reactants) before burning is the same as the mass of the gases and the candle (the products) after burning.

What were Lavoisier's experiments?

Lavoisier wanted to know exactly what happened when substances changed form. To answer this question, he experimented with mercury. He put solid mercury(II) oxide, a red powder, in a sealed container. He found the mass of the reactant in the container. When he heated the container, the mercury(II) oxide changed to a silvery liquid. It also gave off a gas. The silvery liquid was the metal mercury. He then found the mass of the products in the container again. It was the same as the mass before the experiment. ☑

mercury(II) oxide		oxygen	plus	mercury
10g	=	0.3g	+	9.7g

Lavoisier also figured out that the gas produced in the experiment, oxygen, was a part of air. He did this by heating mercury metal with air. He saw that a portion of the air combined with mercury to make mercury(II) oxide. He studied the effect of oxygen on living animals and humans.

Lavoisier did hundreds of experiments in his laboratory. He confirmed that in a chemical reaction, matter is not created or destroyed, but is conserved. This principle is known as the law of conservation of mass. This means that the total starting mass of the reactants of a chemical reaction always equals the total final mass of the products.

Why is Lavoisier called the father of modern chemistry?

Lavoisier's explanation of the law of conservation of mass started modern science. He also was the first to describe a chemical reaction called combustion. These discoveries are why Lavoisier is called the father of modern chemistry.

Reading Check

2. **Identify** What did Lavoisier find about the mass of the container with reactants and the mass of the container with products in his experiment with mercury(II) oxide?

_____

_____

_____

Think it Over

3. Draw Conclusions Imagine that chemists did not use the same rules to name compounds. How might this cause problems for a chemist who was trying to repeat an experiment done by the first chemist?

_____
_____
_____

Picture This

4. Summarize What does the symbol (g) placed next to a compound in a chemical equation mean?

_____
_____

Applying Math

5. Explain What do you notice about the numbers on the left side of the arrow and the numbers on the right side of the arrow in the chemical equation?

_____
_____

Why are names important?

Scientists needed better ways to describe their ideas. He wanted to improve the way elements and compounds were named. Lavoisier knew that if all chemists used the same names for elements and compounds, they could understand one another better. In 1787, Lavoisier and several other scientists wrote the first instructions for naming compounds. Since then, the guidelines have continued to evolve. In 1919, an organization was formed to coordinate guidelines for naming compounds. It is called the International Union of Pure and Applied Chemistry (IUPAC).

Writing Equations

It is important to include all the information when you describe a chemical reaction. What were the reactants? What did you do with them? What happened when they reacted? What were the products? When you answer all of these questions, the description of the reaction can be quite long.

Scientists have a shortcut for describing chemical reactions. A **chemical equation** is a way to describe a chemical reaction using chemical formulas and other symbols. Some of the symbols used in chemical equations are shown in the table.

Symbols Used in Chemical Equations

Symbol	Meaning	Symbol	Meaning
→	produces or forms	(aq)	aqueous, a substance is dissolved in water
+	plus	heat →	the reactants are heated
(s)	solid	light →	the reactants are exposed to light
(l)	liquid	elec. →	an electric current is applied to the reactants
(g)	gas		

Look at this description of a chemical reaction:

Nickel(II) chloride, dissolved in water, plus sodium hydroxide, dissolved in water, produces solid nickel(II) hydroxide plus sodium chloride, dissolved in water.

If you use a chemical equation, the same description is shorter and easier to understand as:

$NiCl_2(aq) + 2NaOH(aq) \rightarrow Ni(OH)_2(s) + 2NaCl(aq)$

362 CHAPTER 21 Chemical Reactions

Unit Managers

Look again at the chemical equation on the previous page. What do the numbers to the left of NaOH and NaCl mean? Remember the law of conservation of mass? Matter is not made or lost in a chemical reaction. Atoms are rearranged, but they are never created or destroyed. The numbers in the equation are called coefficients. A **coefficient** shows the number of units of a substance taking part in a reaction. You can think of coefficients as unit managers.

Suppose you were going to make sandwiches for a picnic. You know that each sandwich needs two slices of bread, one slice of turkey, one slice of cheese, two slices of tomato, and one leaf of lettuce. If you also know how many sandwiches you need to make, you can figure out how much bread, turkey, cheese, tomato, and lettuce you need to buy so you do not have any food left over.

Making sandwiches is like a chemical reaction. The ingredients for the sandwiches are the reactants. The finished sandwiches are the products. The number of units of bread, turkey, cheese, tomato, and lettuce are the coefficients of the reactants. The number of finished sandwiches is the coefficient of the product.

How do chemists use coefficients?

When chemists know the number of units of each reactant, they are able to add the correct amounts of reactants needed for a reaction. The units or coefficients will tell how much product will form. For example, here is the chemical equation from the example on the previous page.

$$NiCl_2(aq) + 2NaOH(aq) \rightarrow Ni(OH)_2(s) + 2NaCl(aq)$$

You can see that one unit of $NiCl_2$ and two units of NaOH produce one unit of $Ni(OH)_2$ and two units of NaCl. The figure below shows you how the coefficients affect the number of molecules in the reaction.

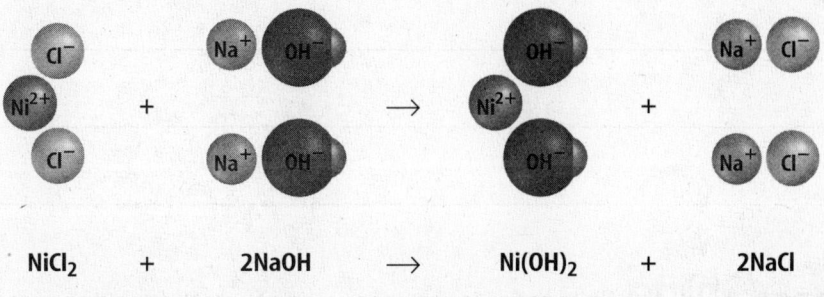

NiCl₂ + 2NaOH → Ni(OH)₂ + 2NaCl

Reading Check

6. **Explain** What does a coefficient show?

Applying Math

7. **Apply** Suppose $NiCl_2$ reacts with NaOH. For each molecule of $NiCl_2$, how many molecules of NaOH are needed?

Picture This

8. **Observe** What does the 2NaOH represent?

After You Read
Mini Glossary

chemical equation: a way to describe a chemical reaction using chemical formulas and other symbols

chemical reaction: a change in which one or more substances are converted into new substances

coefficient: a number that shows how many units of a substance take part in a reaction

products: the new substances that are produced in a chemical reaction

reactants: the substances that react in a chemical reaction

1. Review the terms and their definitions in the Mini Glossary. Write a sentence describing a chemical equation.

   _____

   _____

2. Complete the concept web by writing three ways that Antoine Lavoisier helped to make chemistry into a modern science.

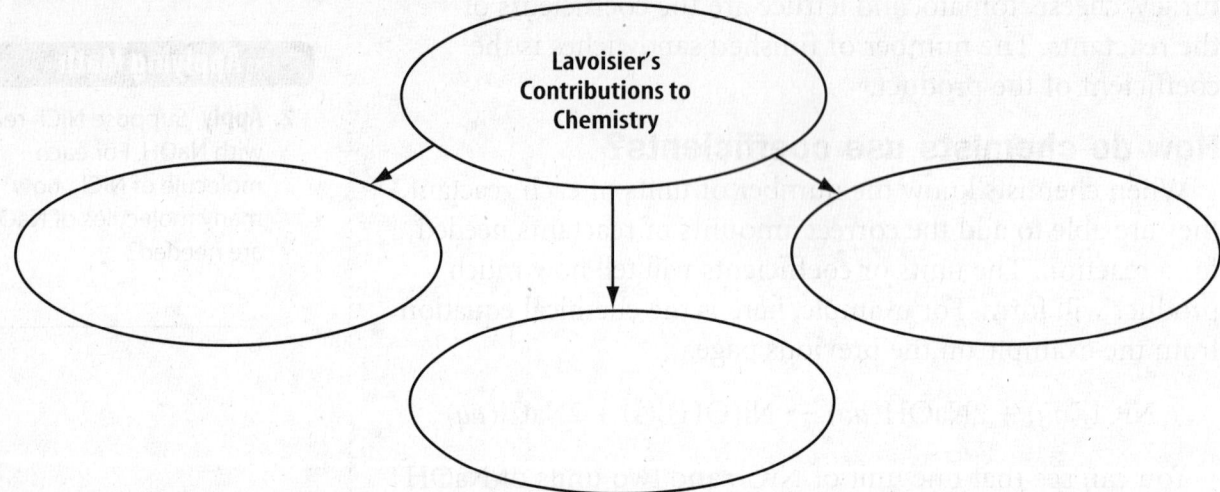

3. You created a quiz with questions about important topics from the section. Which question was the hardest for you to answer? Why do you think this was?

   _____

   _____

   _____

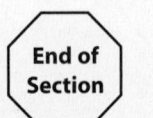

 Visit **gpscience.com** to access your textbook, interactive games, and projects to help you learn more about chemical changes.

364 CHAPTER 21 Chemical Reactions

Chemical Reactions

section 2 Chemical Equations

Before You Read

Do you remember playing on a seesaw when you were younger? What happened if the person on the other side was much larger or much smaller than you?

What You'll Learn
- what a balanced chemical equation is
- how to write a balanced chemical equation

Read to Learn

Balanced Equations

The equation below is for Lavoisier's mercury(II) oxide reaction.

$$HgO(s) \xrightarrow{heat} Hg(l) + O_2(g)$$

How many atoms of mercury (Hg) are on each side of the equation? There is one mercury (Hg) atom on the reactant side and one mercury (Hg) atom on the product side. How many atoms of oxygen (O) are on each side? Notice that there is one oxygen (O) atom on the reactant side, but the product side has two oxygen (O) atoms.

Atoms	HgO	→	Hg	+	O_2
Hg	1		1		
O	1				2

Remember that according to the law of conservation of mass, one oxygen atom cannot become two oxygen atoms. You cannot rewrite HgO as HgO_2. That would make the number of oxygen atoms balance, but HgO and HgO_2 are not the same compound. The formulas in a chemical equation must accurately represent the compounds that react.

Study Coach

Outline As you read the section, create an outline of what you read. Write down the headings that you see in the text. Under the headings, include important information.

FOLDABLES

B Organize Information Make the following Foldable to help you organize information about balanced chemical equations.

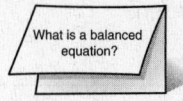

What does a balanced equation show?

A chemical equation must be balanced. Balancing only changes the way a reaction is represented. It does not change what happens in the reaction. To balance a chemical equation, you change the coefficients. A **balanced chemical equation** has the same number of atoms of each element on each side of the equation.

How do you choose coefficients?

You often can find the coefficients to balance an equation just by guessing and checking your guess. In the mercury(II) oxide equation, the number of mercury atoms is balanced. You need to balance the number of oxygen atoms. Try putting a coefficient of 2 in front of HgO on the left side of the equation. This balances the oxygen, but not the mercury.

Atoms	2HgO	→	Hg	+	O₂
Hg	2		1		
O	2				2

To balance the mercury, put a 2 in front of the mercury on the right side of the equation.

Atoms	2HgO	→	2Hg	+	O₂
Hg	2		2		
O	2				2

Now the equation is balanced.

$$2H_gO \rightarrow 2H_g + O_2$$

What are the steps for balancing an equation?

Magnesium burns with a very bright light. Have you ever seen a flare burning at the scene of a traffic accident? The flare probably was made of magnesium. When magnesium burns, it leaves a white powder, magnesium oxide. To write a balanced chemical equation for the burning of magnesium, follow these steps.

Step 1 Write a chemical equation using formulas and symbols. Remember that oxygen is a diatomic molecule, which means that it consists of two oxygen atoms in a covalent bond.

$$Mg(s) + O_2(g) \rightarrow MgO(s)$$

💡 Think it Over

1. **Analyze Results** Why does putting a coefficient of 2 in front of HgO on the left side of the equation balance the oxygen but not the mercury?

   _____

   _____

   _____

💡 Think it Over

2. **Apply** Hydrogen (H), like oxygen, is a diatomic molecule. Write the chemical formula for hydrogen.

   _____

Step 2 Count the atoms in the reactants and products.

Atoms	Mg	+	O_2	→	MgO
Mg	1				1
O			2		1

The magnesium atoms are balanced, but the oxygen atoms are not. So, the equation is not balanced.

Step 3 Choose coefficients to balance the equation. Remember that you cannot change the subscripts of a formula to balance the equation. Instead, try putting the coefficient 2 in front of MgO.

$$Mg(s) + O_2(g) \rightarrow 2MgO(s)$$

Step 4 Check the number of atoms on each side of the equation again. Now, there are two magnesium atoms on the right side of the equation, and only one on the left. You need to put the coefficient 2 in front of Mg to balance the equation.

$$2Mg(s) + O_2(g) \rightarrow 2MgO(s)$$

The above is the balanced chemical equation for the burning of magnesium.

Now you try one on your own. Balance the equation for the following reaction:

$$Fe(s) + Cl_2(g) \rightarrow FeCl_3(s)$$

Write your answers in the margin for what goes in the indicated blanks in the margin.

$$\underline{(a.)}\ Fe + \underline{(b.)}\ Cl_2 \rightarrow \underline{(c.)}\ FeCl_3$$

Applying Math

3. **Apply** When lithium metal is treated with water, hydrogen gas and lithium hydroxide are produced. Balance the following chemical equation that shows this reaction.

   $Li(s) + H_2O(g) \rightarrow LiOH(aq) + H_2(g)$

   _____

   _____

Applying Math

4. **Calculate** What number goes in each indicated blank?

   a. _____

   b. _____

   c. _____

Reading Essentials

After You Read
Mini Glossary

balanced chemical equation: a chemical equation with the same number of atoms of each element on both sides

1. Review the term and its definition in the Mini Glossary. Use the term in a sentence that shows that you understand what it means.

   _____

   _____

2. Complete the flow chart with the steps you follow to balance a chemical equation.

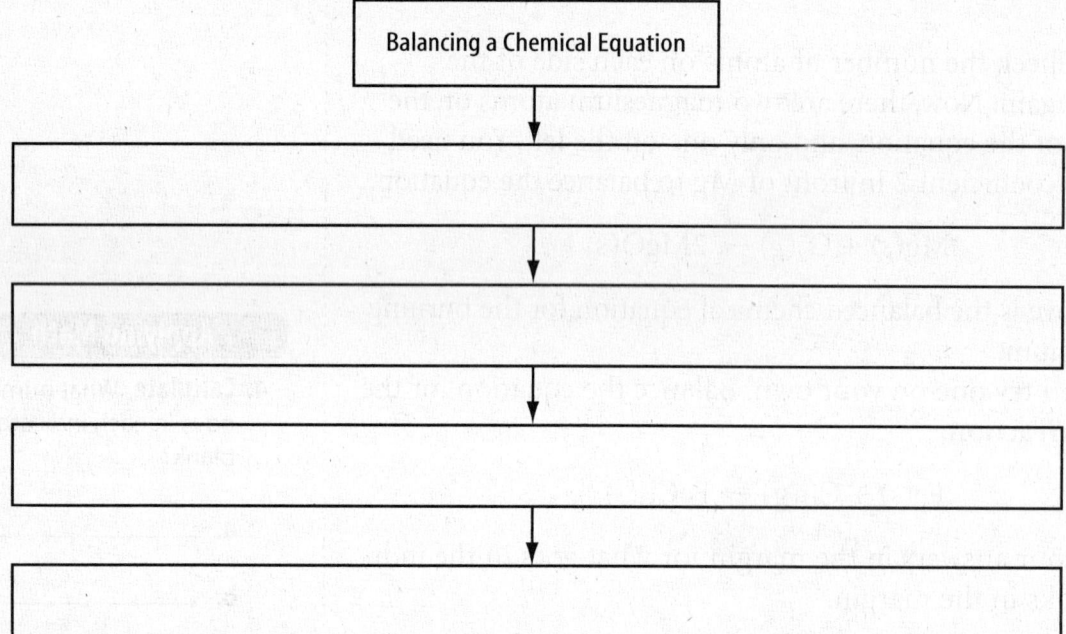

3. You made an outline of the section as you read it. How did this strategy help you learn the material in the section?

   _____

   _____

Science online Visit **gpscience.com** to access your textbook, interactive games, and projects to help you learn more about chemical equations.

368 CHAPTER 21 Chemical Reactions

Chapter 21: Chemical Reactions

section ③ Classifying Chemical Reactions

● Before You Read

Why do you think there are different sections in your school library to organize, or classify, the books?

_____

_____

● Read to Learn

Types of Reactions

Millions of chemical reactions occur every day. Scientists organize, or classify, reactions into five types—combustion, synthesis, decomposition, single displacement, and double displacement. Organizing reactions in this way helps scientists use the knowledge they gain, just as classifying books in the library helps you use the books.

What are combustion reactions?

When you see something burning, you are seeing a combustion reaction. A **combustion reaction** occurs when a substance reacts with oxygen to produce heat and light. Combustion reactions produce one or more products that contain the elements of the reactants. For example, carbon reacts with oxygen to produce carbon dioxide. This reaction describes what happens when coal burns. Many combustion reactions also fit into other types of reactions. For example, the reaction between carbon and oxygen is a combustion reaction and a synthesis reaction.

What are synthesis reactions?

In a **synthesis reaction**, two or more substances combine to form another substance. The general formula for a synthesis reaction is A + B → AB. Substance A reacts with substance B to form substance AB.

What You'll Learn
- what the five kinds of chemical reactions are
- what oxidation and reduction are
- what a redox reaction is
- which metals replace others in compounds

Mark the Text

Find Main Ideas As you read this section, highlight the headings that are questions in one color. Then highlight the answers to those questions in another color.

FOLDABLES

C **Build Vocabulary** Make the following Foldable to help you organize information about chemical reactions. Label the tabs as shown.

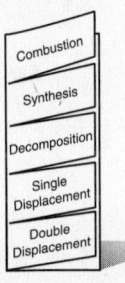

An example of a synthesis reaction is hydrogen burning in oxygen to form water.

$$2H_2(g) + O_2(g) \rightarrow 2H_2O(g)$$

This reaction is used to power some rockets, including the main engines of a space shuttle. Have you ever seen a rusty car or bike? The reaction that causes rust is a synthesis reaction. When iron reacts with oxygen in the presence of water, hydrated iron(II) oxide, or rust, is formed.

Reading Check

1. **Explain** What type of reaction is the formation of rust?

What are decomposition reactions?

A decomposition reaction is the reverse of a synthesis reaction. A **decomposition reaction** occurs when one substance breaks down, or decomposes, into two or more substances. The general formula for a decomposition reaction is $AB \rightarrow A + B$. Most decomposition reactions use heat, light, or electricity. For example, an electric current passed through water produces hydrogen and oxygen. The chemical equation and figure below show this.

$$2H_2O(l) \xrightarrow{\text{elec.}} 2H_2(g) + O_2(g)$$

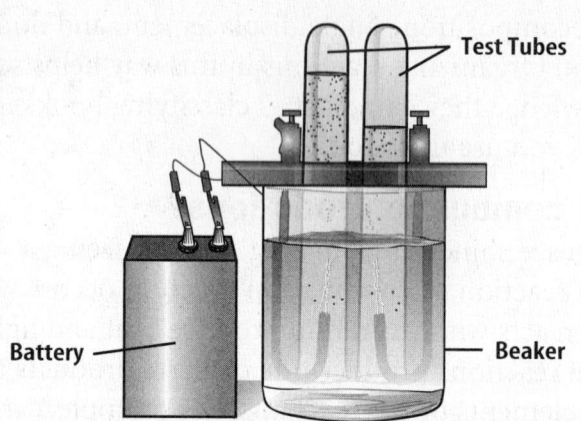

Picture This

2. **Interpret Illustrations** Look at the coefficients of the products in the equation. Which test tube has hydrogen in it, the left one or the right one? How do you know this?

What are single-displacement reactions?

A **single-displacement reaction** happens when one element replaces another element in a compound. The general formula for a single-displacement reaction is $A + BC \rightarrow AC + B$. Atom A displaces, or takes the place of, atom B. A new molecule, AC, forms. A single-displacement reaction occurs when a copper wire is put into a solution of silver nitrate. Copper is a more active metal than silver, so it replaces the silver. A blue copper(II) nitrate solution forms. The silver, which is not soluble, forms on the wire.

$$Cu(s) + 2AgNO_3(aq) \rightarrow Cu(NO_3)_2(aq) + 2Ag(s)$$

What is the activity series?

You can predict which metal will replace another metal in displacement reactions. The diagram lists metals by how reactive they are. The most active metals are at the top of the list. The least active metals are at the bottom. A metal will replace any less active metal. Notice that copper, silver, and gold are the least active metals on the list. That is why these elements often occur in relatively pure deposits.

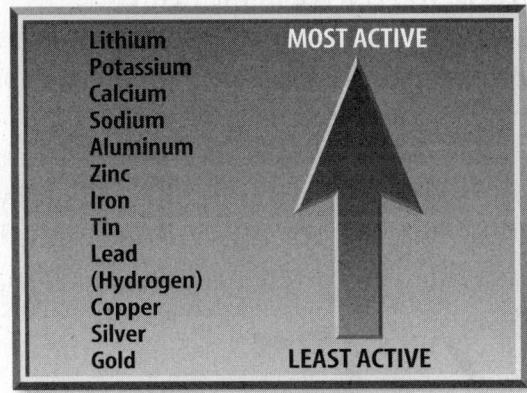

Picture This

3. **Order** Write these elements in order from least active to most active: lead, aluminum, copper, silver, zinc.

_____
_____
_____
_____
_____

What are double-displacement reactions?

In a **double-displacement reaction**, the positive ion of one compound replaces the positive ion of the other compound to form two new compounds. The general formula for a double-displacement reaction is AB + CD → AD + CB. A double-displacement reaction takes place if a precipitate, water, or gas forms when two ionic compounds in solution are combined. A **precipitate** is an insoluble compound (one that cannot be dissolved) that comes out of solution during a double-displacement reaction.

Look at the following example of a double-replacement reaction.

$$Ba(NO_3)_2(aq) + K_2SO_4(aq) \rightarrow BaSO_4(s) + 2KNO_3(aq)$$

The reactants are barium nitrate and potassium sulfate. Both of these compounds are in solutions. The products are solid barium sulfate, which is the precipitate, and potassium nitrate, in solution.

The chemical reactions you learned about in this section are only a few examples. Thousands more reactions of each type happen all around you.

✔ Reading Check

4. **Identify** Which type of reaction causes a precipitate to form?

_____
_____

Reading Essentials 371

Think it Over

5. Draw Conclusions Chlorine has seven electrons in its outer energy level. In reactions, chlorine usually gains an electron. In a redox reaction, would chlorine be oxidized or reduced? Explain.

_____

_____

_____

What are oxidation-reduction reactions?

In many chemical reactions, substances gain or lose electrons. Chemists use two terms to describe gaining or losing electrons. **Oxidation** is a loss of electrons during a chemical reaction. **Reduction** is a gain of electrons during a chemical reaction. Chemical reactions involving electron transfer of this sort often involve oxygen. Oxygen is very reactive. It often pulls electrons from metals. Oxidation-reduction reactions cause metals to corrode, or rust, as you can see in the figure.

During an oxidation-reduction reaction, one substance gains electrons and another loses electrons. The substance that gains electrons becomes more negative, so we say it is reduced. Another substance loses electrons and becomes more positive. We say it is oxidized. Reduction and oxidation always happen together. That is why these reactions are often called redox reactions. Redox stands for *red*uction and *ox*idation.

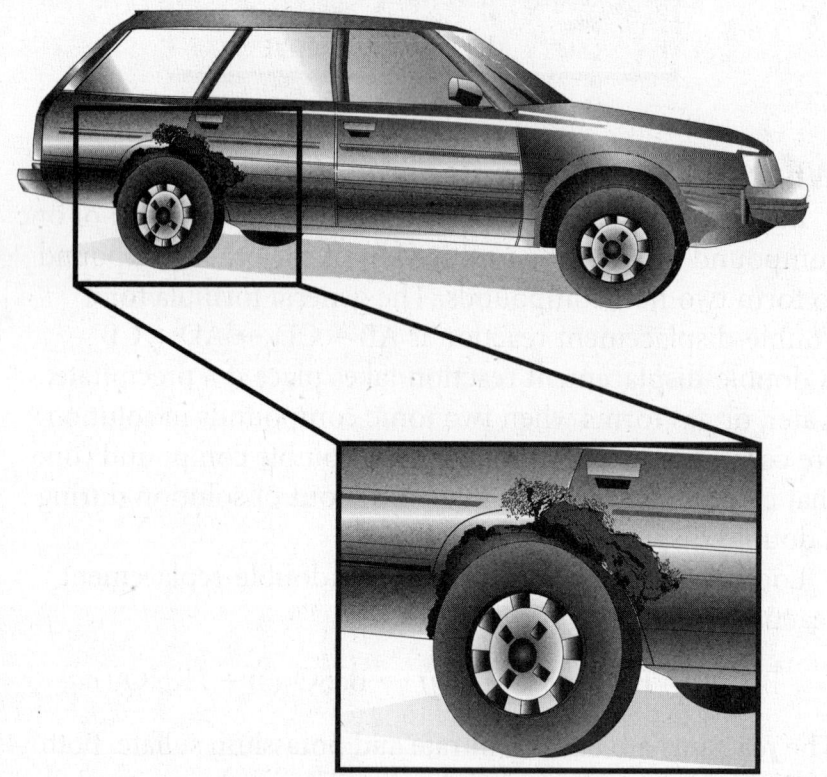

Picture This

6. Explain What has happened to the electrons in the metal of the car in the figure?

_____

_____

_____

After You Read

Mini Glossary

combustion reaction: a reaction in which a substance reacts with oxygen to produce heat and light

decomposition reaction: a reaction in which one substance breaks down, or decomposes, into two or more substances

double-displacement reaction: a reaction in which the positive ion of one compound replaces the positive ion of the other compound to form two new compounds

oxidation: a loss of electrons during a chemical reaction

precipitate: an insoluble compound that comes out of solution during a double-displacement reaction

reduction: a gain of electrons during a chemical reaction

single-displacement reaction: a reaction in which one element replaces another element in a compound

synthesis reaction: a reaction in which two or more substances combine to form another substance

1. Review the terms and their definitions in the Mini Glossary. Choose two terms that are related and write a sentence using both of them.

   _____

   _____

2. Write the letter of the description in Column 2 that matches the reaction in Column 1.

   Column 1

   _____ 1. decomposition reaction

   _____ 2. single-displacement reaction

   _____ 3. synthesis reaction

   _____ 4. combustion reaction

   _____ 5. double-displacement reaction

   Column 2

   a. A + B → AB

   b. AB + CD → AD + CB

   c. burning

   d. A + BC → AC + B

   e. AB → A + B

3. How did highlighting the questions and answers in different colors help you learn the material in this section?

   _____

   _____

 Visit gpscience.com to access your textbook, interactive games, and projects to help you learn more about classifying chemical reactions.

chapter 21

Chemical Reactions

section ④ Chemical Reactions and Energy

What You'll Learn
- energy change sources in chemical reactions
- the difference between exergonic and endergonic reactions
- how catalysts and inhibitors are used

Mark the Text

Underline As you read this section, underline the information you think is important. When you finish reading, look back at what you underlined.

FOLDABLES

D Make a Venn Diagram Make the following Foldable to compare and contrast exergonic and endergonic reactions.

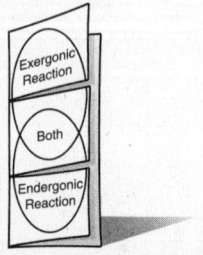

● Before You Read

You have probably seen video of a building being demolished by an explosion. Describe what happened on the lines below.

_____

_____

● Read to Learn

Chemical Reactions—Energy Exchanges

When they are no longer useful, buildings are sometimes demolished with dynamite. A dynamite explosion is an example of a rapid chemical reaction.

Most chemical reactions happen more slowly than a dynamite explosion, but all chemical reactions release or absorb energy. The energy released in a chemical reaction can be in the form of heat, light, sound, or electricity. Wood burns and releases heat and light. A glow stick releases only light.

Chemical bonds are the source of this energy. Most chemical reactions break some chemical bonds in the reactants. It takes energy to break the chemical bonds. That is why many substances need heat to make them react. For products to be produced, new bonds must form. When bonds form, energy is released. The amount of energy required to break the chemical bonds in dynamite is much less than the amount of energy released when new bonds form. The result is a release of energy and sometimes a loud explosion.

More Energy Out

An **exergonic** (ek sur GAH nihk) **reaction** releases energy. In an exergonic reaction, less energy is needed to break the bonds in the reactants than is released when new bonds in the products form. Exergonic reactions give off energy, such as light or heat. An exergonic reaction produces visible light in a glow stick.

What are exothermic reactions?

In some reactions, the energy is given off as heat. Have you ever used a heat pack? Heat packs release energy as heat. An **exothermic** (ek soh THUR mihk) **reaction** releases energy, usually as heat. Burning wood and exploding dynamite are examples of exothermic reactions. Iron rusting is also exothermic. The chemical reaction that produces rust happens so slowly that you can't detect the heat.

How are exothermic reactions used?

Have you turned on a light or used a blow dryer today? The energy you used probably came from exothermic reactions. The power plant the electricity came from probably uses fossil fuels. The carbon in the fossil fuels combines with oxygen to form carbon dioxide gas and energy. This reaction is exothermic.

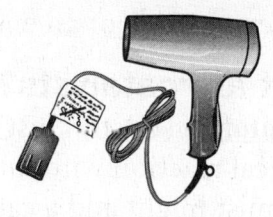

Other substances in fossil fuels also react. Often, the products of these other reactions are pollutants. Sulfur in fossil fuels reacts with oxygen to form sulfur dioxide, which combines with water in the atmosphere to form acid rain.

Picture This
1. **Identify** What type of reactions provides the energy for the items in the figures?

More Energy In

An **endergonic** (en dur GAH nihk) **reaction** absorbs energy. In an endergonic reaction, it takes more energy to break the bonds in the reactants than is released when new bonds in the products form. Endergonic reactions absorb energy such as heat, light, or electricity. ✓

Electricity is often used to supply energy to endergonic reactions. Electricity supplies energy to a reaction that puts a coat of metal onto a surface. This reaction is called electroplating. Electricity also is used to supply energy to separate aluminum metal from its ore.

In the following endergonic reaction, energy from electricity keeps the reaction going.

$$2Al_2O_3(l) \xrightarrow{elec.} 4Al(l) + 3O_2(g)$$

Reading Check
2. **Determine** What happens to energy in an endergonic reaction?

Think it Over

3. Compare and Contrast What is the difference between an endergonic reaction and an endothermic reaction?

_____
_____
_____

FOLDABLES

E Compare and Contrast Make the following Foldable to help you understand how catalysts and inhibitors are similar and different.

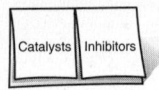

Reading Check

4. Define What is an inhibitor?

_____
_____
_____

What is an endothermic reaction?

An <u>endothermic</u> (en duh THUR mihk) <u>reaction</u> absorbs energy, usually as heat. When an endothermic reaction takes place in a beaker, it can make the beaker feel cold. Physical changes also can be described as endothermic. For example, a salt dissolving in water is an endothermic physical change.

An endothermic reaction is used to make homemade ice cream. In an ice-cream maker, salt is added to a bucket of ice and water. The salt dissolves and absorbs heat. This makes the mixture of salt and water colder. Without salt, the ice would not make the ice cream mixture cold enough to freeze.

Some reactions are extremely endothermic. When barium hydroxide ($BaOH_2$) reacts with ammonium chloride (NH_4Cl) in a beaker of water, it is so endothermic that it causes a drop of water on the outside of the beaker to freeze. Cold packs contain ammonium nitrate crystals and water. They are another example of an endothermic reaction.

What are catalysts?

A <u>catalyst</u> (KA tuh lust) is a substance that speeds up a chemical reaction without being permanently changed itself. A chemist might add a catalyst to a reaction that would be too slow to be useful. When a catalyst is added to a reaction, the mass of the products does not change. The catalyst only speeds up the reaction. A catalyst can be recovered and reused because it is unchanged in the reaction.

Catalysts are used in industry, especially in making plastics. Cars also use catalysts. Palladium and platinum metals are used in the catalytic (ka tuh LIH tihk) converter of a car's exhaust system. The catalysts in the catalytic converter speed up the reaction of unburned fuel with oxygen. This reduces pollution in the exhaust gases that the car emits.

What are inhibitors?

Sometimes, it is helpful to prevent or to slow down chemical reactions. An <u>inhibitor</u> is a substance that slows down a chemical reaction. Like catalysts, inhibitors do not change the amount of product produced in a chemical reaction. Food preservatives called BHT and BHA are inhibitors. They slow down reactions and prevent foods, such as cereals and crackers, from spoiling.

After You Read
Mini Glossary

catalyst: a substance that speeds up a chemical reaction without being permanently changed itself
endergonic reaction: a reaction that absorbs energy
endothermic reaction: a reaction that absorbs energy, usually as heat
exergonic reaction: a reaction that releases energy
exothermic reaction: a reaction that releases energy, usually as heat
inhibitor: a substance that slows down a chemical reaction

1. Review the terms and their definitions in the Mini Glossary. Write a sentence giving an example of the type of reaction that can cause its container to get cold or even freeze water.

   _____
   _____

2. Complete the table. The first row gives examples of different kinds of reactions. Name the type of reaction for each example. Then describe the reaction.

Example of Reaction	Glow Stick	Dynamite Exploding	Electroplating Metals	Cold Pack
Type of reaction	Exergonic reaction		Endergonic reaction	
Description of the reaction		Release energy in the form of heat		

3. What idea was the hardest for you to understand in this section? How would you explain that idea to a friend?

   _____
   _____

Science Online Visit **gpscience.com** to access your textbook, interactive games, and projects to help you learn more about chemical reactions and energy.

Reading Essentials **377**

chapter 22 Solutions

section ❶ How Solutions Form

What You'll Learn
- three types of solutions
- how things dissolve
- the rate solids and gases dissolve

Mark the Text

Locate Information As you read this section, highlight the factors that affect how quickly substances dissolve.

FOLDABLES

Ⓐ Draw and Label Make a half-sheet Foldable like the one below. As you read, draw three diagrams showing solute and solvent molecules before dissolving, during dissolving, and after dissolving. Label your drawings.

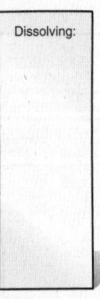

● Before You Read

Have you ever made a drink from a powdered mix? What happened to the powder?

_____

● Read to Learn

What is a solution?

Many people like to watch hummingbirds and put up hummingbird feeders in their yards. They fill the feeders with a red liquid made of water, sugar, and red food coloring. The sweet, colored liquid attracts hummingbirds. To make this food, you add sugar to water and stir. When you stir, the sugar crystals disappear. Next, you add a few drops of red food coloring and stir again. The red color spreads evenly. Why does this happen?

The red sugar-water is a solution. A **solution** is mixture that has the same ingredients, color, density, and even taste mixed evenly throughout. You cannot see the sugar crystals in the solution because they have broken up into molecules. The food coloring also breaks up into molecules. The sugar molecules and the food coloring molecules mix evenly among the water molecules throughout the solution. The figure shows what a sugar-water solution would look like if you could see the separate molecules.

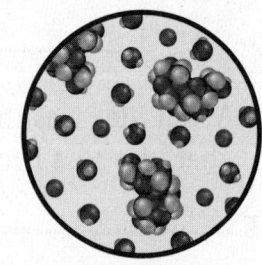

Liquid Solution

Solutes and Solvents

To describe a solution, you could say that one substance dissolves in another. In a solution, the **solute** is the substance that dissolves. The **solvent** is the substance that is doing the dissolving. In sugar-water, the sugar is the solute and the water is the solvent.

378 CHAPTER 22 Solutions

What can dissolve in a liquid?

In a solution made with a liquid and a solid, the solid is the solute and the liquid is the solvent. In salt water, salt is the solute and water is the solvent.

Some solutions are made from a gas dissolved in a liquid. In carbonated soda, carbon dioxide gas is dissolved in water. The gas is the solute and the liquid is the solvent.

Other solutions have a liquid dissolved in another liquid, such as the liquid food coloring in the hummingbird food. The solvent is usually the liquid present in the larger amount.

Are there solutions that do not contain a liquid?

Solutions also can be mixtures of gases or even mixtures of solids. The air you breathe is a solution. Air is a solution of 78 percent nitrogen, 20 percent oxygen, and small amounts of other gases. Look at the figures below. The figure on the left shows different gas molecules in a gas solution.

Some jewelry is made of mixtures of metals. Sterling silver is a mixture of silver and copper. Mixtures of metals are solid solutions. In a solid solution, one metal is the solute and the other metal is the solvent. The two metals are melted together. Solid solutions are known as alloys. Most of the coins you use are made of alloys. Some musical instruments are made of an alloy called brass. Brass is a solid solution of copper and tin. The figure on the right shows atoms of copper and tin in a solid solution brass.

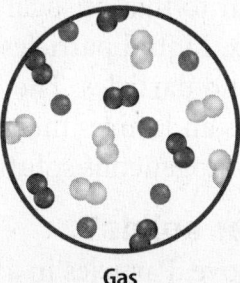
Gas

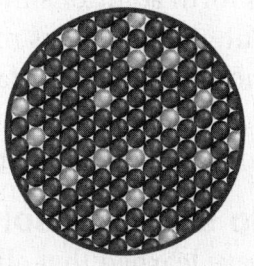
Solid

Picture This

1. **Observe** Look at the spacing of the atoms in the two solutions on this page and the liquid solution on the first page. Which is denser, a solid solution, a gaseous solution, or a liquid solution?

_____

How Substances Dissolve

Fruit drinks and sports drinks are examples of solutions made by dissolving solids in liquids. Both contain sugar and other substances that add color and flavor to the drink. You know that sugar dissolves in water, but how does it happen?

A solid starts to dissolve at its surface. To understand how a water solution forms, you need to remember two things you have learned about water molecules. First, water molecules, like all particles, are always moving. Second, water molecules are polar. **Polar** means that they have a positive end and a negative end. Sugar molecules are also polar.

Reading Check

2. **Explain** Where does a solid start to dissolve?

_____

Picture This

3. Identify Circle a water molecule in the first figure. Mark the positive ends with a plus sign and the negative end with a minus sign.

How does a solid dissolve in a liquid?

The figures below show how sugar dissolves in water. In the figure on the left, water molecules move toward a sugar crystal. The positive ends of sugar molecules attract the negative ends of water molecules.

In the middle figure, water molecules pull sugar molecules into solution. This process continues as layer after layer of sugar molecules move away from the crystal. Finally, the crystal dissolves completely.

The water molecules and sugar molecules spread out and mix evenly. They are now a homogeneous solution. The figure on the right shows sugar molecules surrounded by water molecules in a sugar-water solution. This process happens whenever a liquid solvent dissolves a solid solute.

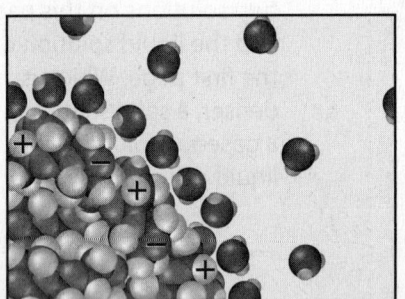 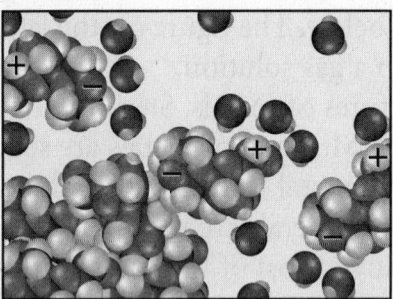 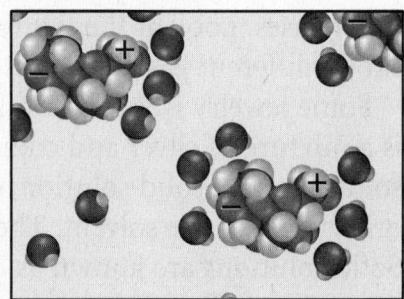

How do liquid and gas solutions form?

Liquids and gases also can form solutions. Liquid and gas particles form solutions in a way similar to that of sugar and water. But, the process is more complex. Liquid particles and gas particles move much faster than solid particles. The movement separates the solute particles and mixes them evenly in the solvent. The result is a homogeneous solution.

How do solids dissolve in other solids?

You have learned that all particles move. Particles in a solid do not move enough to spread out and mix evenly with other solids. Solids must be melted into liquids and then mixed together. In the liquid state, the atoms can move more freely. They spread out and form a homogeneous solution. The atoms stay in solution after cooling. ✓

✓ Reading Check

4. Explain what must be done to solids so they can be mixed together.

_____

_____

Rate of Dissolving

Sometimes a solute dissolves quickly into a solvent. At other times, it dissolves more slowly. There are some things you can do to make a solute dissolve faster. You can stir the solution or heat the solution. If the solute is a solid, you can break it into smaller pieces.

How does stirring speed up dissolving?

Stirring a solution speeds up the dissolving process by making the solvent and solute particles move faster. More solvent particles come into contact with more solute particles. The solid solute dissolves more quickly.

How does breaking up a solid solute speed up dissolving?

Suppose you put a piece of rock candy in your drink to sweeten it. You may have to wait a long time for the candy to dissolve. Now, suppose you crush the rock candy into a powder before adding it to your drink. The small pieces of candy dissolve much more quickly than the chunk of candy. Why?

When you break a solid solute into smaller pieces, you increase its surface area. Remember that the dissolving process takes place at the surface of the solute. More surface area means that more solute comes in contact with the solvent. When the surface area of the solute increases, the solute dissolves more quickly. ✔

How does heating speed up dissolving?

When you make hot chocolate from a mix, you mix the powder into a hot liquid. The sugar in the mix dissolves faster in a hot liquid solvent. Solvent particles move faster when the temperature of the solvent increases. Fast-moving solvent particles have more chances to come in contact with solute particles. The more often they come in contact, the faster the solute particles break loose and dissolve.

Can you combine these methods?

If you use more than one of these methods at the same time, you can speed up the dissolving process even more. Suppose you place a sugar cube in cold water. You know the sugar will dissolve eventually. If you heat the water, the sugar will dissolve at a faster rate. If you heat the water and stir the solution, you increase that rate even more. Finally, if you crush the sugar cube, heat the water, and stir the solution, the sugar will dissolve at the fastest rate. The rate of dissolving increases with each additional method you use.

✔ Reading Check

5. Determine Why does breaking up a solid solute into smaller pieces speed up the dissolving process?

_____
_____
_____

💡 Think it Over

6. Infer What do both stirring and heating do that increases the rate that particles dissolve?

_____
_____
_____

After You Read
Mini Glossary

polar: having a positive area and a negative area
solute: the substance that is dissolved in a solution
solution: a mixture that has the same ingredients, color, and density mixed evenly throughout
solvent: the substance that does the dissolving in a solution

1. Review the terms and definitions in the Mini Glossary. Select one term and write a definition of the term in your own words.
   _____
   _____

2. Complete the outline to help you organize what you learned about solutions.

   Solutions
   I. Solutions
      A. A solution is _____
      B. An example of a solution made with a liquid and a solid is _____
      C. An example of a solution made with a liquid and a gas is _____
      D. An example of a solution made with two solids is _____
   II. Solutes and solvents
      A. A solute is _____
      B. A solvent is _____
   III. How substances dissolve
      A. Water molecules approach the solid solute.
      B. _____
      C. _____
   IV. Rate of dissolving is affected by
      A. _____
      B. _____
      C. _____

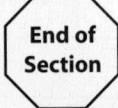

 Visit gpscience.com to access your textbook, interactive games, and projects to help you learn more about how solutions form.

chapter 22 Solutions

section ❷ Solubility and Concentration

● Before You Read

The labels of some containers of fruit juice say "Not from concentrate." What do you think this means?

_____

_____

What You'll Learn
- what solubility is
- about the concentration of solutions
- three types of solutions
- factors that affect gas solutions

● Read to Learn

How much can dissolve?

Suppose you like your lemonade very sweet. You add a teaspoon of sugar to it and stir. The sugar dissolves. If you keep adding sugar to the lemonade, you will reach a point when no more sugar will dissolve in it. The excess sugar crystals fall to the bottom of the glass. They do not go into the solution.

You have reached the solubility of sugar in a given amount of water. **Solubility** (sahl yuh BIH luh tee) is the greatest amount of solute that can dissolve in a specific amount of solvent at a given temperature.

Are the solubilities of all substances the same?

You can dissolve more than 32 g of salt in 113 g of water at 25°C. But, you can only dissolve about 12 g of baking soda in 113 g water. Salt and baking soda have different solubilities in water. The difference in the solubilities of solutes depends on the nature of the solute and the nature of the solvent.

Concentration

Suppose you and a friend are making lemonade. You add one teaspoon of lemon juice to a glass of water. Your friend adds four teaspoons of lemon juice to the same amount of water. Which lemonade has more lemon flavor? Your friend's does. Your friend's lemonade is concentrated, because it has a large amount of solute dissolved in the solvent. Your lemonade is dilute, because it has a small amount of solute dissolved in the solvent.

Mark the Text

Check for Understanding As you read this section, highlight any sentences that you read more than once. After you finish the section, go back and read the highlighted sentences again.

FOLDABLES

B Classify Make a Foldable like the one below. As you read this section, take notes about the different types of solutions.

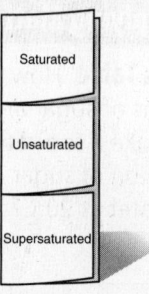

Reading Essentials **383**

How much solute is in a concentrated solution?

Concentrated and dilute are not precise terms. Concentration of solutions can be described precisely, though. One way is to state the concentration as a percentage by volume of the solute. For example, the label on a bottle of orange-flavored drink states that the drink contains 10 percent fruit juice. This means that in 100 mL of the drink, there are 10 mL of juice. Or, to make 100 mL of a 10 percent solution of orange drink, the manufacturer added 10 mL of juice to 90 mL of water. To be sure you are getting the highest concentration of juice, choose a drink that is 100 percent juice.

Types of Solutions

Different solutes have different solubilities. You can use the amount of a solute dissolved to describe three different types of solutions—saturated, unsaturated, and supersaturated.

What is a saturated solution?

The table lists the solubilities of some compounds in 100 g of water at different temperatures. If you add 35 g of copper(II) sulfate, $CuSO_4$, to 100 g of water at 20°C, only 32 g will dissolve. The solution is saturated. A **saturated solution** is a solution that contains all the solute it can hold at a given temperature. If you increase the temperature of the mixture, more copper(II) sulfate can dissolve. As shown in the table, the solubility of solid solutes increases as the temperature of the liquid solvent increases.

Reading Check

1. **Identify** Which of the following does **not** describe a solution according to the amount of solute dissolved? (Circle your answer.)

   a. saturated

   b. soluble

   c. unsaturated

   d. supersaturated

Applying Math

2. **Use a Table** How many grams of sugar are needed to make a saturated solution of sugar in 100 g of water at 20°C?

Solubility of Compounds in g/100 g of Water

Compound	0°C	20°C	100°C
Copper(II) sulfate	23.1	32.0	114
Potassium bromide	53.6	65.3	104
Potassium chloride	28.0	34.0	56.3
Potassium nitrate	13.9	31.6	245
Sodium chlorate	79.6	95.9	204
Sodium chloride	35.7	35.9	39.2
Sucrose (sugar)	179.2	203.9	487.2

What is a solubility curve?

The graph shows some of the information from the table on the previous page. Each line on the graph is a solubility curve for a substance. You can use this to find how much solute dissolves in 100 g of water at a given temperature. To find how much sodium chloride, NaCl, dissolves in 100 g of water at 50°C, find 50°C on the x-axis. Trace a line upward from 50°C to the curve for NaCl. Read the amount shown on the y-axis at that point. About 35 g of NaCl dissolves in 100 g of water at 50°C. As the temperature of the water increases, more solute can dissolve. This is true for most liquid solvents and solid solutes. ✓

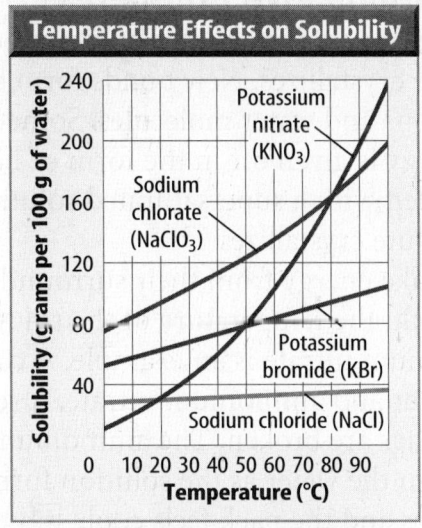

Applying Math

3. **Use a Graph** About how much potassium nitrate is needed to make a saturated solution in 100 g of water at 80°C?

_____

Reading Check

4. **Explain** As the temperature of the water increases, what happens to the amount of solid solute that can be dissolved? (Circle your answer.)

   a. It does not change.

   b. Less can dissolve.

   c. None can dissolve.

   d. More can dissolve.

What is an unsaturated solution?

You learned that 32 g of copper(II) sulfate forms a saturated solution with 100 g of water at 20°C. What if a solution has less than 32 g of copper(II) sulfate? Then it is an **unsaturated solution**, a solution that can dissolve more solute at a given temperature.

The term *unsaturated solution* is not precise. You know exactly how much copper(II) sulfate makes a saturated solution in 100 g of water at 20°C. An unsaturated solution can have any amount of copper(II) sulfate less than 32 g in 100 g of water at 20°C.

How can a solution be supersaturated?

Suppose you make a saturated solution of potassium nitrate (KNO_3) with 100 g of water at 100°C. You add 245 g of (KNO_3) to the water, just as the solubility table shows. You then let the solution cool to 20°C. What happens? Some of the (KNO_3) solute comes out of solution and falls to the bottom of the container. You can see from the table that at 20°C only 31.6 g of (KNO_3) dissolves in 100 g of water. At the lower temperature, the solvent cannot hold as much solute.

Think it Over

5. **Infer** Would a solution made of 30 g of copper(II) sulfate dissolved in 100 g of water at 20°C be a saturated or an unsaturated solution?

_____

_____

Most saturated solutions behave the same way as the potassium nitrate solution when cooled. But some solutions can become supersaturated. A **supersaturated solution** is a solution that has more solute than a saturated solution at the same temperature. For example, if you cool a saturated solution of sodium acetate from 100°C to 20°C, no solute comes out of the solution. This solution is supersaturated. A supersaturated solution is unstable. If a crystal of sodium acetate is dropped into this solution, crystals begin to form. The extra sodium acetate comes out of solution.

When do solutions give off energy?

The supersaturated sodium acetate solution becomes hot as sodium acetate crystallizes. New bonds form between the sodium acetate ions and water molecules. Sometimes when bonds form, energy is given off in the form of heat. Some heat packs are filled with a supersaturated solution that gives off heat as the solute crystallizes.

Some solutes take energy from their surroundings to dissolve. As a result, the temperature of the solution is reduced. Ammonium nitrate is an example. A cold pack has inner bags of water and ammonium nitrate. A solution forms when the inner bags are broken. The ammonium nitrate draws energy from the water as the solution forms. The water temperature drops and the pack feels cool. ✔

Solubility of Gases

Soda is a solution of carbon dioxide gas dissolved in flavored water. When you shake an open bottle of soda, it bubbles. Shaking or stirring a solution of a gas in a liquid allows more gas molecules to reach the surface of the liquid, where they escape from the liquid into the air.

How do pressure and temperature affect a gas dissolved in a liquid?

Increasing the pressure of a gas over a liquid forces more gas to dissolve in the liquid. Soda is bottled under a large amount of pressure. The pressure increases the amount of gas that can dissolve in the soda and also keeps the carbon dioxide gas in solution. When you open a bottle of soda, the pressure is released and bubbles of gas come out of solution.

Cooling a liquid increases the amount of gas that will dissolve in it, the opposite of dissolving amounts of most solids in a liquid. Less gas is able to dissolve in warm liquids. This is why warm soda bubbles up more than cold soda.

Reading Check

6. **Apply** What is happening when a solution gets colder?

_____
_____
_____

Think it Over

7. **Draw Conclusions** How could you make the bubbles come out of solution before you drink a soda?

_____
_____
_____

After You Read

Mini Glossary

saturated solution: a solution that contains all the solute it can hold at a given temperature

solubility: the greatest amount of solute that can dissolve in a specific amount of solvent at a given temperature

supersaturated solution: a solution that has more solute than a saturated solution at the same temperature

unsaturated solution: a solution that can dissolve more solute at a given temperature

1. Review the terms and definitions in the Mini Glossary. Write a sentence using one of the terms that describes a type of solution.

2. Complete the table below to help you organize the information you learned about solubility and types of solutions.

Type of Solution	How is it made?	Does it depend on temperature?	Does it depend on the amount of solute?
Saturated			
Unsaturated			
Supersaturated			

3. As you read this section, you highlighted the sentences that you read more than once. How could you use this strategy if you were studying with a friend?

 Visit **gpscience.com** to access your textbook, interactive games, and projects to help you learn more about solubility and concentration.

Reading Essentials **387**

chapter 22 Solutions

section ❸ Particles in Solution

What You'll Learn
- how some solutes form positively or negatively charged particles
- how some solutions conduct electricity
- how antifreeze works

Mark the Text

Locate Information Highlight every heading in this section that asks a question. Then highlight each answer as you find it.

FOLDABLES

C Summarize Make a Foldable like the one below. As you read the section, write down the most important ideas under each heading.

| Particles With a Charge: | Ionic Solutions: | Effects of Solute Particles: |

● Before You Read

Why do you think many electrical appliances have labels warning the user not to use the appliance near water?

_____

_____

● Read to Learn

Particles with a Charge

There are charged particles in your body that conduct electricity. A particle with a charge is an **ion**. Ions are in the fluids that are in and around all the cells in your body. You could not live without ions. Some ions help nerve cells send messages to other parts of your body. Nerves control how your muscles move each time you blink your eyes or wave your hand.

An **electrolyte** is a compound that produces solutions of ions that conduct electricity in water. Strong electrolytes dissolve completely into ions in solution. Strong electrolytes conduct a strong electric current, because they contain only ions. Sodium chloride is an example of a strong electrolyte.

Weak electrolytes stay mainly as molecules when they dissolve in water. Weak electrolytes produce only a few ions and conduct current weakly. Acetic acid in vinegar is an example of a weak electrolyte.

Nonelectrolytes are substances that do not form ions in water and cannot conduct electricity. Examples of nonelectrolytes are organic molecules like sucrose (sugar) and ethyl alcohol.

How do ionic solutions form?

Ionization Ionic solutions form in two ways. Some electrolytes are molecules made up of neutral atoms. To form ions, these molecules must be broken apart in a way that the atoms take on a charge. This process of forming ions is **ionization.**

388 CHAPTER 22 Solutions

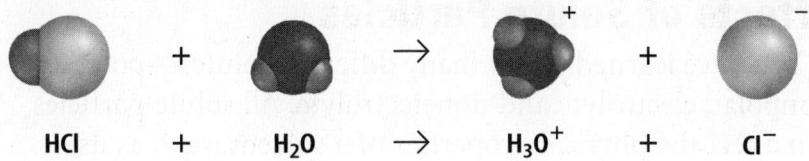

HCl + H₂O → H₃O⁺ + Cl⁻

Picture This

1. **Use a Scientific Illustration** In the figure, circle the positive ion.

Hydrogen chloride is an electrolyte. The above figure shows how hydrogen chloride molecules that are made up of neutral atoms go through the process of ionization when they mix with water molecules. Recall that hydrogen chloride and water are both polar molecules. Water molecules surround the hydrogen chloride molecules and pull them apart into positive hydrogen ions and negative chloride ions, Cl⁻. A symbol for a hydrogen ion in water is H₃O⁺.

Dissociation The second way that ionic solutions form is by the separation of ionic compounds. Some electrolytes, such as sodium chloride, are ionic compounds. Ionic compounds already contain ions that are attracted into solution by the surrounding water molecules. The process in which an ionic solid separates into its positive and negative ions is **dissociation**.

The figure below shows what happens when a crystal of sodium chloride goes through the process of dissociation in water. Water molecules pull the ions from the surface of the crystal. Water molecules surround and separate the sodium ions and chloride ions. The positive ends of the water molecules are attracted to the negative chloride ions. The negative ends of the water molecules are attracted to the positive sodium ions. The sodium ions and chloride ions dissociate from each another. Because the sodium and chloride ions can move freely through the solution, they can conduct an electric current.

Think it Over

2. **Contrast** What is the difference between ionization and dissociation?

_____
_____
_____
_____

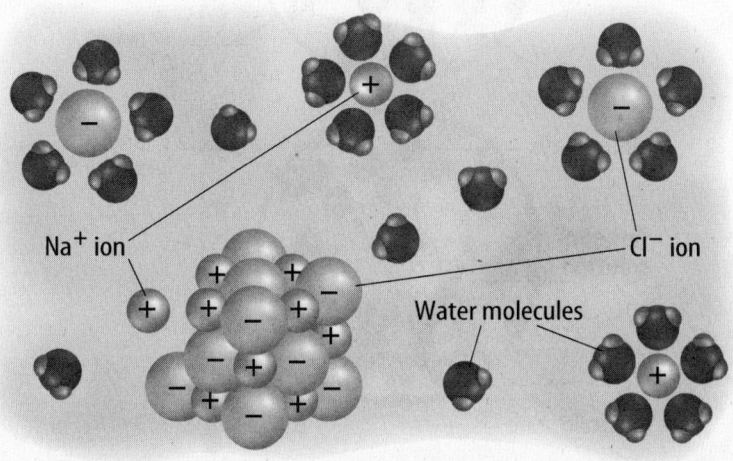

Sodium and Chloride Ions Mixed with Water

Picture This

3. **Identify** In the figure, circle the sodium chloride crystal.

Reading Essentials **389**

Effects of Solute Particles

You have learned about many different solutes—polar and nonpolar, electrolyte and nonelectrolyte. All solute particles can affect the physical properties of a solvent, such as its freezing point and its boiling point. How a solute affects the freezing point or the boiling point of a solvent depends on the number of solute particles in solution. Some of these effects are useful. Adding antifreeze to the water in a car radiator lowers the freezing point of the radiator fluid.

How does antifreeze lower the freezing point?

As a substance freezes, its particles arrange themselves in an orderly pattern. When a solute is added, the solute particles interfere with this pattern, making it harder for the solvent to freeze. To overcome the interference, a lower temperature is needed to freeze the solvent.

The figure shows how water molecules in pure water can move easily from the liquid state to the solid state. Pure water freezes at 0°C. When a solute is added to the water, the solute particles interfere with water molecules as they try to form a crystalline pattern. In a car radiator, antifreeze molecules added to water block the formation of ice crystals. When enough solute particles are present, water cannot freeze at 0°C. The temperature must be lower to overcome the interference of solute particles.

Think it Over

4. Explain how antifreeze keeps water in a car radiator from freezing at 0°C.

Picture This

5. Illustrate In the figure, highlight the particles that keep the water from freezing at its normal temperature.

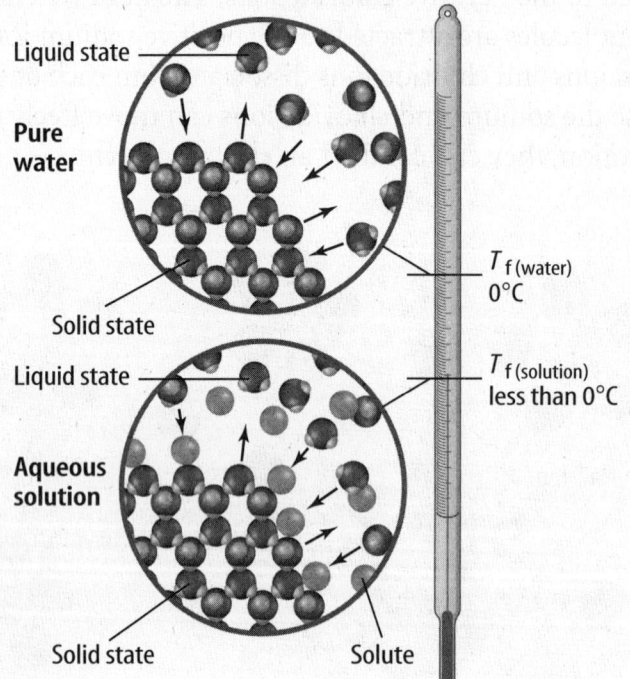

Why can some animals live in cold climates?

Some animals that live in very cold climates have their own kind of antifreeze. Caribou have substances in their bodies that keep their legs from freezing when the temperature falls below 0°C. Fish that live in very cold water also have a natural kind of antifreeze. The antifreeze keeps ice crystals from forming in their tissues. Many insects have a similar antifreeze chemical that protects them from freezing temperatures.

How can the boiling point of water be raised?

You may be surprised to learn that antifreeze raises the boiling point of water. Solute particles interfere with the evaporation of solvent particles. More energy is needed for the solvent particles to escape from the liquid surfaces. The more solute particles in the solution, the higher the boiling point of the solution will be.

How does antifreeze work in a car radiator?

In the figure below, the beaker on the left represents water in a car radiator without antifreeze. As pure water is heated, water molecules move easily to the surface and vaporize. The number of molecules that vaporize depends on the temperature of the water. Water molecules move faster as the temperature rises and more molecules can vaporize. When the pressure of the water vapor equals the pressure of the atmosphere, the water boils.

The beaker on the right shows what happens when a solute, such as antifreeze, is added to the water. The solute particles are distributed evenly throughout the solution, including the surface. Solute particles block part of the surface, so fewer water molecules can reach the surface and vaporize. The solution cannot boil because the vapor pressure of the solution is lower than the vapor pressure of the solvent. Energy must be added to overcome the interference and raise the vapor pressure of the solution to make it boil. The added energy means the solution boils at a temperature higher than the boiling point of the pure water.

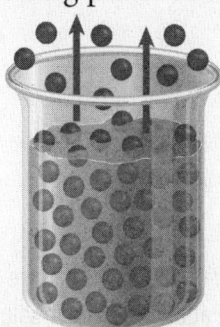 

Think it Over

6. Draw Conclusions What substance would you add to a car radiator during the hot summer months to prevent the radiator from boiling over?

_____

Picture This

7. Draw arrows above the beakers in both figures to show the direction of the atmospheric pressure on the water.

● After You Read
Mini Glossary

dissociation: the process in which an ionic solid separates into its positive and negative ions

electrolyte: a compound that produces solutions of ions that conduct electricity in water

ion: a particle with a charge

ionization: the process of forming ions

nonelectrolyte: substances that do not form ions in water and cannot conduct electricity

1. Review the terms and their definitions in the Mini Glossary. Write a sentence using the terms *ion* and *electrolyte* correctly.

   _____

   _____

2. Write the letter of the term in Column 2 that matches the description or example in Column 1.

   Column 1

   _____ 1. The process used by soldium chloride to break apart into ions

   _____ 2. Sodium chloride

   _____ 3. The process used by hydrogen chloride to break apart into ions

   _____ 4. Na$^+$

   Column 2

   a. ion

   b. electrolyte

   c. ionization

   d. dissociation

3. You highlighted every heading in this section that asks a question and then highlighted each answer as you found it. How would this strategy help you study for a test?

   _____

   _____

   _____

 Visit **gpscience.com** to access your textbook, interactive games, and projects to help you learn more about particles in solution.

Solutions

section 4 Dissolving Without Water

Before You Read

You probably have gotten stains on your clothes that didn't rinse out with just water. Write about a stain that you could not remove with water.

_____

_____

What You'll Learn
- what solutes do not dissolve in water
- how polar and nonpolar solvents work in water
- how to choose the right solvent for cleaning

Read to Learn

When Water Won't Work

Water often is called the universal solvent because it dissolves so many things. But, there are some things that water cannot dissolve, such as a salad dressing with vinegar and oil. The water in vinegar cannot dissolve the oil. Why not?

Water molecules have positive and negative ends. The charged areas of water molecules help water dissolve polar solutes. Some substances are nonpolar. A **nonpolar** material does not have positive and negative areas. Nonpolar substances are not attracted to polar substances, including water. Most nonpolar substances do not dissolve in water.

How do nonpolar solutes behave?

Think again about the example of a vinegar-and-oil salad dressing. If you do not mix the dressing, it separates into two layers. The bottom layer is vinegar, which is a solution of acetic acid in water. The top layer is salad oil.

Most salad oils are made of large molecules of carbon and hydrogen atoms called hydrocarbons. Hydrocarbon molecules share the electrons in their bonds in a nearly equal way. So, hydrocarbon molecules do not have separate positive and negative ends. The nonpolar oil molecules are not attracted to the polar water molecules in the vinegar solution. The water molecules in the salad dressing cannot dissolve the nonpolar oil molecules. That is why you have to shake a vinegar-and-oil dressing to mix it before using it.

Study Coach

Create a Quiz As you read this section, write five questions that could be used on a quiz. Write different types of questions, such as multiple choice and fill-in-the-blank. Be sure to answer the questions after you write them.

FOLDABLES

D Classify Make a Foldable like the one below. As you read this section, write down the different nonpolar solvents and what they can be used for.

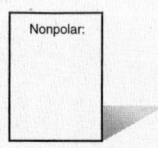

Reading Essentials

Why are alcohols special?

Molecules of some substances have a polar end and a nonpolar end. These substances can form solutions with polar and nonpolar solutes. Alcohols are examples of this kind of substance. The polar end of an alcohol molecule dissolves polar solutes. The nonpolar end dissolves nonpolar solutes.

The figure shows a molecule of the alcohol ethanol and its formula. The –OH group is polar. The other part of the molecule, $-C_2H_5$, is nonpolar. Ethanol can dissolve iodine, a nonpolar substance. It also can dissolve water, a polar substance.

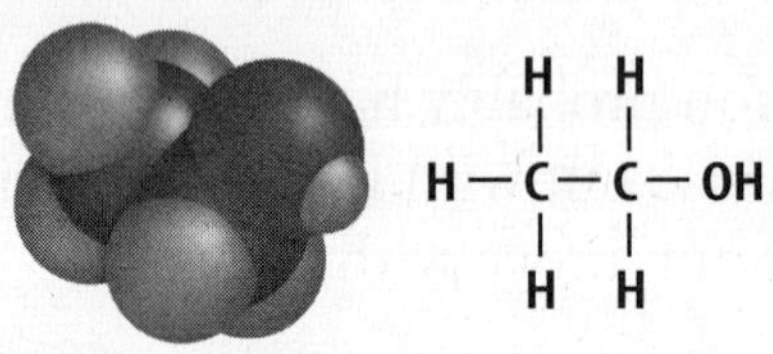

Picture This

1. **Determine** Circle the nonpolar part of the ethanol molecule.

Useful Nonpolar Molecules

You may have nonpolar solvents around your house. Mineral oil is a useful nonpolar solvent. It can be used to remove candle wax from candleholders. Mineral oil can also help remove bubble gum from some surfaces. Mineral oil, candle wax, and bubble gum are all nonpolar substances.

Oil-based paints contain pigments dissolved in oils. Oils are nonpolar molecules. To remove wet paint or to make it thinner, you must use a nonpolar solvent, such as turpentine.

Another useful nonpolar substance is the gasoline used in cars and lawnmowers. Gasoline is a solution of different hydrocarbons. Remember that hydrocarbons are nonpolar substances.

Dry cleaners use nonpolar solvents to remove oil and grease stains. The word *dry* means that no water is used in the process. Molecules of a nonpolar solute can slip easily among molecules of a nonpolar solvent. Dry cleaning can remove stains from clothes that cannot be removed when the clothes are washed in water.

Think it Over

2. **Interpret** Explain in your own words what the phrase "like dissolves like" means.

_____
_____
_____

"Like dissolves like" is a useful rule to remember. It means that polar solvents dissolve polar solutes and nonpolar solvents dissolve nonpolar solutes.

When are nonpolar solvents not helpful?

Nonpolar solvents can be dangerous when they are not used carefully. Many nonpolar solvents are flammable, which means that they burn easily. Some nonpolar solvents are toxic, which means they can harm you if they touch your skin or if you breathe their vapors. Never use nonpolar solvents in a closed room. These solvents evaporate more readily than water. You must always have fresh air coming into the room so you do not breathe in the vapors of a nonpolar solvent.

How does soap work?

You have natural oils on your skin and hair. These oils keep your skin and hair from drying out. They also attract and hold dirt. Dirt forms a nonpolar mixture when it combines with oils. Water won't wash this nonpolar mixture away, because water is a polar solvent. You need a substance with both polar and nonpolar properties. You need to use soap.

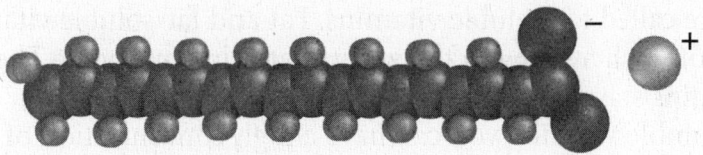

Picture This

3. Illustrate Circle the part of the soap molecule that is ionic.

_____
_____
_____

Look at the soap molecule in the figure above. Soaps start out as large fatty acid molecules. Fatty acids have long hydrocarbon ends that are nonpolar. A carboxylic acid group, –COOH, is at the other end.

To make soap, the hydrogen atom of the carboxylic acid group is removed. Without the hydrogen atom, the end has a negative charge. The negative end attracts a positive ion of sodium or potassium. The finished soap molecule has a nonpolar end and an ionic end. The figure below shows how soap helps oil and water combine. The ionic end of a soap molecule dissolves in water. The nonpolar end dissolves in oily dirt. Together the two ends of a soap molecule remove dirt so it can be rinsed away.

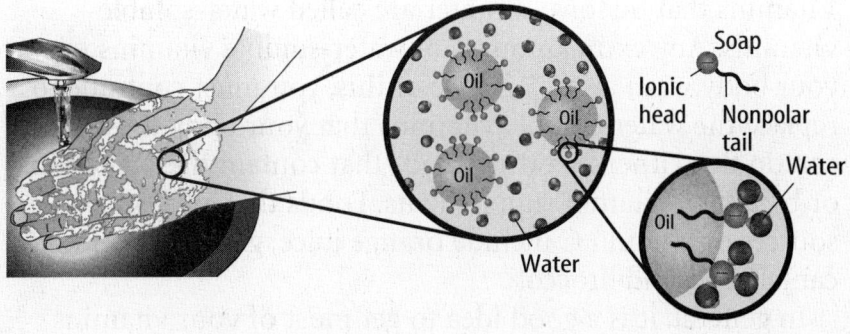

Picture This

4. Explain Why can't you wash away oily dirt with only water?

_____
_____
_____

Reading Check

5. Identify Name three foods that are good sources of vitamin A.

_____

_____

_____

Reading Check

6. Determine Name a vitamin that is made of polar molecules.

_____

Think it Over

7. Describe What is the difference between fat-soluble vitamins and water-soluble vitamins?

_____

_____

_____

Polarity and Vitamins

You know that your body needs vitamins to stay healthy. Vitamin A is an important vitamin. Good sources of Vitamin A include liver, lettuce, cheese, eggs, carrots, sweet potatoes, and milk.

Vitamin A is nonpolar and can dissolve in fat in your body because fat is also a nonpolar substance. Fat also dissolves vitamin D, vitamin E, and vitamin K. Vitamins that dissolve in fat are called fat-soluble vitamins. Fat and fat-soluble vitamins do not wash away with the water that is in your body. The vitamins stay in your tissues. If you take too much of a fat-soluble vitamin, you can have a high concentration of the vitamin in your body. Some fat-soluble vitamins can be very harmful at high concentrations. You should not take large amounts of fat-soluble vitamins unless your doctor tells you to.

Polar Vitamins Other vitamins, such as vitamins B and C, are polar. You can see the structure of vitamin C in the figure to the right. Notice it has many carbon-to-carbon bonds. This might make you think vitamin C is nonpolar. But vitamin C also has oxygen-to-hydrogen bonds that resemble the bonds in water. These oxygen-to-hydrogen bonds make vitamin C a polar molecule.

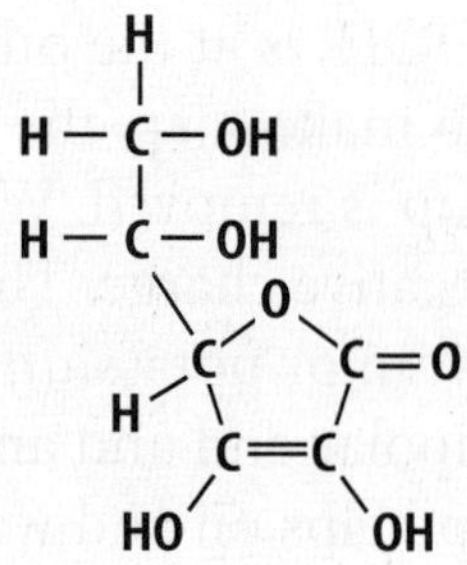

Polar vitamins dissolve in the water that is in your body. Vitamins that dissolve in water are called water-soluble vitamins. Any extra amounts of water-soluble vitamins in your body wash away. Because of this, you must continue to replace the water-soluble vitamins that your body uses. You can do this either by eating foods that contain these vitamins or by taking vitamin supplements. Foods that are good sources of vitamin C include orange juice, green peppers, cantaloupe, and broccoli.

In general, it is a good idea to get most of your vitamins from eating a healthful diet. That way you will not risk a harmful overdose.

After You Read

Mini Glossary

nonpolar: a material that does not have a positive area and a negative area

1. Review the term and its definition in the Mini Glossary. Write a sentence describing how a nonpolar liquid behaves when added to water.

   _____

   _____

2. Write the name of each substance in the box that contains the kind of solvent that can dissolve it.

   vitamin C   bubble gum   salad oil   candle wax   vinegar

Nonpolar solvent	Polar solvent

3. You created a quiz based on what you have learned in this section, then answered the questions. What kind of question best helped you learn the material?

   _____

   _____

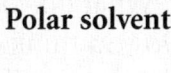 Visit gpscience.com to access your textbook, interactive games, and projects to help you learn more about dissolving without water.

chapter 23 Acids, Bases, and Salts

section ❶ Acids and Bases

What You'll Learn
- how acids and bases are similar and different
- formulas and uses of common acids and bases

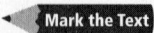

Underline As you read, underline any words or sentences you think might be important to remember. When you finish reading, look back at what you underlined to make sure you understand it.

FOLDABLES

A **Compare and Contrast** Make the following Foldable to help you understand acids and bases.

Describe	Properties	Common Examples
Acids		
Bases		

● Before You Read

Think of a food or drink you like that tastes sour. What do you think gives it a sour taste?

_____

● Read to Learn

Acids

What do you think about when you hear the word *acid*? Do you think of a substance that can burn your skin or put a hole in metal? Many sour foods contain acids. Some acids are dangerous to handle. Others are safe, and some are good to eat.

What are the properties of acids?

An **acid** is a substance that produces hydrogen ions, H^+, in a water solution. When an acid dissolves in water, some of the hydrogen atoms are released as hydrogen ions, H^+. It is the ability to produce these H^+ ions that gives acids their characteristic properties. When an acid dissolves in water, H^+ ions interact with water molecules to form hydronium ions. A **hydronium ion** (hi DROH nee um • I ahn), H_3O^+, is a combination of an H^+ ion and a water molecule.

Acids have some common properties. All acids taste sour. But you should never taste a substance to see if it is acidic. Some acids can burn you. Acids are corrosive, which means they seem to eat away some metals. When an acid reacts with a metal, hydrogen gas and metallic compounds form. Acids also cause indicators to change color. An **indicator** is an organic compound that changes color in acid and base. Litmus paper is an indicator that turns red in acids.

What are some common acids?

Many foods contain acids. Citrus fruits contain citric acid. Lactic acid is found in yogurt and buttermilk. Pickled foods contain vinegar, also known as acetic acid. Your stomach uses hydrochloric acid to help it digest food.

398 CHAPTER 23 Acids, Bases, and Salts

Common Acids and Their Uses		
Name, Formula	**Use**	**Other Information**
Acetic acid, CH_3COOH	Food preservation and preparation	When in solution with water, it is known as vinegar.
Acetylsalicylic acid, $HOOC-C_6H_4-OOCCH_3$	Pain relief, fever relief, to reduce inflammation	Known as aspirin
Ascorbic acid, $H_2C_6H_6O_6$	Antioxidant, vitamin	Called vitamin C
Carbonic acid, H_2CO_3	Carbonated drinks	Involved in cave, stalactite, and stalagmite formation and acid rain
Hydrochloric acid, HCl	Digestion as gastric juice in stomach, to clean steel in a process called pickling	Commonly called muriatic acid
Nitric acid, HNO_3	To make fertilizers	Colorless, yet yellows when exposed to light
Phosphoric acid, H_3PO_4	To make detergents, fertilizers and soft drinks	Slightly sour but pleasant taste, detergents containing phosphates cause water pollution
Sulfuric acid, H_2SO_4	Car batteries, to manufacture fertilizers and other chemicals	Dehydrating agent, causes burns by removing water from cells

Some common acids and their uses are listed in the table. Many of these acids are important in making products, such as fertilizer. Remember that many acids can burn your skin.

Bases

A <u>base</u> is any substance that forms hydroxide ions, OH^-, in a water solution. A base is also any substance that accepts H^+ ions from acids.

Unlike acids, not many foods are bases. Egg whites and baking powder are two foods that are basic. Some medicines, such as antacids, are basic. A common base is soap. A characteristic of bases is that they feel slippery, like soapy water. Many cleaning products contain bases. Bases are important in industry, also. For example, sodium hydroxide is a base that is used to separate cellulose fibers from wood pulp to make paper.

Picture This

1. **Identify** Look at the acids listed in the table above. What is the first element in the chemical formula of most acids?

_____

What are the properties of bases?

Bases are the opposites of acids. While bases and acids do share some features, bases also have their own properties. When they are not dissolved in water, many bases are solids in the form of crystals. In solution, bases feel slippery and taste bitter. Like strong acids, strong bases are corrosive. Bases can burn you. Never taste or touch a substance to see if it is basic. Bases also cause indicators to change color. Litmus paper is an indicator that turns blue in bases.

What are some common bases?

The table below lists some common bases and their uses. You may have used many common bases found in cleaning products and not even known it.

Picture This
2. Identify Look at the bases listed in the table. What molecule do most of the bases have in their chemical formulas?

Common Bases and Their Uses		
Name, Formula	**Use**	**Other Information**
Aluminum hydroxide, $Al(OH)_3$	Color-fast fabrics, antacid, water purification	Sticky gel that collects suspended clay and dirt particles on its surface
Calcium hydroxide, $Ca(OH)_2$	Leather-making, mortar and plaster, lessen acidity of soil	Called caustic lime
Magnesium hydroxide, $Mg(OH)_2$	Laxative, antacid	Called milk of magnesia when in water
Sodium hydroxide, $NaOH$	To make soap, oven cleaner, drain cleaner, textiles, and paper	Called lye and caustic soda; generates heat (exothermic) when combined with water, reacts with metals to form hydrogen
Ammonia, NH_3	Cleaners, fertilizer, to make rayon and nylon	Irritating odor that is damaging to nasal passages and lungs

Reading Check
3. Explain Why is water the main solvent for acidic and basic solutions?

Solutions of Acids and Bases

Many products that contain acids and bases are solutions. Water is the main solvent for these solutions because water molecules have polarity. Remember, polarity means a molecule has a slight positive charge on one end and a slight negative charge on the other end.

What happens when acids dissolve in water?

Remember that an acid produces hydrogen ions (H⁺) in water. When an acid dissolves in water, the negative ends of nearby water molecules attract the positive hydrogen in the acid. The acid separates into ions, which is called dissociation. What is left after dissociation are negative ions and positive H⁺ ions. The H⁺ ions combine with water molecules to form hydronium ions (H₃O⁺). Therefore, an acid compound produces hydronium ions when dissolved in water. The figure shows the dissociation of hydrogen chloride (HCl) into hydronium and chloride ions.

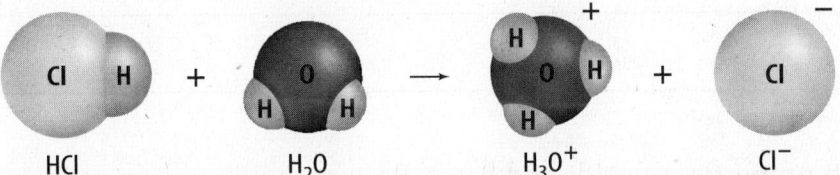

HCl   H₂O   H₃O⁺   Cl⁻

Picture This
4. Identify When HCl dissociates, what ions does it produce?

_____

_____

What happens when bases dissolve in water?

Bases form hydroxide ions (OH⁻) in water. In the table of bases on the previous page, you can see that most bases have –OH in their formulas. When bases dissolve in water, the positive ends of nearby water molecules attract the OH⁻ ions in the base. The base dissociates. What is left after dissociation are positive ions and negative OH⁻ ions. Unlike acid dissociation, the OH⁻ ions do not combine with water molecules. The figure shows the dissociation of sodium hydroxide (NaOH).

$$NaOH + H_2O \rightarrow Na^+ + OH^- + H_2O$$

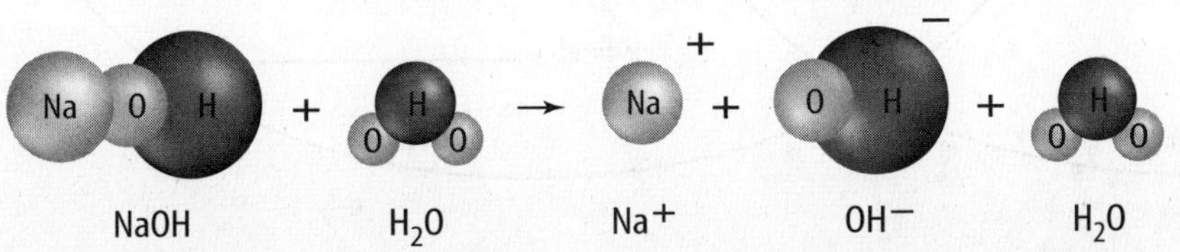

NaOH   H₂O   Na⁺   OH⁻   H₂O

Picture This
5. Identify When NaOH dissociates, it produces a molecule of water, H₂O and two ions. What ions does it produce?

_____

_____

How is ammonia different from other bases?

Ammonia, NH₃, is a base that does not contain –OH. In water, ammonia actually dissociates water molecules. An ammonia molecule attracts a hydrogen ion from a water molecule to form an ammonium ion, NH₄⁺. The rest of the water molecule is a hydroxide ion, OH⁻. Ammonia is found in many household cleaners. You should never use products containing ammonia with other cleaners that contain chlorine (sodium hypochlorite), such as bathroom bowl cleaners and bleach. Ammonia reacts with sodium hypochlorite and produces toxic gases that can severely damage lung tissue and cause death.

● After You Read
Mini Glossary

acid: a substance that produces hydrogen ions, H⁺, in a water solution

base: any substance that forms hydroxide ions, OH⁻, in a water solution or a substance that accepts H⁺ ions from acids

hydronium ion: a combination of a water molecule and a hydrogen ion

indicator: an organic compound that changes color in acid and base

1. Review the terms and their definitions in the Mini Glossary. Write a sentence using the term for the substance that makes many foods taste sour.

   _____

   _____

2. Fill in the Venn diagram with properties of acids and bases. Be sure to put any properties that acids and bases have in common in the part where the ovals overlap.

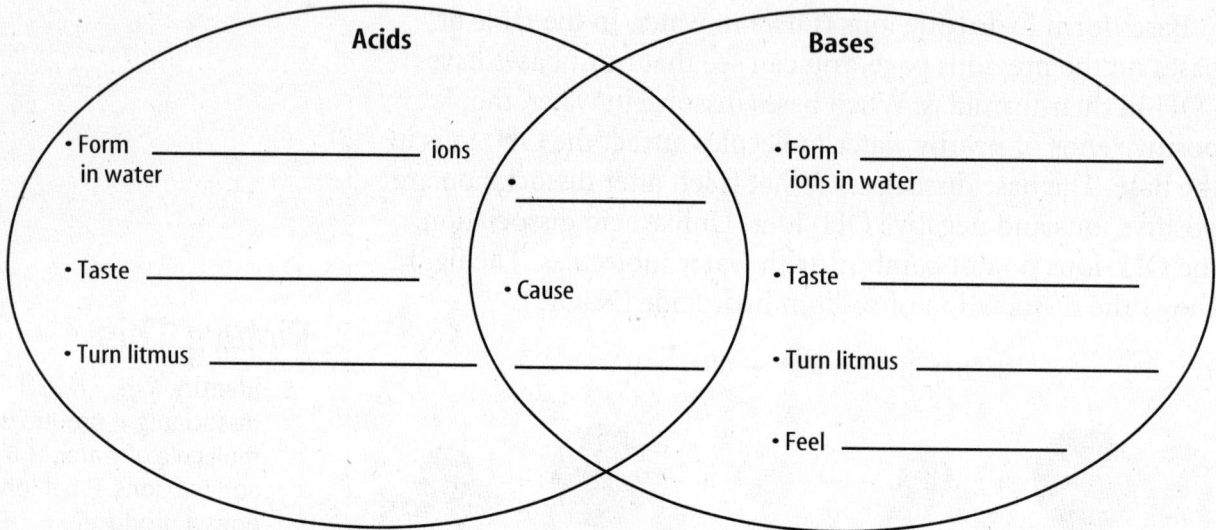

3. You underlined words and sentences in this section that you thought would be important to remember. How did this help you learn about the topics in the section?

   _____

   _____

   _____

 Visit **gpscience.com** to access your textbook, interactive games, and projects to help you learn more about acids and bases.

402 CHAPTER 23 Acids, Bases, and Salts

Acids, Bases, and Salts

section ❷ Strength of Acids and Bases

● Before You Read

Describe the difference between a food that is very sour and a food that is just a little sour.

_____

_____

What You'll Learn
- what makes acids and bases strong or weak
- how strength and concentration are alike and different

● Read to Learn

Strong and Weak Acids and Bases

Some acids, like the sulfuric acid in car batteries, can burn your skin. Some acids are stronger than others. The strength of an acid or a base depends on how many acid molecules dissociate into ions in water. A **strong acid** is one in which almost all the acid molecules dissociate in water.

Hydrochloric acid (HCl), nitric acid (HNO_3), and sulfuric acid (H_2SO_4), are examples of strong acids. A **weak acid** is one in which only a small number of the acid molecules dissociate in water. Acetic acid (CH_3COOH) and carbonic acid (H_2CO_3) are weak acids.

Ions in solution can conduct electric current. Acids and bases can carry an electric current because they dissociate into ions. The more ions in a solution, the more electric current the solution can conduct. Strong acids have many ions in solution, so they conduct current well. Weak acids have few ions in solution and do not conduct current as well. ☑

Study Coach

Identifying the Main Point Write down the main point of each paragraph as you read this section. After you read, look over the main points to make sure you understand them.

How do you write chemical equations for acid dissociation?

An equation for the dissociation of a strong acid uses a single arrow that points toward the ions that form.

$$HCl(g) + H_2O(l) \rightarrow H_3O^+(aq) + Cl^-(aq)$$

Reading Check

1. **Identify** Which conducts an electric current best, a strong acid or a weak one?

_____

FOLDABLES

B Compare and Contrast
Make the following Foldables to understand how the strength and concentration of acids and bases are the same and different.

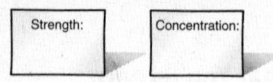

Picture This

2. **Describe** In the figure, the solution of acetic acid is more concentrated than the solution of hydrochloric acid. Does this mean acetic acid is stronger than hydrochloric acid? Explain.

_____
_____
_____
_____

Weak Acids In weak acids, like acetic acid, only some of the acid dissociates. An equation for the dissociation of a weak acid uses double arrows pointing in opposite directions.

$$CH_3COOH(l) + H_2O(l) \rightleftarrows H_3O^+(aq) + CH_3COO^-(aq)$$

What are strong and weak bases?

A **strong base** is one that dissociates completely in solution. Sodium hydroxide (NaOH) is an example of a strong base. An equation for the dissociation of a strong base uses a single arrow that points toward the ions that form.

Strong base: $NaOH(s) \rightarrow Na^+(aq) + OH^-(aq)$

A **weak base** is one that does not dissociate completely in solution. Ammonia (NH_3) is a weak base. An equation for the dissociation of a weak base uses double arrows pointing in opposite directions.

Weak base: $NH_3(aq) + H_2O \rightleftarrows NH_4^+(aq) + OH^-(aq)$

How are strength and concentration described?

The words *strong* and *weak* refer to how easy it is for the acid or base to dissociate in solution. Strong acids and bases dissociate completely. Weak acids and bases dissociate only partially.

The words *dilute* and *concentrated* tell how much acid or base is dissolved in the solution. A dilute solution means there is a small amount of acid or base in the solution. A concentrated solution means there is a large amount of acid or base in the solution.

You can have dilute solutions of strong acids and bases. You also can have concentrated solutions of weak acids and bases. The figure on the left does not have as many particles in the solution. But, all of the particles are ions. This is a dilute solution of a strong acid. The next figure has more particles in solution, but, not all of the particles are ions. Some of them are still acetic acid. This is a concentrated solution of a weak acid.

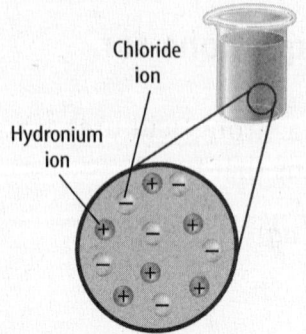

This is a dilute solution of HCl.

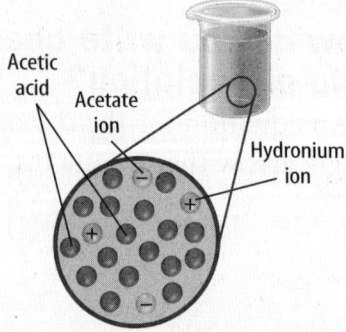

This is a concentrated solution of acetic acid.

404 CHAPTER 23 Acids, Bases, and Salts

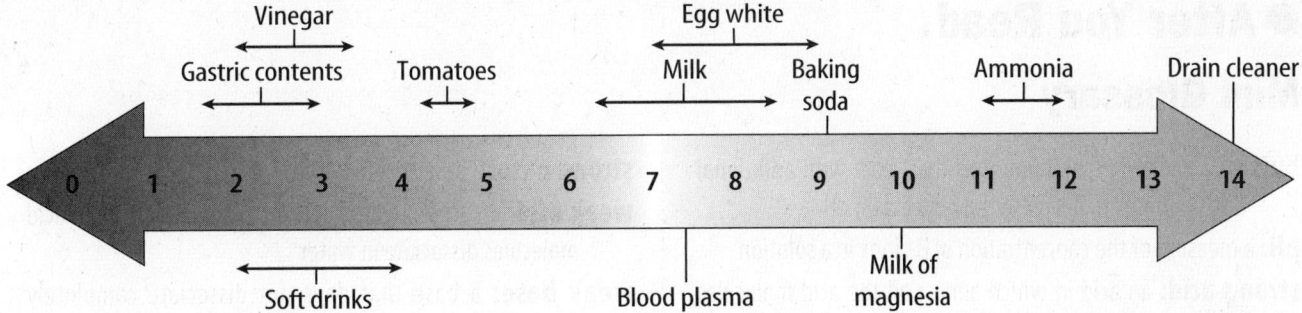

pH of a Solution

If you have a swimming pool or keep tropical fish, you know that the pH of the water must be kept at certain levels. The **pH** of a solution is a measure of the concentration of H^+ ions in the solution. The pH of a solution is measured on a scale ranging from 0 to 14. The greater the concentration of H^+ ions is, the lower the pH and the more acidic the solution.

The figure shows a pH scale and the pH of some common acid and base substances. Solutions with a pH lower than 7 are acidic. Solutions with a pH greater than 7 are basic. The greater the pH value, the more basic the solution is. A solution with a pH of 7 is called neutral. In a neutral solution, concentrations of H^+ and OH^- ions are equal. Pure water at 25°C has a pH of 7.

You can measure pH with universal indicator paper. The paper changes color when H_3O^+ or OH^- ions are present in solution. The color of the paper is then compared to colors in a chart to find the pH. You also can find pH with an electronic pH meter. The meter has electrodes that determine the pH when they are placed in a solution.

Why is pH of blood important?

The pH of your blood must remain between 7.0 and 7.8. If your blood pH goes outside of this range, enzymes do not work. Enzymes are protein molecules that act as catalysts for many of your body's reactions. Why doesn't your blood pH change when you eat acidic foods? Your blood contains compounds called buffers that allow small amounts of acids or bases to be absorbed without changing the pH.

Buffers are solutions containing ions that react with additional acids or bases to decrease their effects on pH. The buffers help your blood stay at an almost constant pH of 7.4. One buffer system in the blood involves bicarbonate ions, HCO_3^-.

Applying Math

3. **Use a Scale** Which is more acidic, a pH of 9 or a pH of 3?

FOLDABLES

C Organize Information Make the following Foldable to help you organize information about pH. Write important information about acids under the first tab and important information about bases under the second tab.

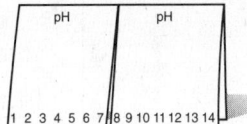

Reading Check

4. **Explain** What are the compounds called that allow you to eat acidic foods without causing a pH change in your blood?

After You Read

Mini Glossary

buffer: a solution containing ions that react with additional acids or bases to decrease their effects on pH

pH: a measure of the concentration of H+ ions in a solution

strong acid: an acid in which almost all the acid molecules dissociate in water

strong base: a base that dissociates completely in solution

weak acid: an acid in which only a small number of the acid molecules dissociate in water

weak base: a base that does not dissociate completely in solution

1. Review the terms and their definitions in the Mini Glossary. Choose two related terms and use them together in a sentence.

   _____
   _____

2. Complete the table to describe acids and bases.

	Understanding Acids and Bases
Property	**Description**
Strength	• Strong acids and bases almost completely dissociate in solution. • Weak acids and bases _____
Concentration	• Concentrated acids and bases _____ • Dilute acids and bases have few particles in solution.
pH	• Acids have a pH lower than 7. • Bases have a pH _____

3. As you read this section, you wrote down the main points of the section. How did these help you understand the section better?

   _____
   _____

 Visit **gpscience.com** to access your textbook, interactive games, and projects to help you learn more about the strength of acids and bases.

Chapter 23: Acids, Bases, and Salts

section ❸ Salts

● Before You Read

What does the word *salt* mean to you? How do you use salt?

_____

_____

What You'll Learn
- identify a neutralization reaction
- about salt and how it forms
- about soaps and detergents

● Read to Learn

Neutralization

<u>Neutralization</u> is a chemical reaction between an acid and a base that happens in a water solution. For example, NaOH neutralizes HCl. Hydronium ions from the acid combine with hydroxide ions from the base, producing neutral water.

$$H_3O^+(aq) + OH^- \rightarrow 2H_2O(l)$$

Antacids are medicines that contain bases or other compounds that neutralize the HCl in your stomach. One of these antacids is sodium bicarbonate—$NaHCO_3$.

$$HCl(aq) + NaHCO_3(s) \rightarrow NaCl(aq) + CO_2(g) + H_2O(l)$$

How is salt formed?

HCl, shown above, is being neutralized by NaOH. But only half of the ions are shown. The other ions react to form a salt. A <u>salt</u> is a compound formed when the negative ions from an acid combine with the positive ions from a base. When HCl reacts with NaOH, the salt formed is sodium chloride.

$$Na^+(aq) + Cl^-(aq) \rightarrow NaCl(aq)$$

Study Coach

Make Flash Cards As you read the section, look for the main points in each paragraph. Write each main point on a note card. Use the note cards as flash cards to help you remember the main points.

FOLDABLES

Ⓐ Organize Information Make the following Foldable to help you organize information about how salts form.

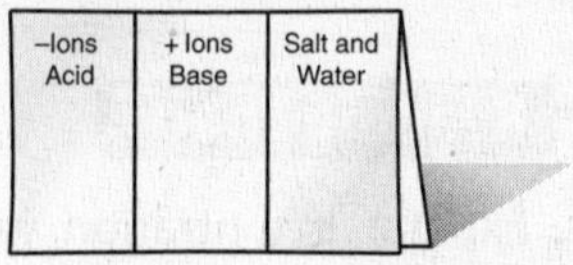

−Ions Acid	+Ions Base	Salt and Water

Reading Essentials 407

How do acid-base reaction equations look?

This general equation shows acid-base reactions in water:

acid + base → salt + water

The equation for the reaction between HCl and NaOH is:

$HCl(aq) + NaOH(aq) \rightarrow NaCl(aq) + H_2O(l)$

Salts

Salt is necessary for many organisms. Some common salts and their uses are shown in the table. Most salts are made up of a positive metal ion and an ion with a negative charge, such as Cl^- or CO_3^{2-}. Ammonium salts contain the ammonium ion, NH_4^+, instead of a metal.

Some Common Salts and Their Uses		
Name, Formula	Common Name	Uses
Sodium chloride, NaCl	Salt	Food, manufacture of chemicals
Sodium hydrogen carbonate, $NaHCO_3$	Sodium bicarbonate Baking soda	Food, antacids
Potassium nitrate, KNO_3	Saltpeter	Fertilizers
Ammonium chloride, NH_4Cl	Sal ammoniac	Dry-cell batteries

Titration

You can find the concentration of an acid or base by titration. In **titration** (ti TRAY shun), a solution of known concentration is used to find the concentration of another solution.

To do a titration on an acid solution, measure the volume of the solution of unknown concentration. Add a few drops of an indicator to the solution. Phenolphthalein (fee nul THAY leen) is an indicator that has no color in an acid but turns pink in a base.

Since the unknown solution is an acid, slowly and carefully add a base solution of known concentration to the acid-and-indicator mixture. Toward the end of the titration, add base solution drop by drop until one drop turns the solution pink and the color stays. The completed titration is the end point. The acid is neutralized. Use the volume of base that was added to calculate the concentration of the acid solution.

Picture This

1. Identify Which salt is used to make fertilizers?

Reading Check

2. Explain In a titration, what does it mean when the color changes in the unknown solution?

What are indicators?

Many natural substances are acid–base indicators. The indicator litmus comes from a combination of a fungus and an algae called lichen. The flowers of the hydrangea plant are indicators. The flowers are blue when the pH of the soil is acidic. The flowers are pink when the pH of the soil is basic. Red cabbage juice is also an indicator. The juice is deep red at pH 1, lavender at pH 7, and yellow-green at pH 10.

Soaps and Detergents

At the supermarket, you will see many different kinds of soaps and detergents. Some are liquids and some are solids.

What are soaps?

Soaps are organic salts. They have a nonpolar organic chain of carbon atoms on one end of the molecule. On the other end, they have either a sodium or potassium salt of a carboxylic (kar bahk SIHL ihk) acid—COOH group. The figure below shows a soap molecule.

Soaps clean well for two reasons. Look at figure on the left below. The long, nonpolar, hydrocarbon tail mixes with oils and dirt so they can be removed easily. The polar head attracts water molecules. The figure on the right shows how the attraction between the soap and water molecules helps to wash away the dirt and oil linked with the soap.

> **Reading Check**
>
> **3. List** Name two acid-base indicators that exist in nature. p-
>
> _____
>
> _____

> **Picture This**
>
> **4. Describe** Which part of a soap molecule attracts dirt and oil? Which part attracts water molecules?
>
> _____
>
> _____

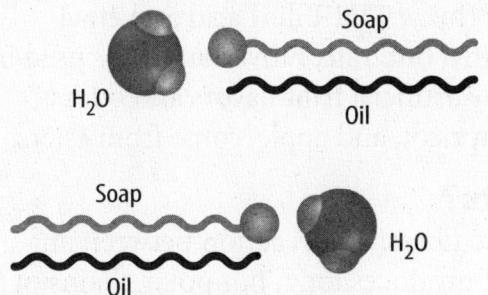

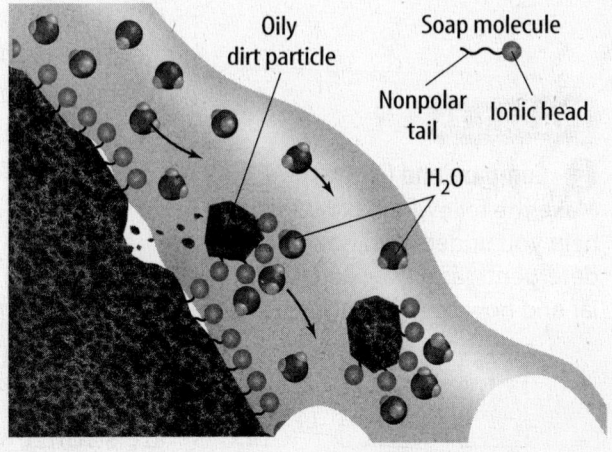

What are commercial soaps and detergents?

A long-chain fatty acid is reacted with sodium or potassium hydroxide to make simple soaps. The fatty acids used to make soaps are natural oils, such as canola, palm, and coconut. A problem with soaps is that tap water often contains metal ions such as calcium, magnesium, and iron. When this occurs, it is known as hard water. These metal ions can react with soap, replacing the sodium or potassium ions. This replacement makes soap insoluble in water. The ions separate out of solution in the form of soap scum.

Detergents are like soap, but they are made from petroleum molecules instead of from natural fatty acids. They also might have a sulfonic acid group at the end, instead of a carboxylic acid group. Detergents also react with metal ions in hard water, but the products are more soluble. So, they don't leave as much soap scum. Detergents have additional ingredients to increase sudsing and to improve cleaning in hard water. Detergents do not break down as easily as soap. Some can cause foam in water treatment plants and in streams.

Reading Check
5. Define What is hard water?

Versatile Esters

Think of esters as organic versions of salts. Like salts, esters are made from acids, and water is a product of the reaction. But, instead of reacting with bases, acids react with alcohols that have a hydroxyl group to produce esters. Esters are used to make soap, flavorings, perfumes, and fibers for clothing.

How are esters used for flavor?

Many drinks taste like real fruit but do not contain any fruit, only artificial flavors. Most likely, the artificial flavors come from esters. The reaction to prepare esters involves removing a molecule of water from an acid and an alcohol. Often, this reaction is helped by adding concentrated sulfuric acid. A reaction between butyric (byew TIHR ihk) acid and ethyl alcohol forms the ester ethyl butyrate. Ethyl butyrate is used in pineapple flavoring. Many artificial fruit flavors and odors, such as banana, orange, apricot, and apple, come from esters.

What are polyesters?

Polyesters are synthetic fibers. The reaction between an organic acid and alcohol produces long, nonpolar chains of many, or poly, esters. The chains of esters are closely packed together, making them strong. Polyester fibers are used alone or woven or knitted with natural fibers to make fabrics that are strong, water repellant, and colorfast.

FOLDABLES

E Compare and Contrast Make the following Foldables to help you understand how soaps, detergents, and esters are similar and how they are different.

Soaps/Detergents: Esters:

After You Read

Mini Glossary

neutralization: a chemical reaction between an acid and a base that happens in a water solution

salt: a compound formed when the negative ions from an acid combine with the positive ions from a base

soap: an organic salt

titration: a process in which a solution of known concentration is used to find the concentration of another solution

1. Review the terms and their definitions in the Mini Glossary. Write a sentence about the types of solutions used with titration.

   _____

   _____

2. The graphic below shows the set-up for a titration. The beaker contains a known amount an acid of unknown concentration. In the boxes, write what must be added to the beaker and the pipette to do a titration. Then, describe the procedure.

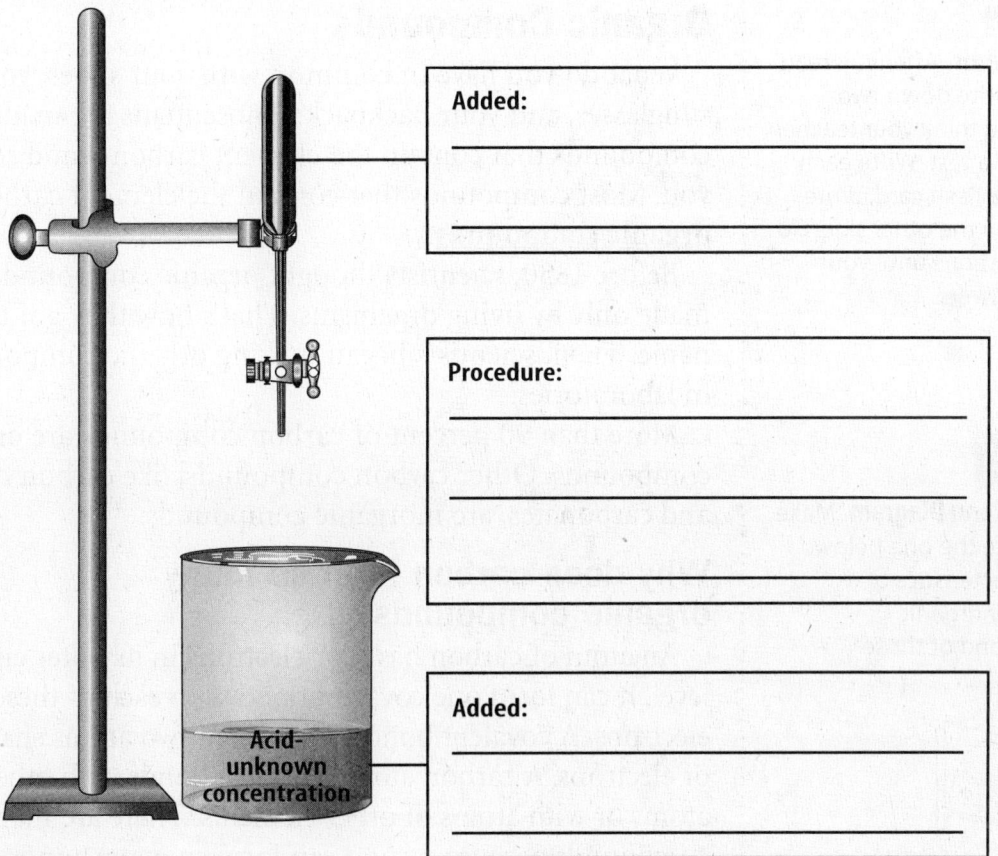

 Visit **gpscience.com** to access your textbook, interactive games, and projects to help you learn more about salts.

chapter 24 Organic Compounds

section ❶ Simple Organic Compounds

What You'll Learn
- about organic and inorganic carbon compounds
- difference between saturated and unsaturated hydrocarbons
- identify isomers

Study Coach

Make Flash Cards After you read each page, write down two questions you think your teacher might ask on a test. Write each question on a flash card. Write the answer on the other side. Go over the questions until you know the answers.

FOLDABLES

Ⓐ **Make a Venn Diagram** Make a Foldable like the one below. List the characteristics of organic and inorganic compounds and of those they share.

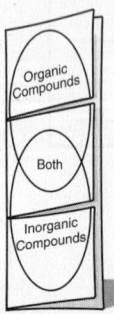

● Before You Read

Many people think organic means that something doesn't contain chemicals. Write what you think organic means. Give an example of something you think is organic.

_____

_____

● Read to Learn

Organic Compounds

What do you have in common with your shoes, your sunglasses, and your backpack? Those items are made of compounds that contain the element carbon—and so are you. Most compounds that contain the element carbon are **organic compounds**.

Before 1830, scientists thought organic compounds were made only by living organisms. That's how they got their name. Then, scientists began making organic compounds in laboratories.

More than 90 percent of carbon compounds are organic compounds. Other carbon compounds, like carbon dioxide and carbonates, are inorganic compounds.

Why does carbon form so many organic compounds?

An atom of carbon has four electrons in its outer energy level. It can form one covalent bond with each of these electrons. A covalent bond forms when two atoms share a pair of electrons. A carbon atom can form bonds with other carbon atoms or with atoms of other elements. There are many carbon compounds because carbon can form so many bonds. Some carbon compounds are small, like the ones used as fuels. Other carbon compounds are complex, like the ones found in medicines and plastics. But the structure of the carbon atom is not the only reason that carbon can form so many compounds.

412 CHAPTER 24 Organic Compounds

Straight Chain   **Branched Chain** 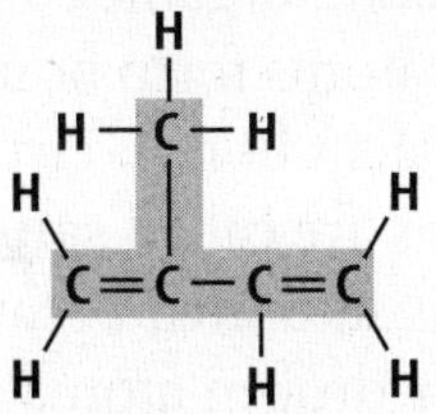   **Cyclic Chain**

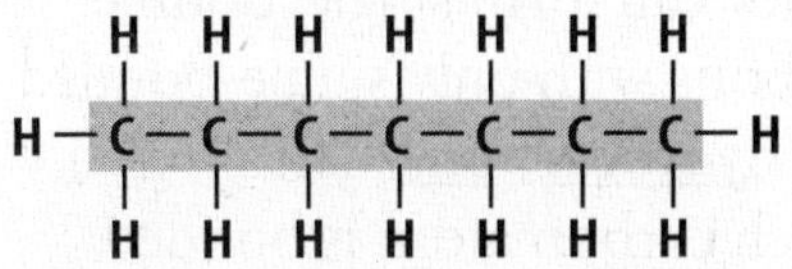

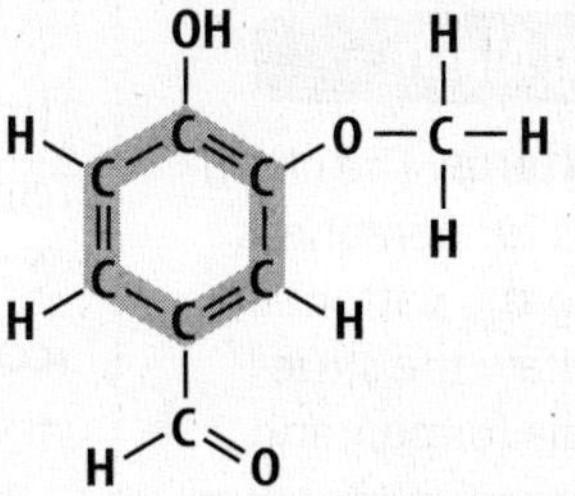

How can carbon atoms arrange themselves?

Another reason carbon atoms can form so many organic compounds is that carbon atoms can bond together in different ways—chains, branched chains, and rings. Look at the figure above. The first structure shows carbon atoms bonded together in a straight chain. This compound is heptane. Heptane is an organic compound in gasoline. In the second structure, the carbon atoms bond together in a branched chain. This compound is isoprene, an organic compound in natural rubber. Look at the third structure. See how some of the carbon atoms form a ring? This ring is called a cyclic chain. The compound is vanillin, which is an organic compound found in vanilla flavoring. Carbon also can form single, double, or triple covalent bonds.

Hydrocarbons

There are many organic compounds that contain only carbon and hydrogen. A compound that is made of only carbon and hydrogen atoms is called a **hydrocarbon**. A furnace, stove, or water heater may burn natural gas. Natural gas contains the hydrocarbon methane. The chemical formula for methane is CH_4.

The figure shows two other ways to represent the formula of a compound. The first is a space-filling model of a methane molecule. It shows the relative size and arrangement of the atoms in the molecule. The second item in the figure is a structural formula for a methane molecule. Structural formulas use lines to show the bonds between atoms. Each line represents a covalent bond. The structural formula of methane shows one carbon atom bonded to four hydrogen atoms. Chemists usually use chemical formulas or structural formulas when they write about chemical reactions.

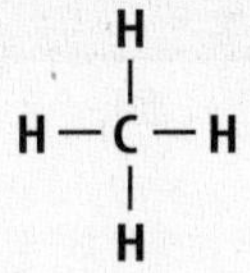

Picture This

1. **Conclude** Why is the structure of isoprene called a branched chain?

   _____
   _____
   _____

Picture This

2. **Interpret a Scientific Illustration** Circle the hydrogen atoms in the space-filling model and in the structural formula of methane.

Reading Essentials 413

Applying Math

3. **Finding Patterns** The chemical formula for pentane is C_5H_{12}. Using the pattern in the table, draw the structural formula for pentane.

Single Bonds

To understand organic compounds you must understand how carbon atoms form bonds. They can form single, double, or triple covalent bonds. Hydrocarbons with only single-bonded carbon atoms are called **saturated hydrocarbons.** The compound is saturated because each carbon atom is bonded to as many hydrogen atoms as possible.

Some Hydrocarbons

Name	Methane	Ethane	Propane	Butane
Chemical Formula	CH_4	C_2H_6	C_3H_8	C_4H_{10}
Structural Formula				

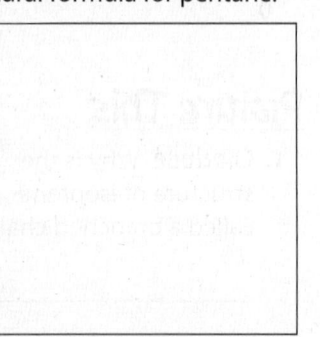

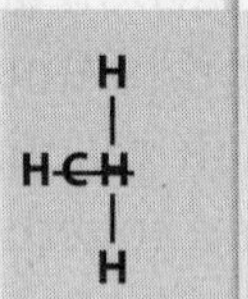

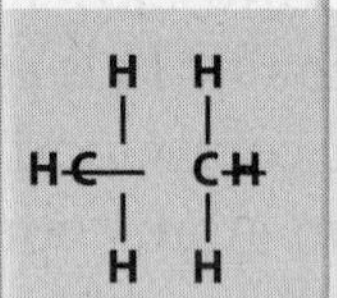

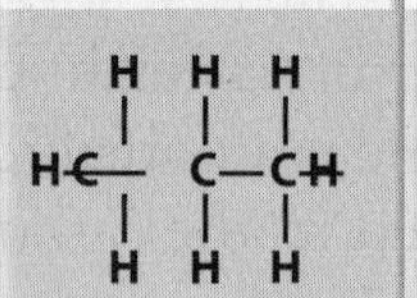

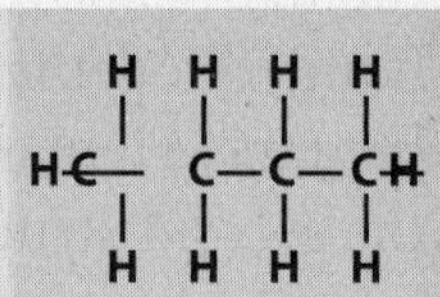

The table lists four saturated hydrocarbons. Look at the structural formulas. You can see how each carbon atom looks like a link in a straight chain connected by single covalent bonds.

Boiling Points of Hydrocarbons The graph below shows the boiling points of five saturated hydrocarbons. Methane has one carbon atom and a boiling point of about –160°C. Ethane has two carbon atoms and a boiling point of about –90°C. Propane has two carbon atoms. What is propane's boiling point? Do you see a pattern? The boiling points of saturated hydrocarbons increase as the number of carbon atoms in the chain increases.

Boiling Points of Hydrocarbons

Applying Math

4. **Use Graphs** What is the approximate boiling point of propane?

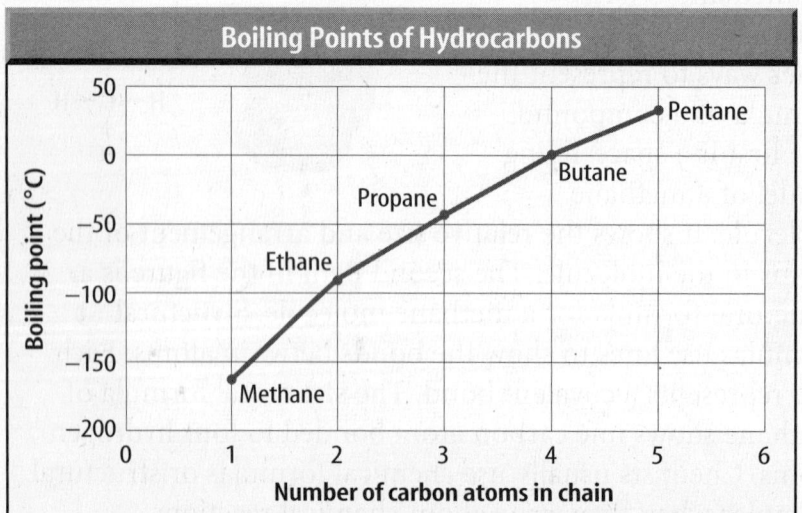

Can different hydrocarbons have the same formula?

If you have cooked on a camping stove, you may have used the hydrocarbon butane. The chemical formula of butane is C_4H_{10}. Another hydrocarbon called isobutane also has the chemical formula C_4H_{10}. How can two different molecules have the same chemical formula but different names?

Look at the structural figures below. The first one shows a molecule of butane. In butane, the carbon atoms form a straight chain. But in isobutane, the second structural figure, the chain of carbon atoms is branched. The arrangement of carbon atoms changes the shape of each molecule. The arrangement of carbon atoms also often affects each molecule's physical properties. Butane and isobutane are isomers.

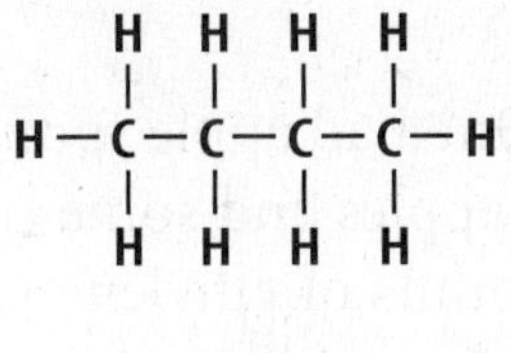

Butane
C_4H_{10}

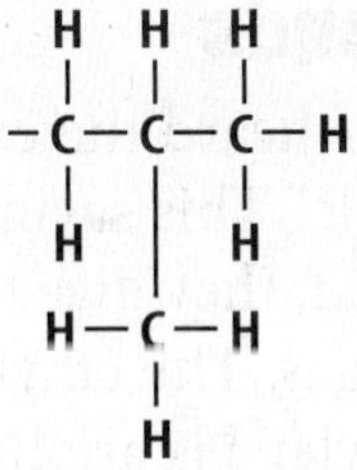

Isobutane
C_4H_{10}

Picture This
5. Illustrate Circle the isomer of butane that is in the form of a straight chain.

What are isomers?

<u>Isomers</u> are compounds that have the same chemical formula, but have different molecular structures and shapes. Thousands of hydrocarbons are isomers. Butane and isobutane are two of them.

Look at the table below. It lists properties for butane and isobutane. The melting point and boiling point of isobutane are lower than the melting point and boiling point of butane. Generally, the more branches an isomer has, the lower its melting point and boiling point are.

Reading Check
6. Identify Name two isomers.

_____

_____

Properties of Butane Isomers		
Property	Butane	Isobutane
Description	Colorless gas	Colorless gas
Density	0.60 kg/L	0.603 kg/L
Melting point	−135°C	−145°C
Boiling point	−0.5°C	−10.2°C

Applying Math
7. Use Negative Numbers Is the boiling point of isobutane higher or lower than the boiling point of butane? By how much?

_____

Picture This

8. Draw and Label In the space below, draw a picture of your right hand and its mirror image. Label one "right hand" and the other "mirror image."

Reading Check

9. Apply Fill in the blanks to complete the sentence: Unsaturated hydrocarbons are compounds that have at least one (a.) _____ bond or (b.) _____ bond.

(a.) _____
(b.) _____

Are there other kinds of isomers?

There are many other kinds of isomers. Not all isomers are in the form of a straight chain molecule and a branched chain molecule. Some isomers differ only slightly in the way their atoms are arranged. Some isomers form what are called right-handed and left-handed molecules.

Try this: Hold your right hand in front of a mirror. What do you see? You see an image that looks like it is your left hand. This is a mirror image. The same idea describes right-handed and left-handed molecules. They are mirror images of each other. Everything else about their structure is the same. Right-handed and left-handed isomers have nearly identical physical and chemical properties.

Multiple Bonds

Have you ever heard the expression, "One bad apple spoils the whole barrel?" This saying is true. As apples and some other fruits ripen, they give off small amounts of ethylene (ETH uh leen) gas. The ethylene makes other fruit ripen or spoil faster. Ethylene is another name for the hydrocarbon ethene (eh THEEN). The chemical formula for ethene is C_2H_4. The figure below shows the structural formula for an ethene molecule. You can see that ethene has one double bond. The two carbon atoms in the double bond share two pairs of electrons.

The figure also shows the structural formula for a molecule of ethyne (eh THIHN). The ethyne molecule has a triple bond. In a triple bond, two carbon atoms share three pairs of electrons. Another name for ethyne is acetylene. Acetylene is used in welding torches. Ethene and ethyne are unsaturated hydrocarbons. An **unsaturated hydrocarbon** is a hydrocarbon that has at least one double bond or triple bond. The compounds are unsaturated because each carbon atom is not bonded to as many hydrogen atoms as possible.

The last three letters in the name of a hydrocarbon tell what type of bond is in the molecule. Compounds that end in –ane have only single bonds. Compounds that end in –ene have at least one double bond. Those that end in –yne have at least one triple bond.

Ethene   Ethyne

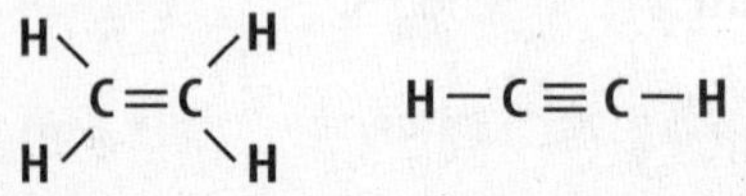

$$\begin{matrix} H \\ \\ H \end{matrix} \! C \! = \! C \! \begin{matrix} H \\ \\ H \end{matrix} \qquad H - C \equiv C - H$$

416 CHAPTER 24 Organic Compounds

After You Read
Mini Glossary

hydrocarbon: a compound that is made of only carbon and hydrogen atoms

isomers: compounds that have the same chemical formula but have different molecular structures and shapes

organic compound: a compound that contains the element carbon

saturated hydrocarbon: a hydrocarbon made of only single-bonded carbon atoms

unsaturated hydrocarbon: a hydrocarbon that has at least one double bond or triple bond

1. Review the terms and their definitions in the Mini Glossary. Write a sentence using a term that could apply to compounds with only single bonds.

   _____

   _____

2. Complete the flow chart by filling in the type of compound and its characteristics.

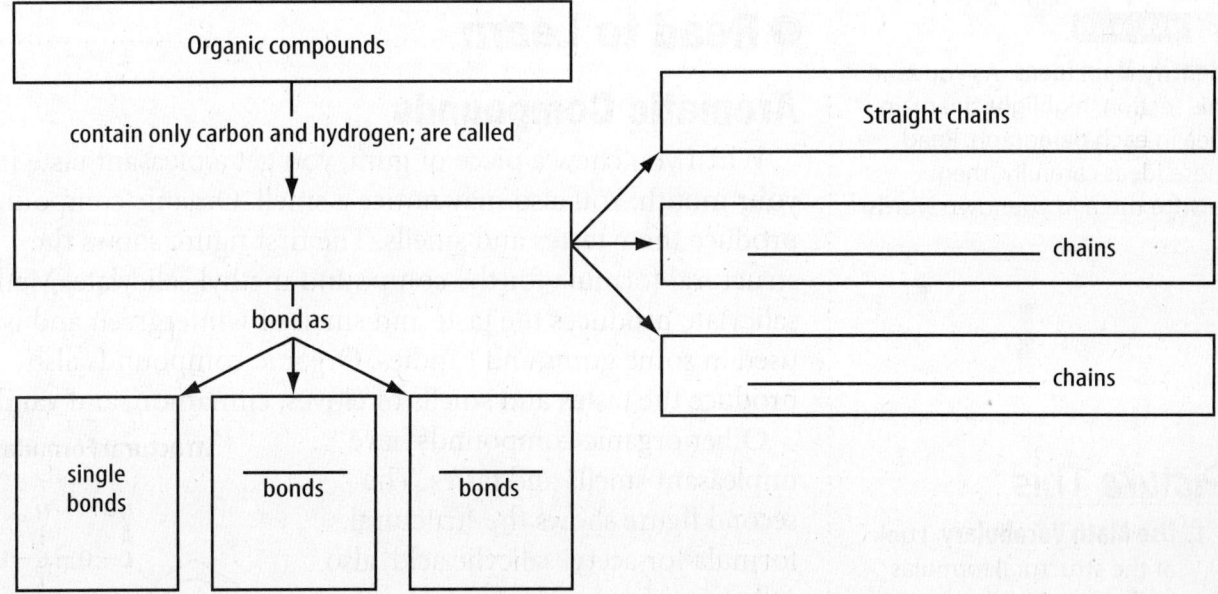

3. You made flash cards of questions your teacher might ask on a test. Was this a good strategy for learning the information? Why or why not?

   _____

   _____

 Visit **gpscience.com** to access your textbook, interactive games, and projects to help you learn more about simple organic compounds.

Reading Essentials **417**

chapter 24
Organic Compounds

section ❷ Other Organic Compounds

What You'll Learn
- what aromatic compounds are
- what alcohols and acids are
- some organic compounds you use every day

● Before You Read

How do you think gum is made to taste and smell like mint or cinnamon? Does the gum have pieces of mint or cinnamon in it? Write your thoughts below.

_____

_____

_____

Mark the Text

Identify Main Ideas As you read this section, highlight the main idea in each paragraph. Read these ideas carefully, then rewrite them in your own words.

● Read to Learn

Aromatic Compounds

When you chew a piece of gum, you get a pleasant taste in your mouth. You also may notice a smell. Organic compounds produce these tastes and smells. The first figure shows the structural formula for the compound methyl salicylate. Methyl salicylate produces the taste and smell of wintergreen and is used in some gums and candies. Organic compounds also produce the tastes and smells of cloves, cinnamon, and vanilla.

Other organic compounds have unpleasant smells and tastes. The second figure shows the structural formula for acetyl salicylic acid, also called aspirin. Aspirin has a sour taste.

These are aromatic compounds. You might think they get that name because they are smelly. Some of them are, but look at the two structural formulas. What do they have in common? They both have a ring in their structural formula. To chemists, an **aromatic compound** is a compound that contains a benzene structure having a ring made of six carbon atoms.

Structural Formulas

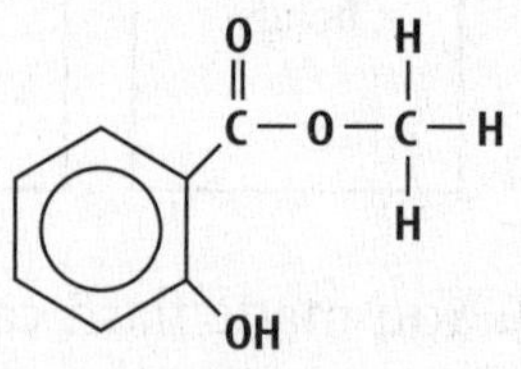

Methyl Salicylate

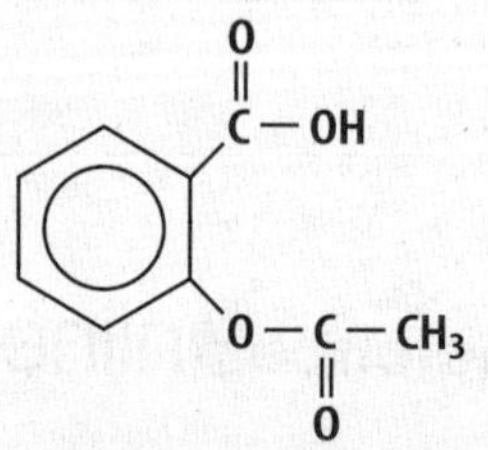

Acetyl Salicylic Acid

Picture This

1. **Use Math Vocabulary** Look at the structural formulas in the figures. What is the name for a geometric figure with six sides?

_____

Why is benzene stable?

Benzene has six carbon atoms bonded together in the form of a ring. The chemical formula for benzene is C_6H_6. Look at the structural formula of benzene.

The benzene ring is made of six carbon atoms bonded together by three double bonds and three single bonds. The single and double bonds alternate around the ring. All six of the carbon atoms in the ring equally share the electrons that form these bonds. This sharing of electrons makes a benzene molecule very stable. The symbol for benzene is a circle inside a hexagon. Many organic compounds contain a benzene ring.

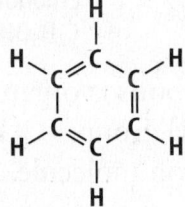
Structural Formula

Benzene Symbol

Are there other ring structures?

Organic compounds can contain more than one ring structure. For example, some moth crystals are made of naphthalene (NAF thuh leen). Naphthalene is an organic compound with two ring structures fused together. The figure shows the symbol and chemical formula for naphthalene. Many known compounds contain three or more rings fused together.

 **Naphthalene** $C_{10}H_8$

Substituted Hydrocarbons

A cook sometimes substitutes one ingredient for another in a recipe. In a similar way, chemists can change molecules by substituting some of their "ingredients." Chemists change hydrocarbons into other compounds with different physical and chemical properties. They may add a double or triple bond. They may substitute different atoms or groups of atoms in a hydrocarbon molecule. A **substituted hydrocarbon** is a hydrocarbon that has one or more of its hydrogen atoms replaced by atoms or groups of atoms of other elements. Chemists first decide what properties they want the new compound to have. Then they choose atoms, groups of atoms, or types of bonds that will give those properties. ✓

FOLDABLES

B Classify Make a Foldable like the one below. As you read the section, list the characteristics and examples of aromatic compounds and substituted hydrocarbons.

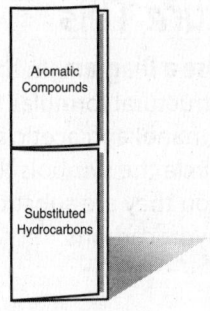

Picture This

2. **Explain** In the structural formula in the figure, what do the segments between the carbon atoms represent?

_____
_____

Reading Check

3. **Determine** Name one way chemists can change hydrocarbons into substituted hydrocarbons.

_____
_____

Reading Essentials **419**

What are some substituted hydrocarbons?

Alcohols are one kind of substituted hydrocarbon. They are an important group of organic compounds. They are used as solvents and disinfectants and as pieces to make larger molecules. An **alcohol** forms when a hydroxyl group, –OH, replaces one or more hydrogen atoms in a hydrocarbon. A hydroxyl group contains an oxygen atom and a hydrogen atom.

The first figure shows the structural and chemical formula for ethanol. Notice that the ethanol molecule has –OH on one end. This tells you that ethanol is an alcohol. When the sugar in grains or fruit ferments, it produces ethanol.

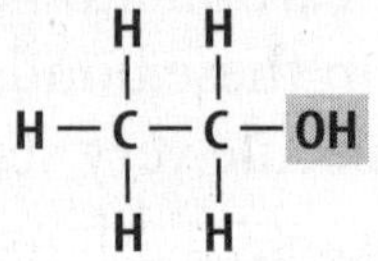

Ethanol
C_2H_5OH

Picture This

4. Use a Diagram In the structural formulas for ethanol and acetic acid, circle the symbols that tell you they are substituted hydrocarbons.

Another group of substituted hydrocarbons is organic acids. Organic acids form when a carboxyl group, –COOH, attaches to a carbon atom of a hydrocarbon molecule. A carboxyl group contains one carbon atom, two oxygen atoms, and one hydrogen atom.

The second figure shows the structural and chemical formula for acetic acid. Acetic acid is an organic acid found in vinegar. Other organic acids include vinegar, citric acid, found in fruits such as oranges, and lactic acid, found in sour milk.

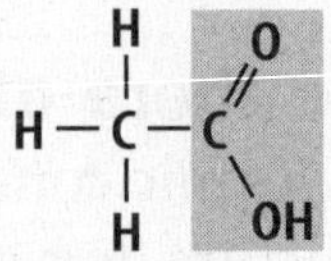

Acetic acid
CH_3COOH

✓ Reading Check

5. Identify What is the symbol for a carboxyl group?

_____

What other elements can be added to hydrocarbons?

Hydrogen and oxygen are not the only atoms added to hydrocarbons to make substituted hydrocarbons. Chlorine atoms are used as well. When four chlorine atoms replace four hydrogen atoms on a molecule of ethene, they form a molecule of tetrachloroethene (teh truh klor uh eth EEN). This is a solvent used in dry cleaning. The third figure shows the structural and chemical formula for tetrachloroethene.

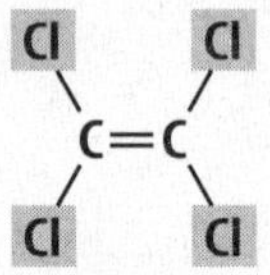

Tetrachloroethene
C_2Cl_4

When four fluorine atoms replace four hydrogen atoms in ethylene, they form a compound that can be made into a black, shiny material used for nonstick cookware. Nitrogen, bromine, and sulfur atoms also are used in substituted hydrocarbons. Compounds called thiols are formed when sulfur replaces the oxygen in the –OH group of an alcohol. Thiols also are called mercaptans. Most mercaptans smell very bad. In fact, there is a mercaptan in skunk spray.

Picture This

6. Infer Look at the diagram. The prefix *tetra-* means "four." How do you suppose tetrachloroethene got its name?

_____

_____

420 CHAPTER 24 Organic Compounds

After You Read

Mini Glossary

alcohol: a compound that forms when a hydroxyl group, –OH, replaces one or more hydrogen atoms in a hydrocarbon

aromatic compound: a compound that contains a benzene structure having a ring made of six carbon atoms

substituted hydrocarbon: a hydrocarbon that has one or more of its hydrogen atoms replaced by atoms or groups of atoms of other elements

1. Review the terms and definitions in the Mini Glossary. On the lines below, write a sentence that tells how the terms substituted hydrocarbon and alcohol are related.

   _____

   _____

2. Complete the outline to help you organize what you learned about other organic compounds.

   Other Organic Compounds

   I. Aromatic Compounds

      A. An aromatic compound is _____

         _____.

      B. A _____ contains six atoms bonded together in a ring.

      C. _____ can contain two or more fused rings.

   II. Substituted Hydrocarbons

      A. A substituted hydrocarbon is _____

         _____

         _____.

      B. In _____, the –OH group is substituted for a _____.

      C. Organic acids contain the group _____.

      D. Substituted hydrocarbons can contain other elements, including _____, _____, and _____.

Science Online Visit gpscience.com to access your textbook, interactive games, and projects to help you learn more about other organic compounds.

chapter 24 Organic Compounds

section ❸ Petroleum—A Source of Carbon Compounds

What You'll Learn
- how carbon compounds are obtained from petroleum
- how carbon compounds form long chains of molecules
- what polymers are

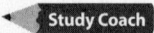 **Study Coach**

Make an Outline Make an outline covering the main ideas about petroleum and carbon compounds.

● Before You Read

Plastics are made from petroleum. Look around you. List and describe two plastic products that have different properties.

_____

_____

● Read to Learn

What is petroleum?

Do you carry a comb in your pocket or purse? It is probably made of plastic. Where did that plastic come from? It probably came from petroleum. Petroleum is a dark, flammable liquid that is often called crude oil. Petroleum exists deep within Earth. Coal, natural gas, and petroleum are all called fossil fuels because they form from fossilized material.

How can thick, dark, liquid petroleum be turned into a hard object like a comb? To answer that question, you have to understand more about petroleum. Petroleum is a mixture of thousands of carbon compounds. First, the petroleum is brought up from its sources underground. Oil wells are used to pump petroleum, or crude oil, to Earth's surface.

Next, chemists and engineers separate the oil mixture into fractions containing compounds with similar boiling points. Within a fraction, boiling points may range more than 100°C. Fractional distillation is the process of separating petroleum into fractions. Fractional distillation happens at petroleum refineries in metal towers called fractionating towers.

What is a fractionating tower?

A fractionating tower can be as tall as 35 m. Inside the tower, metal plates are arranged like floors inside a building. The plates have many small holes to let vapors pass through. Pipes are connected to the outside of the tower at different levels. The tower uses fractional distillation to separate crude oil into fractions.

1. **Draw Conclusions** What do we call the process of separating petroleum into fractions?

_____

_____

How does fractional distillation work?

The figure below shows how a fractioning tower works. Crude oil is pumped into the bottom of the tower. The crude oil is heated to more than 350°C. At 350°C, most of the hydrocarbons in the oil mixture turn into vapor and start to rise. The vapors rise up inside the tower. Vapors of the fractions with the highest boiling points only reach the lowest plates in the tower before they condense. Condensed vapor turns into liquid. The first liquid fractions drain out of the tower through pipes connected to its sides and are collected.

The rest of the vapors continue to rise inside the tower. Vapors of fractions with lower boiling points condense when they reach plates in the middle part of tower. These liquid fractions also drain off and are collected.

Vapors of fractions with the lowest boiling points rise to the highest plates before they condense. Some vapors never condense and are collected as gases at the top of the tower. The figure lists some fractions and the temperature at which they condense.

Think it Over

2. Explain What property do fractions that condense at the lowest part of a fractioning tower have?

_____

_____

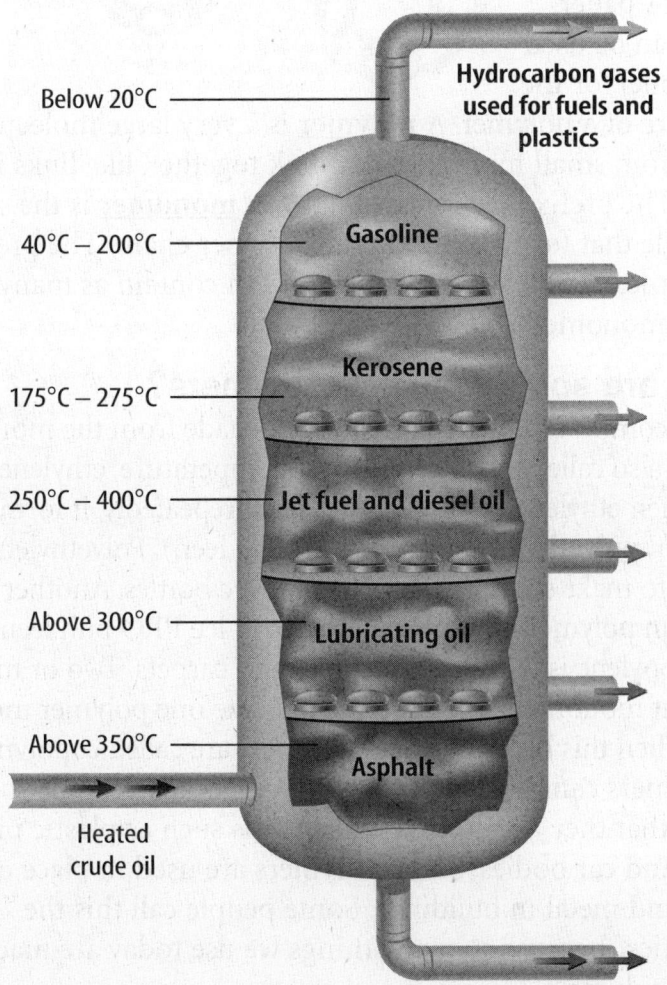

Picture This

3. Interpret a Figure What do the temperatures in the figure represent?

_____

_____

Applying Math

4. Use Data Which fractions condense at a temperature of 300°C or greater?

_____

_____

_____

_____

FOLDABLES

C Classify Make half-sheet Foldables like the ones below. Use one to explain what petroleum is and then list things that are made from petroleum. On the other tell how monomers and polymers are related.

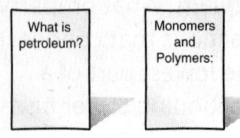

Picture This

5. Distinguish Use one color highlighter to circle a monomer in the figure. Use another color to circle the polymer. In the box below, make a key to show what each color represents.

☐ monomer

☐ polymer

Think it Over

6. List Name three things made of polymers.

_____

_____

_____

Uses for Petroleum Compounds

Some petroleum fractions are used for fuels. Butane and propane are some of the lightest fractions. They are taken from the top of the fractioning tower. Molecules of propane have three carbon atoms. Molecules of butane have four carbon atoms.

The molecules of fractions that condense on the upper plates of the tower are larger and heavier than butane and propane. They have five to ten carbon atoms in their molecules. These compounds are used for gasoline and solvents. Fractions that condense on lower plates have 12 to 18 carbon atoms in their molecules. They are kerosene and jet fuel. Most of the bottom fractions are lubricating oil. Whatever is left over is used to make asphalt to pave roads.

Polymers

Have you ever made a paper chain like the one in the figure? A paper chain can be used as a model for the

structure of a polymer. A **polymer** is a very large molecule made from small molecules that link together, like links in a chain. The prefix *poly-* means many. A **monomer** is the small molecule that forms a link in the polymer chain. The prefix *mono-* means one. A polymer chain can contain as many as 10,000 monomers.

What are some common polymers?

One common polymer, or plastic, is made from the monomer ethene, also called ethylene. At room temperature, ethylene is a gas. When ethylene combines with itself repeatedly, it forms a polymer, **polyethylene** (pah lee EH thuh leen). Polyethylene is used to make shopping bags and plastic bottles. Another common polymer is polypropylene (pah lee PRO puh leen). Polypropylene is used to make glues and carpets. Two or more different monomers can combine to make one polymer molecule. When this happens, the monomers are called copolymers.

Polymers can be light and flexible. They also can be so strong that they are used to make items such as plastic pipes, boats, and car bodies. Some polymers are used in place of wood and metal in buildings. Some people call this the "Age of Plastics" because so many things we use today are made of plastic polymers.

What determines the properties of polymer materials?

The properties of a polymer depend on which monomers are used to make them. Polymers can have branches in their chains, like hydrocarbons. Also, like hydrocarbons, the amount of branching and the shape of the polymer affect its properties.

Sometimes the same polymer can take two completely different forms. Polystyrene (pah lee STI reen) is an example. Polystyrene is made from the monomer styrene. Clear CD cases are made from polystyrene. When carbon dioxide gas is blown into melted polystyrene as it is molded, it becomes lightweight and opaque. It is used to make foam cups and packing materials. Bubbles remain in the polymer when it cools, so it is a good insulator. ☑

Polymers can be spun into thread. Fabrics made from strong polymer thread can be used for products that are treated roughly, such as suitcases and backpacks. Polymer materials are used to make bulletproof vests. Polymer fibers also can be made to stretch and then return to their original shape. These polymers are used to make exercise clothing.

What are some other petroleum products?

Other petroleum products are made by separating individual compounds from the petroleum fractions. These compounds can be changed into substituted hydrocarbons, as you learned in the previous section. Chemists use these hydrocarbons to make medicines like aspirin, insecticides, printing ink, and flavorings. Dyes made from petroleum have replaced dyes made from plants and other natural sources.

Are there problems with polymers?

Disposing of the many things made of polymers is a problem in this "Age of Plastics." Polymers do not decompose. A piece of plastic will remain a piece of plastic almost forever.

What can we do with all this plastic? One answer is to recycle. Recycling reuses clean plastics to make new products. Another answer is a process called depolymerization. **Depolymerization** uses heat or chemicals to break long polymer chains into monomer fragments. Then the monomers can be used to make other polymers. Today, depolymerization is expensive because a different process must be used to depolymerize each different polymer. There is still much research to do to make depolymerization practical.

Reading Check

7. **Apply** Which is made from polystyrene? (Circle your answer.)

   a. foam cups

   b. aspirin

   c. ink

   d. fabric

Think it Over

8. **List** Name two other petroleum products that are not polymers.

   _____

   _____

After You Read
Mini Glossary

depolymerization: a process that uses heat or chemicals to break long polymer chains into monomer fragments

monomer: the small molecule that forms a link in the polymer chain

polyethylene: a polymer formed when ethylene combines with itself repeatedly

polymer: a very large molecule made from small molecules that link together

1. Review the terms and definitions in the Mini Glossary. Write a sentence using the terms *polymer* and *polyethylene*.

   _____

   _____

   _____

2. Complete the flow chart to help you organize what you learned about petroleum and its compounds.

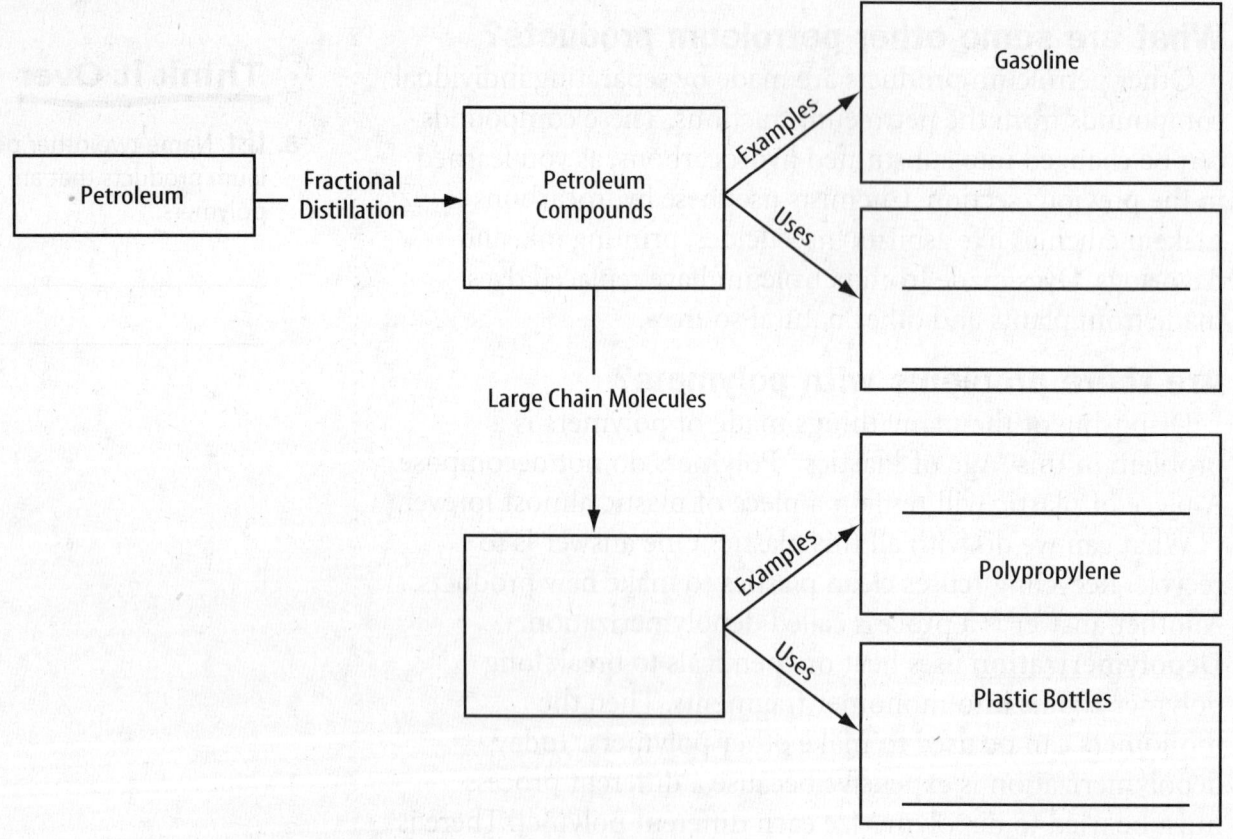

End of Section

Science Online Visit gpscience.com to access your textbook, interactive games, and projects to help you learn more about petroleum as a source of carbon compounds.

426   CHAPTER 24   Organic Compounds

Organic Compounds

section ❹ Biological Compounds

● Before You Read

Red meat, poultry, fish, dairy products, some beans, and some vegetables contain protein. List three things you have eaten in the past several days that contain protein.

What You'll Learn
- about proteins, nucleic acids, carbohydrates, and lipids
- polymers in food
- biological polymers

● Read to Learn

Biological Polymers

Many of the important biological compounds in your body are biological polymers. Biological polymers are huge molecules, just like the polymers you learned about in Section 3. Biological polymers also are made of monomers, but they are larger than monomers in other polymers.

Proteins

<u>Proteins</u> are large organic polymers formed from organic monomers called amino acids. There are only 20 different amino acids. But they can be arranged in so many ways that they make millions of different proteins. Many of the tissues in your body are made of proteins, including muscles, tendons, hair, and fingernails.

What are amino acids?

Amino acids are the monomers that combine to form proteins. The figure shows the structures of two amino acids, glycine and cysteine. Both molecules have an amine group, $-NH_2$. Both also have the group that all organic acids have, the carboxyl group, $-COOH$. All amino acids have an amine and a carboxyl group.

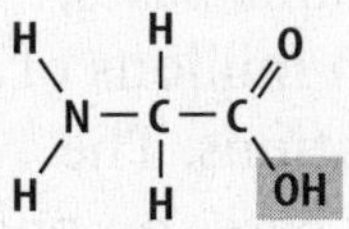

Mark the Text

Locate Information As you read about biological compounds in this section, draw lines from the words or phrases in the text to the corresponding parts of the figures.

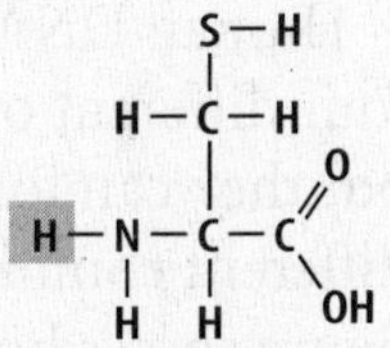

Picture This

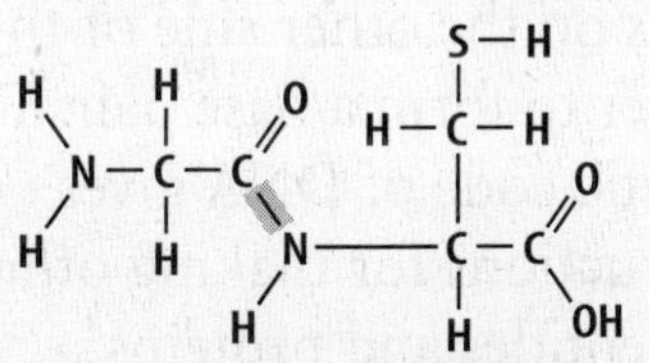

1. **Compare** Highlight the amine group in glycine and cysteine in one color. Highlight the carboxyl group in another color.

Reading Essentials **427**

FOLDABLES

A **Classify** Make a Foldable like the one below. As you read this section, write how your body uses each of these compounds.

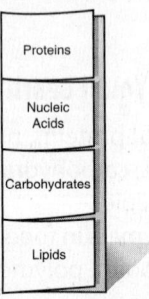

Reading Check

2. Infer What kind of organic polymer is DNA?

_____
_____

Picture This

3. Draw To show that there are two bases connected at the slash, draw a slash across each box indicating the base pairs.

The amine group of one amino acid can combine with the carboxyl group of another amino acid. The compound they form is a peptide. The bond joining the two amino acids is a peptide bond. When a peptide contains about 50 or more amino acids, the molecule is called a protein.

What do proteins look like?

Because a protein molecule is such a long chain, it twists around itself. Scientists can identify each protein by the way it twists. Many foods, including meat, dairy products, and some vegetables, contain proteins. When you eat these foods, your body breaks down the large protein molecules into their amino acid monomers. Then, your body uses the amino acids to make new proteins for muscles, blood, and other body tissues.

Nucleic Acids

Nucleic acids are another group of organic polymers that are essential for life. They control the activities and reproduction of cells. One nucleic acid in your body is deoxyribonucleic (dee AHK sih ri boh noo klay ihk) acid, or DNA. Deoxyribonucleic acid is in the nuclei of cells, where it codes and stores genetic information. This information is the genetic code.

What is DNA made of?

Monomers called nucleotides make up DNA. Nucleotides contain an organic base, a sugar, and a phosphoric acid unit. There are two chains of nucleotides in DNA. These two chains twist around each other and form something like a twisted ladder. This structure is called the double helix. *Helix* means "spiral."

Human DNA contains only four different organic bases, but they can form millions of different combinations. The figure shows how bases on one side of the ladder link with bases on the other side of the ladder to form a base pair. The genetic code of DNA gives instructions for making other nucleotides and proteins needed by the body.

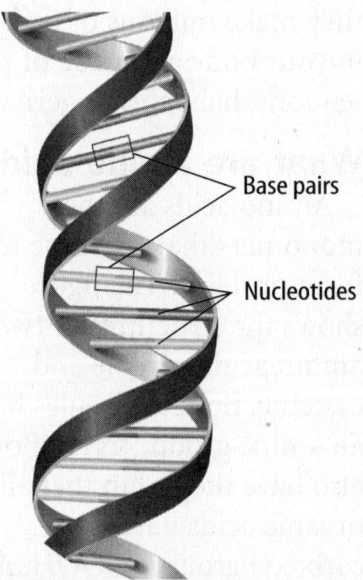

Base pairs

Nucleotides

428 CHAPTER 24 Organic Compounds

What is DNA fingerprinting?

Each molecule of DNA in your body has more than 5 million base pairs. Your DNA is different from the DNA of every other person in the world, unless you have an identical twin. Identical twins have exactly the same DNA.

DNA can be used to solve crimes. Scientists can remove DNA from hair, blood, or saliva left at a crime scene. They break up the DNA polymer into its nucleotides. Then scientists use radioactive and X-ray methods to make a picture of the pattern formed by the nucleotides. They compare the DNA pattern from the crime scene to the DNA pattern of a suspect. If the patterns match, they can link the suspect to the crime scene.

Carbohydrates

<u>Carbohydrates</u> are organic compounds made of carbon, hydrogen, and oxygen that have twice as many hydrogen atoms as oxygen atoms. Carbohydrates include sugars and starches. Foods like bread and pasta contain carbohydrates.

What are sugars?

Sugars are one of the main groups of carbohydrates. The figure shows the structural and chemical formulas of sucrose and glucose. Sucrose is also known as table sugar. Your body breaks sucrose down into two simpler sugars, fructose and glucose. Fruit contains the sugar fructose.

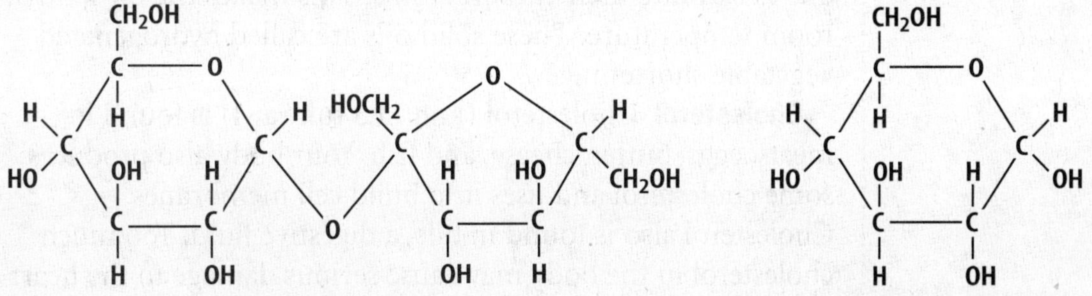

Picture This

4. Determine Look at the structural formula of glucose. How many hydrogen atoms does it contain? _____ How many oxygen atoms? _____ How do these numbers confirm that glucose is a carbohydrate?

_____
_____
_____
_____

Sucrose $C_{12}H_{22}O_{11}$

Glucose $C_6H_{12}O_6$

Glucose is found in your blood. Glucose also is in fruit and honey. You get a quick boost of energy after eating a food rich in sugar.

What is a starch?

A starch is a carbohydrate that is also a polymer. It is made of monomers of the sugar glucose. Your body breaks down starch molecules into monomers of glucose and other sugars. These sugars release energy into your cells.

Reading Essentials **429**

Think it Over

5. Summarize Why would marathon runners eat a lot of carbohydrates before a race?

_____

_____

Reading Check

6. Determine What are lipids made of? (Circle your answer.)

a. oxygen

b. hydrogen and oxygen

c. carbon and hydrogen

d. oxygen, carbon, and hydrogen

Athletes use starches to give them energy that lasts a long time. The energy from starches can be stored in liver and muscle cells in a compound called glycogen (GLI kuh jun). During a long race, the body releases the stored energy and the runner gets a fresh burst of power.

Lipids

Lipids are organic compounds like fats and oils. Lipids include animal fats, such as butter, and vegetable oils, such as corn oil. Lipids are made of carbon, hydrogen, and oxygen, just as carbohydrates are. But lipids contain these elements in different proportions. For example, lipids have fewer oxygen atoms than carbohydrates have. Also, lipids contain carboxyl groups and carbohydrates do not.

What are some lipids in your diet?

Fats and oils are similar in structure to hydrocarbons. If they have only single bonds between carbon atoms, they are saturated fats. Unsaturated fats have one or more double bonds. An unsaturated fat that has only one double bond is monounsaturated. An unsaturated fat that has two or more double bonds is polyunsaturated.

Fats are lipids that come from animals. Fats usually are saturated and solid at room temperature. Oils are lipids that come from plants. Oils are unsaturated and usually liquid at room temperature. Sometimes hydrogen is added to vegetable oils to saturate their carbon atoms. This makes the oil solid at room temperature. These solid oils are called hydrogenated vegetable shortenings.

Cholesterol Cholesterol (kuh LES tuh rawl) is found in meats, eggs, butter, cheese, and fish. Your body also produces some cholesterol and uses it to build cell membranes. Cholesterol also is found in bile, a digestive fluid. Too much cholesterol in the body may cause serious damage to the heart and blood vessels.

Have you heard that eating too much fat can be unhealthy? Eating too many foods with high amounts of saturated fats and cholesterol may cause heart disease. Some unsaturated fats may protect the heart from some diseases. The body may be more likely to change saturated fats into substances that can block the arteries going to the heart. It is not wise, though, to stop eating all fats. A balanced diet contains some fats as well as proteins and carbohydrates.

After You Read

Mini Glossary

carbohydrate: an organic compound made of carbon, hydrogen, and oxygen that has twice as many hydrogen atoms as oxygen atoms

deoxyribonucleic acid: a nucleic acid found in the nuclei of cells where it codes and stores genetic information

lipid: an organic compound like fats and oils

nucleic acid: one of a group of organic polymers that are essential for life

protein: a large organic polymer formed from organic monomers called amino acids

1. Review the terms and definitions in the Mini Glossary. Circle the three terms that describe compounds that are in the foods you eat.

2. Write the letter of the term in **Column 2** that matches the example in **Column 1**.

Column 1

_____ 1. Made of amino acids

_____ 2. Form peptide bonds

_____ 3. DNA is one of these

_____ 4. Gives your body a quick boost of energy

_____ 5. Long-lasting fuel that can be stored in the body

_____ 6. Fats and oils

Column 2

a. lipid

b. nucleic acid

c. protein

d. starch

e. amino acids

f. sugar

3. You drew a line from the words or phrases in the text to the corresponding part of each figure. How would this strategy help you if you were studying for a test?

_____

_____

 Visit gpscience.com to access your textbook, interactive games, and projects to help you learn more about biological compounds.

chapter 25 — New Materials Through Chemistry

section ❶ Materials with a Past

What You'll Learn
- the uses of alloys
- the properties of alloys
- how the use of alloys depends on their properties

Study Coach

Outline Create an outline of the section as you read. Include the main ideas and definitions that you read about.

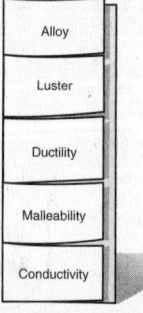

A Build Vocabulary Make the following Foldable to make sure you understand the vocabulary terms in this section.

Before You Read

What are some metal objects you have used? How do metals make your life easier?

_____

_____

Read to Learn

Alloys

For ages, people have searched for things to make their lives easier. Ancient people used stone tools. Later, methods for processing metals were discovered. Today, scientists are still finding ways to blend metals, or make alloys, to make better metal products. An **alloy** is a mixture of elements that has the properties of metal. Have you ever seen a pewter mug? Pewter is an alloy made of tin, copper, and antimony. Pewter looks and feels like metal. Alloys sometimes have improved properties over the metals they are made of. Alloys can be lighter, stronger, harder, and more durable than the elements they are made of.

How were alloys discovered?

Historians believe the ancient Sumerians discovered bronze in about 3500 B.C. The Sumerians lived in what is now Iraq. By accident, copper and tin ore found in rocks the Sumerians used to keep their fires from spreading probably melted. The alloy of copper and tin is bronze. Sumerians then used the bronze to make tools and other objects. Bronze was the main alloy used to make tools for about 2,000 years. Historians call this time period the Bronze Age.

Properties of Metals and Alloys

What are the properties of metals and alloys? **Luster** is the ability to reflect light or have a shiny appearance. **Ductility** (duk TIH luh tee) is the ability to be pulled into wires.

432 CHAPTER 25 New Materials Through Chemistry

Two other properties are malleability and conductivity. **Malleability** (mal yuh BIH luh tee) is the ability to be hammered or rolled into sheets—like aluminum foil. **Conductivity** is the ability to carry heat and electrical charges easily. Copper has good ductility and conductivity so it is often used for electrical wires.

Which properties of an alloy are most important?

The properties of an alloy that are most important depend on how the alloy is used. For example, most gold jewelry is made from an alloy of gold and copper. Pure gold is a bright, expensive metal that is soft and bends easily. Copper is not expensive and it is harder than gold. Gold and copper are melted, mixed, and allowed to cool, forming an alloy.

The properties of the alloy depend on how much of each metal is used. A ring made with a larger percentage of gold would be more expensive, but it would bend more easily. A ring made with a smaller percentage of gold would be less expensive, but it would not bend as easily.

How do you choose an alloy?

A gold and copper alloy would probably not be used to make a drill bit. Why? A drill bit must be much harder than jewelry. The alloys used to make tools, jewelry, or other metal objects depend on how the object will be used. In some tools, such as drill bits, hardness is very important. In jewelry, value and luster are important. In electrical wires, flexibility and conductivity are important. The uses of some alloys and the elements they contain are listed in the table below.

✔ Reading Check

1. **Conclude** Which metal is softer, gold or copper?

Picture This

2. **Interpret Data** Which metal is used more than any other in the alloys listed in the table?

Common Alloys		
Name	**Composition**	**Use**
Bronze	copper, tin	jewelry, marine hardware
Brass	copper, zinc	hardware, musical instruments
Sterling silver	silver, copper	tableware
Pewter	tin, copper, antimony	tableware
Solder	lead, tin	plumbing
Wrought iron	iron, carbon	porch railings, fences, sculpture

Uses of Alloys

As you saw in the table on the previous page, alloys are used in many objects. In fact, most metal objects you see around you are made of alloys. Strong alloys are used to make machinery and building materials. The alloys used in car and airplane bodies must be light and strong and resist corrosion.

Alloys are even used in the human body. Steel pins and screws are used to repair broken bones. Steel plates are used to repair damage to the skull. The metal parts used are made from surgical steel. Surgical steel has properties that allow it to resist tissue rejection. Tooth fillings usually are made from silver and mercury.

What is steel and how is it used?

Steel is an alloy of iron, carbon, and other elements. The properties of steel alloys depend on how much carbon and other elements are used in the alloy. The alloys have different properties and different uses. Steel alloys can be very strong and often are used in large buildings and bridges. Ship hulls, bedsprings, and many automobile parts are made from steel. Stainless steel is an alloy that resists corrosion. It is used in surgical instruments and cooking utensils.

New Alloys

Steel is not the only common alloy. Aluminum is a common metal used to make soda cans and cooking foil. Aluminum alloys are strong and light and are used to build large airplanes. Alloys of the metal titanium also are used on aircraft because of their strength and light weight. Strong materials last longer and lighter airplanes are less expensive to fly.

What alloys are used in space?

The original heat shield on the space shuttle was made from ceramic tiles. These tiles resist the large amount of heat generated when a shuttle reenters Earth's atmosphere. The ceramic tiles often crack and break because of the heat and stress of reentry into Earth's atmosphere. Fixing the broken tiles is expensive and time consuming.

Tiles made from a new titanium alloy are being developed. These extremely heat resistant new tiles may be used on space shuttle heat shields. The titanium alloy tiles are larger and easier to attach than the ceramic tiles. The tiles could make the heat shield easier and less expensive to maintain.

Reading Check

3. **List** What are the two main elements in the alloy steel?

Think it Over

4. **Describe** What are some properties of aluminum and titanium alloys that make them useful for building aircraft?

After You Read

Mini Glossary

alloy: a mixture of elements that has the properties of a metal

conductivity: the property of metals and alloys that allows heat and electrical charges to pass through the material easily

ductility: the ability to be pulled into wires

luster: the ability to reflect light or have a shiny appearance

malleability: the ability to be hammered or rolled into sheets

1. Review the terms and their definitions in the Mini Glossary. Write a sentence using one of the terms that describes a property of alloys.

   _____

   _____

2. In the graphic organizer below, list the properties of alloys that might be used to make the listed items. Also give an example of a type of an alloy that has those properties.

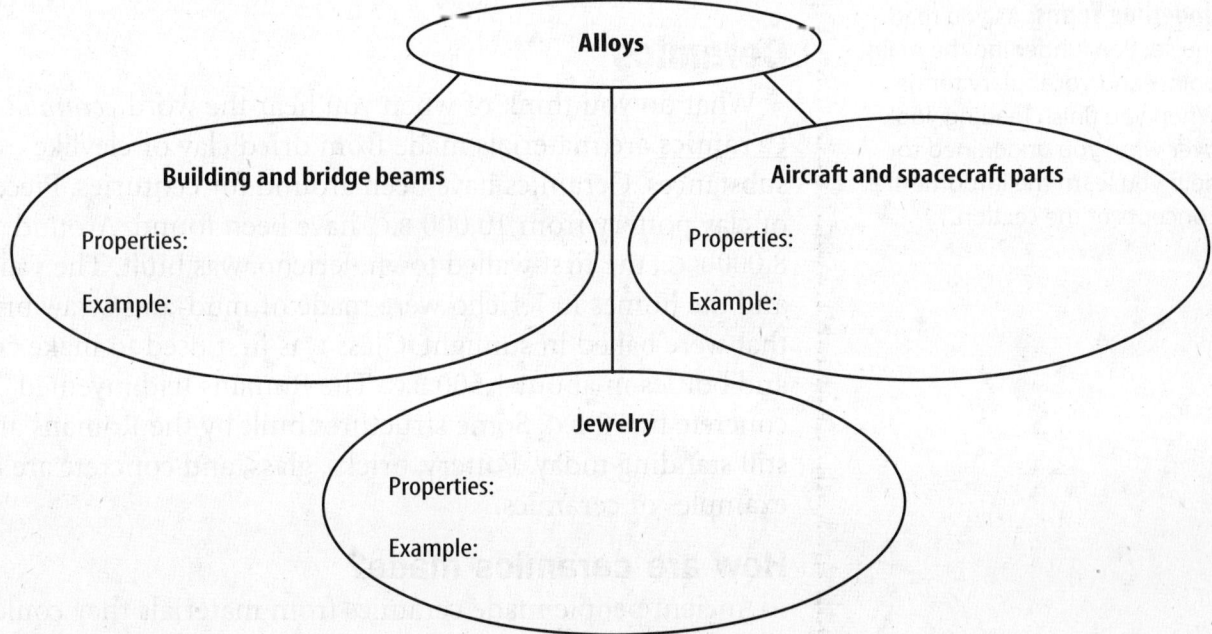

3. You made an outline as you read this section. What did you use for the major categories in your outline?

   _____

 Visit **gpscience.com** to access your textbook, interactive games, and projects to help you learn more about alloys.

Reading Essentials **435**

New Materials Through Chemistry

section ❷ Versatile Materials

What You'll Learn
- properties of ceramics
- how ceramic materials are used
- what semiconductors are

● Before You Read

Think about eating a meal at home. You might see metal cooking pans on the stove and metal silverware on the table. What other materials might you see around you?

___

___

Mark the Text

Underline Terms As you read the section, underline the main points and vocabulary terms. When you finish reading, look over what you underlined to help you learn the important concepts of the section.

● Read to Learn

Ceramics

What do you think of when you hear the word *ceramic*? **Ceramics** are materials made from dried clay or claylike substances. Ceramics have been around for centuries. Pieces of clay pottery from 10,000 B.C. have been found. Around 8,000 B.C., the first walled town, Jericho, was built. The wall and the homes in Jericho were made of mud-and-straw bricks that were baked in sunlight. Glass was first used to make cups and bottles in about 1,500 B.C. The Romans had invented concrete by 50 B.C. Some structures built by the Romans are still standing today. Pottery, bricks, glass, and concrete are all examples of ceramics.

How are ceramics made?

Ancient people made ceramics from materials they could find easily. These were clay, sand, and crystalline rock called feldspar. These materials are still used today to make some ceramics. Newer ceramics might contain metallic compounds and carbon, nitrogen, or sulfur.

Why are ceramics heated?

Ceramic objects are made by molding the processed raw materials into a shape. The ceramic then is heated to a temperature between 1,000°C and 1,700°C. This process, called firing, shrinks the spaces between particles. Ceramic objects are smaller after firing.

436 CHAPTER 25 New Materials Through Chemistry

Firing The figures show how firing makes the spaces between the ceramic particles shrink. This makes ceramics strong. They also are resistant to heat. For these reasons, ceramic tiles are used on the heat shield of the space shuttle. But ceramics, even those on the shuttle, are fragile. They will break if dropped or if their temperature changes too quickly.

Not Fired   Fired

 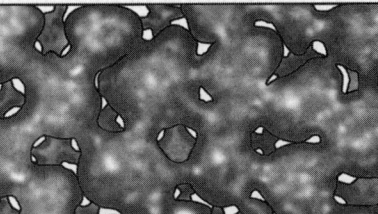

How are ceramics used?

Ceramics have some useful properties. They do not react with oxygen, water, acids, bases, salts, or strong solvents. This makes ceramics useful where they might come into contact with these materials. For example, tableware and cookware are often made of ceramics because foods contain acids, salts, and water. Ceramics also make good insulators. You may have seen the ceramic insulators used to connect electric wires to posts. The insulators keep the electric current flowing through the wires instead of through the post and into the ground.

Like alloys, ceramics can be customized, or made to have certain properties. Ceramics are customized by using different raw materials or different processes. The ceramic material in a coffee cup does not need to be the same as the ceramic material in the space shuttle tiles.

What are some new uses for ceramics?

Some new ceramics have been created with properties that most ceramics do not have. For example, most ceramics are insulators. But ceramics made with chromium dioxide are good conductors and some ceramics made with copper are even superconductors. Transparent, conductive ceramics are used to make airplane windshields that stay free of snow and ice.

Ceramics have medical uses in the human body. The properties of ceramics make them resistant to body fluids, which can damage many materials. Surgeons use ceramics to repair or replace joints, such as hips, knees, shoulders, elbows, fingers, and wrists. Dentists use ceramics to repair and replace teeth and to make braces.

Picture This

1. **Describe** Look at the figures showing ceramic material before and after firing. Why is the ceramic denser after firing?

   _____

   _____

Reading Check

2. **Explain** Why are ceramic insulators used to connect electric wires to posts?

   _____

   _____

Reading Essentials **437**

FOLDABLES

B Organize Information Make the following Foldable to help you organize what you learn about semiconductors.

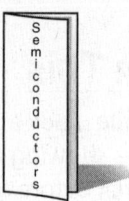

✓ Reading Check

3. Describe how the conductivity of a semiconductor can be changed.

_____
_____

Picture This

4. Explain In this figure, what is different about the element that was added to the silicon crystal?

_____
_____

Semiconductors

Another group of useful materials is semiconductors. <u>Semiconductors</u> are materials that do not conduct electricity as well as metals but do conduct better than nonmetals, and their conductivity can be controlled. These materials make computers and other electronic devices work.

Look at the periodic table in this book. The elements along the stair-step line between metals and nonmetals are the metalloids. Some metalloids are semiconductors. Recall that metals are good conductors of electricity. Nonmetals are poor conductors of electricity. Semiconductors lie between the two. The metalloid elements silicon (Si) and germanium (Ge) are common semiconductors.

How is the conductivity of a semiconductor changed?

Adding other elements to some metalloids can change their conductivity. For example, you can increase the conductivity of silicon crystals by replacing some silicon atoms with atoms of other elements. Atoms that commonly are added to silicon are arsenic (As) and gallium (Ga). The figure below shows how the conductivity of silicon increases when another element is added. ✓

The added atoms are called impurities. If the impurities have fewer electrons than silicon atoms, the silicon crystals will have areas with fewer electrons. These areas are called holes. Electrons now can move from hole to hole across the crystal. This increases the conductivity of the material. Adding just one atom of an impurity to a million silicon atoms changes the conductivity of the silicon a great deal. By controlling the type and number of atoms added, the conductivity of silicon can vary over a wide range.

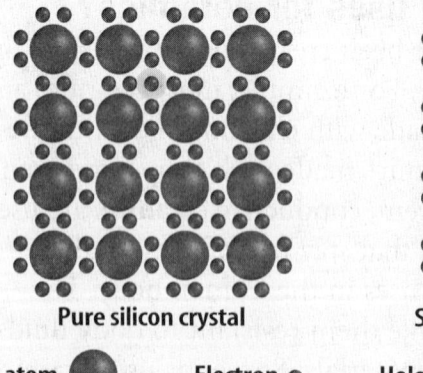
Pure silicon crystal

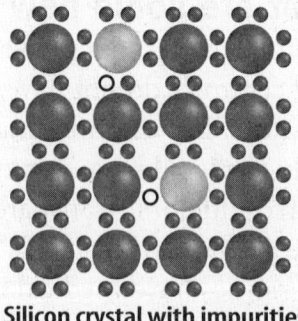
Silicon crystal with impurities

Si atom ●   Electron •   Hole ○   Other element (impurity)

What is doping?

Doping is the process of adding impurities to semiconductors to change their conductivity. If doping increases the number of electrons in a crystal, the semiconductor is called an *n-type* semiconductor. If doping reduces the number of electrons in a crystal, the semiconductor is called a *p-type* semiconductor.

What are integrated circuits?

Electronic devices such as transistors and diodes are made by placing n-type and p-type semiconductor crystals next to each other. Transistors and diodes control the flow of electrons in electric circuits of devices like radios and computers. The figure shows how electrons flow from the arsenic-doped silicon n-type semiconductor to the gallium-doped silicon p-type semiconductor.

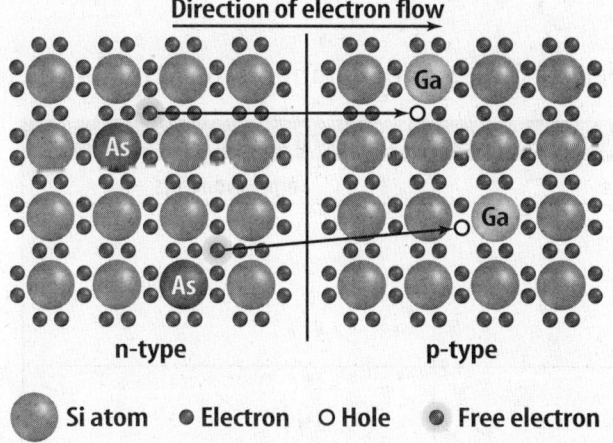

Picture This
5. Classify What is the chemical symbol of the element used in the p-type semiconductor shown in the figure?

_____

In the 1960s, scientists developed very small transistors and other devices. At the same time, the integrated circuit was developed. An **integrated circuit** contains many tiny semiconducting devices. Integrated circuits are sometimes called microchips. The integrated circuit was a great advance in technology. The small electronic devices we use today, such as radios, TVs, cell phones, calculators, and especially computers, would not be possible without integrated circuits.

Integrated circuits are being made smaller and smaller. It takes less time for electric current to travel through smaller circuits, which makes the computers that use them even faster. When computers were first invented in the 1940s, they filled entire rooms. They needed thousands of electric circuits to process information. Today, these thousands of circuits are all found on tiny integrated circuits. In about 60 years, computers have gone from being the size of buildings to being small enough to fit in your hand.

Reading Check
6. Identify What is another name for an integrated circuit?

_____
_____

After You Read
Mini Glossary

ceramics: materials made from dried clay or a clay-like substance

doping: the process of adding impurities to semiconductors to change their conductivity

integrated circuit: a circuit that contains many tiny semiconducting devices

semiconductor: a material that does not conduct electricity as well as metals but conducts better than nonmetals, and its conductivity can be controlled

1. Review the terms and their definitions in the Mini Glossary. Write a sentence that shows you understand why manufacturers use doping when making some semiconductors.

   _____

   _____

2. Complete the table to describe the properties and uses of ceramics and semiconductors.

Properties and Uses of Ceramics and Semiconductors		
	Ceramics	**Semiconductors**
Materials they are made of		
Properties		
Uses		

3. How are semiconductors similar to traffic lights?

   _____

   _____

 Visit **gpscience.com** to access your textbook, interactive games, and projects to help you learn more about ceramics and semiconductors.

440 CHAPTER 25 New Materials Through Chemistry

Chapter 25: New Materials Through Chemistry

section 3 Polymers and Composites

Before You Read

What was the last plastic item you used and how did you use it?

_____

_____

What You'll Learn
- what a polymer is and the kinds of polymers
- composites and why they are used

Read to Learn

Polymers

Polymers are natural or manufactured substances that are made of long chains of molecules with small, simple, repeating units. A **monomer** is one specific molecule that is the repeated unit in polymer chains. For example, polypropylene polymer chains might have 50,000 to 200,000 monomers in each chain. Some examples of manufactured polymers are shown in the table.

Mark the Text

Highlight Main Points As you read this section, highlight the vocabulary words and main points. After you read, look over the sentences you highlighted.

Common Polymers	
Polymer	**Uses**
Polyethylene	bottles, garment bags
Polyvinyl chloride (PVC)	pipe, bottles, compact discs, computer housing
Polypropylene	rope, luggage, carpet, film
Polystyrene	toys, packaging, egg cartons, flotation devices

FOLDABLES

C Compare and Contrast Make the following Foldable to help you understand how synthetic and natural polymers are similar and different.

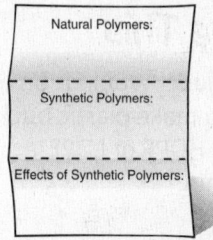

You might know some of the polymers in the table as plastics. Not all polymers are plastics, and not all polymers are manufactured. Proteins, cellulose, and nucleic acids are polymers found in living things. In this section, you will learn mostly about synthetic polymers. **Synthetic** means not naturally occurring but manufactured in laboratories or chemical plants.

Reading Essentials **441**

How were synthetic polymers invented?

Humans have used natural polymers for centuries. The ancient Egyptians used polymers in natural resins to help preserve their dead. The first synthetic polymers were made while scientists were trying to improve natural polymers. In 1839, an American inventor named Charles Goodyear found a way to improve natural rubber. He heated it with sulfur. The result was a rubber that did not break when it was cold or get too soft when it was hot. In the 1860s, John Hyatt invented celluloid, a polymer that was used to make billiard balls, umbrella handles, and toys. These early polymers were not as useful as today's polymers. However, they were better than many natural polymers. Now there are many synthetic polymers. They are often divided into groups, such as plastics, synthetic fibers, adhesives, surface coatings, and synthetic rubbers.

What are synthetic polymers made of?

Today, synthetic polymers are made from fossil fuels such as oil, coal, and natural gas. Fossil fuels are called hydrocarbons because they are compounds mostly of hydrogen and carbon. Since synthetic polymers are made from hydrocarbons, synthetic polymers contain mostly hydrogen and carbon.

Why do synthetic polymers have different properties?

The properties of polymers depend on the type of monomers used to make them. The way the monomers are arranged in the chains also affects the properties. There are many different types of synthetic polymers because there are many ways to arrange the chains.

For example, the monomer ethylene is used in several different types of polymers. Polyethylene can be high density or low density, depending on how the molecules are attached to the monomer. Low-density polyethylene (LDPE) and high-density polyethylene (HDPE) differ mostly by the way the monomers are arranged in the polymer chains. The table shows the properties of LDPE and HDPE and some of the ways they are used.

Material	Properties	Typical uses
Low-density polyethlene (LDPE)	Flexible, tough, chemical resistant, low density	Shopping bags, milk jugs
High-density polyethylene (HDPE)	Firm, strong, less translucent, higher density	Toys, outdoor furniture

Reading Check

1. **Recall** What was one use for celluloid?

Think it Over

2. **Recognize Cause and Effect** Why could using fewer plastics help save natural resources?

Picture This

3. **Conclude** Would it be better to make plastic buckets from HDPE or LDPE? Explain.

Plastics One of the groups of synthetic polymers is plastics. Plastics are used in many products because they have useful properties. For example, they are lightweight, strong, impact resistant, waterproof, moldable, chemical resistant, and inexpensive. There are many types of plastics. Some are clear, some are flexible, and some are hard. The properties of plastics depend on the polymers they are made from.

Synthetic Fibers Like plastics, synthetic fibers can be made with almost any properties. Nylon, polyester, acrylic, and polypropylene are examples of synthetic fibers. Nylon is often used in wind- and water-resistant clothing, such as jackets and windbreakers. Polyester is used to make many fabrics. It is often blended with natural fibers, such as cotton. Polyester fibers also are used to stuff pillows and quilts. Polyurethane is the foam used in mattresses and pillows.

Aramids are a family of nylons with special properties. Some aramids are fireproof. They are used to make protective clothing for firefighters, military pilots, and race car drivers. Another aramid fiber is used to make bulletproof vests, race car survival cells, puncture-resistant gloves, and motorcycle clothing. These aramids are light enough for clothing, yet stronger than steel.

Adhesives Adhesives are products that stick things together. Glue is an adhesive. Synthetic polymers are used to make adhesives that can be modified for different uses.

Content cement is a synthetic polymer adhesive. It is used in the manufacture of car parts, furniture, and leather goods. Contact cement sticks instantly. The bond gets stronger after the cement dries. Structural adhesives are used in construction projects. One structural adhesive is silicon. Silicon adhesive is used to seal the spaces around windows and doors to prevent heat loss. Orthodontists use adhesives to attach braces to teeth. One such adhesive bonds after it is exposed to ultraviolet light.

Surface Coatings and Elastic Polymers Many products that are used to coat and protect surfaces are synthetic polymers. Polyurethane is a polymer that is used to protect wood. Many paints also contain synthetic polymers. Synthetic rubber is an elastic polymer. It is used to make tires, gaskets, belts, and hoses. The soles of some shoes are made from this synthetic elastic polymer.

Reading Check

4. Identify Name a synthetic fiber used to stuff pillows.

_____

_____

Think it Over

5. Apply Suppose water was leaking around a shower door. What kind of synthetic polymer might be used to seal the leak?

_____

_____

How do synthetic fibers copy nature?

Spinning long fibers into threads and fabrics is not a new idea. Spiders spun fibers for their webs long before humans copied the idea. Nylon fiber is another idea borrowed from nature. The silkworm produces a fiber that is woven into fabric called silk. Silk is used for blouses and other clothing. Can you imagine how long it takes for a worm to produce enough silk for one blouse? Nylon was produced in a laboratory as a possible substitute for silk.

Composites

The properties of a synthetic polymer can be changed by adding other materials. A **composite** is a mixture of two or more materials—one embedded or layered in the other. Composite materials of plastic have many uses. Some boat and car bodies are made of a composite called fiberglass. Fiberglass is made from small fibers or threads of glass embedded in a plastic. The glass fibers make the plastic stronger. Fiberglass is a strong, lightweight composite. Glass and other fibers are often added to plastics to make them stronger, more flexible, and less brittle.

How are composites used in spacecraft and aircraft?

Some satellites are made from composite materials. These satellites are stronger and weigh less than satellites made of aluminum. Lighter satellites are less expensive to launch.

The figure below shows that aircraft contain composite materials. Just as in satellites, the composites used in aircraft are lighter and stronger than the metals once used. The weight of some commercial aircraft is reduced by more than 2,600 kg by using composite materials and new alloys. Lighter airplanes cost less to operate because they use less fuel.

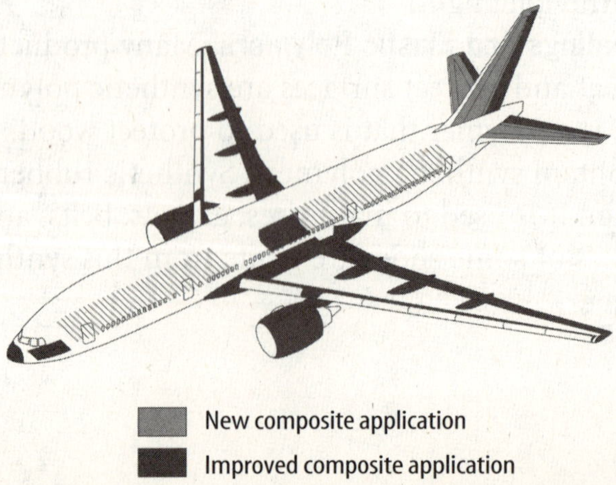

▪ New composite application
▪ Improved composite application

✓ **Reading Check**

6. Reflect What effect do the glass fibers have on the plastic they are layered in?

_____
_____
_____

Picture This

7. Observe Look at the figure. In what areas of the aircraft are composites used most often?

_____
_____
_____

444 CHAPTER 25 New Materials Through Chemistry

After You Read

Mini Glossary

composite: a mixture of two or more materials—one embedded or layered in the other

monomer: one specific molecule that is the repeated unit in polymer chains

polymer: a natural or manufactured substance that is made of long chains of molecules with small, simple, repeating units

synthetic: not naturally occurring, but manufactured in laboratories or chemical plants

1. Review the terms and their definitions in the Mini Glossary. Describe fiberglass using at least one of the terms in the Mini Glossary.

   _____
   _____

2. Fill in the Venn diagram below to compare polymers and composites. Give examples of both polymers and composites.

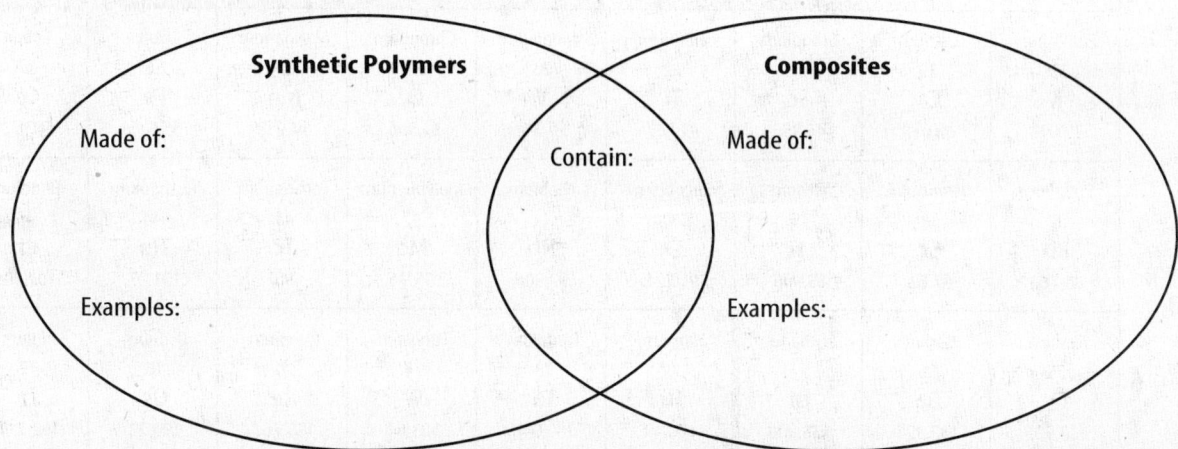

3. As you read this section, you highlighted vocabulary words and main points. After you read, you looked again at the sentences you highlighted. How could you use your highlighting to help you remember the material in this section?

   _____
   _____

Visit **gpscience.com** to access your textbook, interactive games, and projects to help you learn more about polymers and composites.

Reading Essentials 445

PERIODIC TABLE OF THE ELEMENTS

Columns of elements are called groups. Elements in the same group have similar chemical properties.

Element — Hydrogen
Atomic number — 1
Symbol — H
Atomic mass — 1.008
State of matter

- Gas
- Liquid
- Solid
- Synthetic

The first three symbols tell you the state of matter of the element at room temperature. The fourth symbol identifies elements that are not present in significant amounts on Earth. Useful amounts are made synthetically.

Period	1	2	3	4	5	6	7	8	9
1	Hydrogen 1 **H** 1.008 (gas)								
2	Lithium 3 **Li** 6.941	Beryllium 4 **Be** 9.012							
3	Sodium 11 **Na** 22.990	Magnesium 12 **Mg** 24.305							
4	Potassium 19 **K** 39.098	Calcium 20 **Ca** 40.078	Scandium 21 **Sc** 44.956	Titanium 22 **Ti** 47.867	Vanadium 23 **V** 50.942	Chromium 24 **Cr** 51.996	Manganese 25 **Mn** 54.938	Iron 26 **Fe** 55.845	Cobalt 27 **Co** 58.933
5	Rubidium 37 **Rb** 85.468	Strontium 38 **Sr** 87.62	Yttrium 39 **Y** 88.906	Zirconium 40 **Zr** 91.224	Niobium 41 **Nb** 92.906	Molybdenum 42 **Mo** 95.94	Technetium 43 **Tc** (98) (synthetic)	Ruthenium 44 **Ru** 101.07	Rhodium 45 **Rh** 102.906
6	Cesium 55 **Cs** 132.905	Barium 56 **Ba** 137.327	Lanthanum 57 **La** 138.906	Hafnium 72 **Hf** 178.49	Tantalum 73 **Ta** 180.948	Tungsten 74 **W** 183.84	Rhenium 75 **Re** 186.207	Osmium 76 **Os** 190.23	Iridium 77 **Ir** 192.217
7	Francium 87 **Fr** (223)	Radium 88 **Ra** (226)	Actinium 89 **Ac** (227)	Rutherfordium 104 **Rf** (261) (synthetic)	Dubnium 105 **Db** (262) (synthetic)	Seaborgium 106 **Sg** (266) (synthetic)	Bohrium 107 **Bh** (264) (synthetic)	Hassium 108 **Hs** (277) (synthetic)	Meitnerium 109 **Mt** (268) (synthetic)

The number in parentheses is the mass number of the longest-lived isotope for that element.

Rows of elements are called periods. Atomic number increases across a period.

The arrow shows where these elements would fit into the periodic table. They are moved to the bottom of the table to save space.

Lanthanide series

Cerium 58 **Ce** 140.116	Praseodymium 59 **Pr** 140.908	Neodymium 60 **Nd** 144.24	Promethium 61 **Pm** (145) (synthetic)	Samarium 62 **Sm** 150.36

Actinide series

Thorium 90 **Th** 232.038	Protactinium 91 **Pa** 231.036	Uranium 92 **U** 238.029	Neptunium 93 **Np** (237) (synthetic)	Plutonium 94 **Pu** (244) (synthetic)